Computer Technology and the Law in Canada

J. Fraser Mann
LL.B., Ph.D.
of the Ontario Bar

Carswell
Toronto • Calgary • Vancouver
1987

Canadian Cataloguing in Publication Data

Mann, J. Fraser
Computer technology and the law in Canada

Includes index.
ISBN 0-459-39850-4

1. Computers — Law and legislation — Canada.
2. Computer industry — Law and legislation — Canada.
I. Title.

KE452.C6M35 1987 343.71′0999 C87-093001-X

To my Parents

Preface

In the late 1980s the field of computer law has come of age. There has been a proliferation of articles, seminars, conferences, reporting services and trade publications devoted to the topic. Computer law is a recognized field of study in several Canadian and many American law schools. Major law firms have set up departments or sections specializing in computer and technology-related matters. National and international legal associations have established computer law committees. Law students and young lawyers are looking to the field of computer law to provide new career opportunities.

There is no doubt that computer law represents one of the most difficult, yet at the same time one of the most intellectually challenging, areas of law today. It is hoped that this book will be of some assistance in clarifying the nature and scope of computer law in Canada and the legal issues that may arise in any dealings with computer technology. At the same time, however, no claim is made that this book is exhaustive in covering this field of law.

I am indebted to various colleagues at Borden & Elliot, including Christopher Bredt, Timothy Buckley, John Hylton, Gary Maavara, Robert Russell and James Shearn, for reviewing and commenting on various portions of this book. In addition, Tony Pallotta and Murray Vince provided assistance by way of research in the preparation of chapters 7 and 8.

I would also like to thank Adrian Wijeyewickrema for his assistance in preparing the section on computer technology.

I am particularly grateful to David T. Tetreault who is principally responsible for the final two chapters of the book dealing with taxation aspects of computer transactions, as well as my colleagues in Borden & Elliot's Tax Department for their review and comments on these chapters.

I would also like to thank members of the Executive Committee of Borden & Elliot for their support in carrying out this project.

A special word of appreciation is extended to my secretary, Marilyn Batuszkin, and to Laura Hunter and other staff members of Borden & Elliot's Word Processing Centre, for their patience and diligence in typing, reviewing and proof-reading numerous drafts of this work. I would also like to thank Geralyn Christmas, Senior Legal Editor of Carswell Legal Publications, for her continuing advice and assistance, and Dianne Overholt and other personnel at Carswell for their work in converting the manuscript into book form.

It goes without saying, of course, that any errors or omissions in the book are the sole responsibility of the author.

This book attempts to state the law as of December 31, 1986.

February, 1987 J. Fraser Mann

Table of Contents

Part III
COMPUTERS, CRIMINAL LAW AND EVIDENCE

Table of Cases

Part I

COMPUTER TECHNOLOGY AND THE LAW IN CANADA

1

Computers and the Law: An Overview

1. INTRODUCTION: THE INFORMATION SOCIETY

It is virtually a truism to say that we live in an information society. Computers play a pivotal role in virtually all economic and social relationships. Businesses of all sizes are dependent upon computers to perform a variety of functions, such as accounting, personnel management, database management, word processing and inter-office and remote communications. The integration of communications facilities with computer systems enables information to be transmitted, processed and stored without regard to national or provincial boundaries or to the distance between the data processing facility and the location of the user.

Computers play an integral role in the functioning of virtually every type and size of business. Manufacturing industries require sophisticated computer systems to operate assembly lines, to run equipment and to monitor operating conditions within plants or factories. Wholesale and retail industries require complex accounting systems and computerized point-of-sale devices to process orders, control inventory and issue invoices. The design and manufacture of new products are facilitated by CAD/CAM ("computer-aided design/computer-aided manufacturing") systems whereby the various components of new products are designed, analyzed and tested by the use of computerized models in place of prototypes.

Computers also play a key role in the financial and service industries. Banking, insurance, brokerage and other financial industries require specialized computer packages and services as well as access to high-speed communications facilities to transfer information or funds, to monitor the status of insurance claims or the performance of securities on the stock markets, or otherwise to acquire, process and disseminate information. Airlines, travel agencies and hotels are heavy users of computer/communications facilities to control reservations and to monitor movements of passengers. Accounting firms, lawyers and other professionals are dependent on accounting, word processing and other computer systems to provide services to their clients.

At the consumer level, a wide variety of products, such as motor vehicles, electrical appliances and electronic equipment incorporate computer programs embodied in silicon chips. The rapid decrease in the size and price of computational power has resulted in millions of consumers acquiring personal computers to be used with mass-distributed software packages for such purposes as the control of household budgets, word processing, data base management, home video games or gaining access to external data banks.

In the area of government, a myriad of institutions collect seemingly endless amounts of information about private citizens as well as statistical data, economic analyses and other information required for the functioning of modern government. Institutions required to administer social programs, income tax legislation and various regulatory schemes are all highly dependent on the use of computers to carry out their responsibilities.

2. THE MEANING OF COMPUTER LAW

Computer law may be described as that field of law that deals generally with the legal aspects of the relationship between society and computer technology. More specifically, computer law refers to those areas of public and private law concerned with the regulation of computer technology and the development, distribution and use of computer products and services.

The widespread use of computer technology affects virtually all substantive areas of law either directly or indirectly. In some cases, the legal issues which arise with respect to computers may not be unique to that technology. For example, many of the contractual issues relating to the acquisition of computer equipment are common to the acquisition of any type of tangible property. In other cases, however, the computer itself is the impetus of the creation of a new legal problem. For example, the creation of computer software, with its unique technological aspects, presents new questions as to the interpretation and application of traditional intellectual property laws.

The various parts of this text are intended to review six major areas of Canadian law relating to computer technology. These areas are as follows.

(a) Constitutional Aspects of Computer Technology

In a federal state such as Canada, legislative jurisdiction is divided between the federal and provincial governments. The growing use of computer technology raises new issues as to the powers of each level of government, and the nature and scope of the fundamental rights of citizens. Some of these issues are considered in the following chapter of this text.

(b) Intellectual Property Laws

The various statutory and common law causes of actions relating to protection of intellectual property are of crucial importance in determining the rights and obligations of the creators, distributors and users of computer programs and computerized information. Part II will review in detail three such areas of law, namely, copyright, trade secrecy and patent law, as they apply to computer programs and data and related subject matter. Other evolving forms of legal protection such as industrial design and chip protection legislation are also considered.

(c) Criminal Law and Evidence

Computer technology provides new opportunities for criminal elements to interfere with private property and personal rights of citizens. At the same time, computer systems and computerized information may be unlawfully destroyed, interfered with or otherwise become the subject of criminal activities. The computerization of virtually all businesses also creates new legal problems concerning the admissibility of computer-produced documents and the proof of facts in criminal and civil actions. Part III will review some of the legal issues in the area of criminal law and evidence as they relate to computer technology.

(d) Privacy, Data Protection and Transborder Data Flow

The growth of computer data banks in the public and private sectors, and the potential for cross-linking of such data banks, raise important issues as to the rights of privacy and the measures that can be taken to protect personal information. At the same time, the technological possibility of transmitting large amounts of information world-wide virtually instantaneously raises issues as to political sovereignty and economic decision-making which may have a major impact on the lives of citizens within any country. Part IV will review the existing Canadian laws as well as some international legal developments relating to these issues. In addition, some of the legal issues relating to the establishment of electronic funds transfer systems will be considered.

(e) Contracting for Computer Products and Services

A wide variety of contractual arrangements may be made between the owners, distributors and users of computer products and services. Although many of the contractual issues are common to other commercial arrangements, legal practitioners must be aware of the unique problems that arise and some measures that may be taken to protect their clients' rights. Part V will review a number of common contracts that may be made relating to the supply and use of computer products and services.

(f) Taxation Issues Relating to Computer Technology Transactions

The final part considers some of the income and commodity tax issues relating to the transfer of various types of computer products. The tax implications of the servicing and development of computer products are also considered.

The foregoing have been selected as being the major areas of law that touch most directly on the growing computerization of modern Canadian society. However, computers affect virtually every area of law, and accordingly no claim is made that this text represents an exhaustive overview of the field of computer law in Canada. Moreover, the issues considered in each part are sufficiently complex that they cannot be considered exhaustively in a single work. Computer law is also rapidly changing and many of the legal issues have only recently been considered by the legislators and the courts of Canada. Accordingly, although in many case the legal principles that are to be applied have evolved over many decades, the ultimate resolution of these issues remains a matter of considerable uncertainty.

With the foregoing qualifications as to the intended scope of this text, the following section of this chapter will provide a rudimentary overview of the nature of computer technology as an introduction to the various substantive areas of law.

3. AN INTRODUCTION TO COMPUTER TECHNOLOGY

(a) Introduction

This section will provide a general overview of the basic elements and terminology of computer technology. While not intended to be exhaustive, it is hoped that such an overview will provide a background to understand the legal implications of the development, supply and use of computer technology.

A computer is "a device capable of converting information by applying prescribed logical and arithmetic processes to the information".[1] Specifically, a computer executes instructions to convert information from one form to another at high speed and without human intervention.

Computers may be divided into two basic types: analog and digital. An analog computer is a hard-wired combination of electrical components

---

1 D. Bender, "An Introduction to Computer Technology", in *Computers and the Law, An Introductory Handbook*, 3rd ed., R.P. Bigelow, ed. (C.C.H., 1981) at 10.

arranged in such a manner as to accept an input, "the problem", and transform it into an output, "the solution". Its flexibility is limited as changing its "program" involves rewiring its configuration. An analog computer uses an analog representation of data, that is, "a representation of the value of a variable by a physical quantity that is considered to be continuously variable."[2] The variable of interest in an analog computer may be represented by electrical characteristics, such as voltages.[3]

A digital computer, on the other hand, represents data by discrete numbers, and functions in terms of "how many" rather than "how much".[4] The structure of a digital computer is "binary" or "two-state", so that all data processing is performed based on the "on/off" position of switches within the computer (that is, according to whether or not it is transmitting electrical current). The two states of an electrical element may be represented by the digits "1" and "0", so that the state of a set of elements can be represented by a sequence of digits, for example, "1001".[5]

A single set of "on/off" elements (and of the information which it represents) is commonly referred to as a "bit" (or binary digit). A combination of several bits, which represents one character and which can be operated on as a unit, is a "byte". A "word" is a sequence of bytes which may represent a single program instruction. A digital computer may be classified according to the number of bits of information which it is capable of processing as a single unit (for example, 8-bit, 16-bit or 32-bit).

Analog computers were the first types of computers to be used, and have been supplanted in most applications by digital computers. Accordingly, the following discussion will focus primarily on the elements and operation of digital computers.

(b) The Elements of a Computer

Computer systems consist of five basic elements. These are: input (for passing data and instructions to the computer); memory or storage (used to retain information input into the computer and that generated by the computer); control (to manipulate data in main memory and on peripheral devices such as disc drives and to control the use of the logic element); logic (interfacing closely with the control element to perform arithmetical and logical operations); and output (to produce an external

---

2 *American National Dictionary for Information Processing Systems*, Computer and Business Equipment Manufacturers Association (American National Standards Institute, 1982) at 5.

3 Bender, *op. cit.*, at 10.

4 *Ibid.*

5 H. Brett & L. Perry, eds., *The Legal Protection of Computer Software* (ESC Publishing, 1981) at 4.

manifestation of the results of the data processing operation).

(i) *Input*

The input element of a computer consists of a device whereby data are "read in" to the computer. The most common form of input device is a typewriter-style keyboard through which information is typed directly into the computer. Another form of keyboard input device is a keypunch which produces a series of punched cards which are then read into the computer. Two other common forms of input devices are magnetic disc drives and magnetic tape drives. Other forms of input devices are optical character readers capable of reading typewritten or graphic information, voice input devices and magnetic strip readers on credit cards.

Input devices may be an integral part of the hardware components of a computer or may be connected to the rest of the computer by cable. Alternatively, input devices may be situated in remote locations and information may be transmitted to the computer by telecommunications facilities such as microwave transmission, satellite, light pulses transmitted along optic fibres or directly-wired computer data transmission lines.

(ii) *Memory*

Memory refers to any device forming part of a computer system which can be used for information storage. Any such device may contain two categories of stored information, namely, programs (instructions to the computer) and data (the information on which programs operate to achieve results).[6]

The two types of memory devices in a computer are main (also known as central or internal) and secondary or auxiliary. Main memory is rapidly accessible and contains the operating systems programs (that is, to control the main functions of the computer), commonly used applications programs (that is, to perform specific tasks) and all data being processed at any given point in time. Instructions and data from main memory can be loaded directly into registers from which the instructions can be executed or the data can be operated upon.[7] Access time for main memory is generally

6 The term "data" is also used to mean any physical representation of information and as such may include representations of programs as well as other information; however, in accordance with the common terminology of the computer industry, the word "data" will be used in this text, unless otherwise indicated, to refer to information or representations of information on which programs operate rather than the programs themselves.

7 *Vocabulary of Data Processing*, International Organization for Standardization, International Standards Organization, cited in Rosenberg, *Dictionary of Computers, Data Processing and Telecommunications* (Wiley & Sons, 1984) at 306.

measured in nanoseconds (billionths of a second) or microseconds (millionths of a second).[8]

Less commonly used data are stored in secondary or external memory devices which are accessible only through input-output channels. Access time for secondary storage is measured in milliseconds (thousandths of a second) or slower. Auxiliary or secondary storage is largely electromechanical whereas main memory is usually electronic.[9]

Memory may also be categorized by mode of access. "Random" or direct access is the facility to obtain data from or to enter data into a storage device independent of the location of the information being acquired. Serial or sequential access, on the other hand, requires that each record be searched in turn based on a reference to data which have been previously accessed. Accordingly, access time in the latter case is a direct function of a particular record's location in memory. The term "random access memory" ("RAM") is also commonly used to refer to "read-write memory", that is, that form of memory which may have new information stored on it each time the computer is used.

The main memory of a computer is usually of a direct access type and is generally comprised of either a semi-conductor or core storage device. A semi-conductor device is a circuit ordinarily embedded in silicon that either will or will not conduct electricity depending on its state. In the case of core storage, each bit is stored in a piece of iron-based ceramic, magnetized in one of two ways depending on the direction of electrical current. Direct access devices may also take the form of magnetic discs or diskettes (commonly called floppy discs) which are usually part of the auxiliary memory of a computer system. Discs and diskettes both resemble phonograph records, and have metallic recording surfaces on which information is recorded in concentric circles or tracks. They differ in that hard discs are generally of much greater capacity (and are consequently more common on mainframe and minicomputers) while floppy diskettes are non-rigid in construction, with smaller memory capacity and are usually restricted to microcomputers. Other forms of direct access memory are magnetic drums, magnetic data cell devices and magnetic bubbles (which use magnetic domains in crystalline sheets).[10] Examples of sequential access memory devices are magnetic tape and magnetic cassette tape drives.

Memory devices may also be distinguished according to whether the contents may be modified or altered, that is, whether they are erasable or non-erasable. Read-only or fixed memory ("ROM") is a device in which contents are either inherently non-erasable or non-erasable under partic-

8 Bender, *op. cit.*, at 12.
9 Bender, *op. cit.*, at 12-13.
10 Bender, *op. cit.*, at 14.

ular conditions such as when controlled by a lockout feature. In other words, the contents of a ROM can only be read. A ROM generally takes the form of a semi-conductor chip which is incorporated as part of the electronic circuitry of a computer. Similar memory devices are "Programmable Read Only Memory" ("PROM's") which can be programmed after manufacture but which cannot be subsequently altered, and "Erasable Programmable Read Only Memory" ("EPROM's") which can be programmed after manufacture but whose contents can subsequently be erased so that the chip can be re-programmed. All such forms of memory should be contrasted with read-write memory or RAM devices, most of which are "volatile" and lose the information stored thereon when the computer is turned off.

Operating systems and commonly used language compilers may be stored in ROM in main random access memory. Statistical and research data bases which are not meant to be altered may be stored in ROM devices in auxiliary memory. Information in erasable memory devices, on the other hand, may be removed from magnetic storage by changing the storage medium to an unmagnetized state or a predetermined magnetized state.

(iii) *Logic Hardware*

Logic hardware devices are part of a computer's central processing unit ("CPU") and contain the mechanisms for performing arithmetic or logical operations on data to achieve a certain result. The logic hardware of a computer simulates arithmetic operations by symbolic logic (or "Boolean algebra"). Boolean algebra is the construction of mathematical and logical expressions using the two digits 1 and 0, being the only digits used in binary or base two arithmetic. All early computers operated on this binary code system as each 1 or 0 could be represented by a core which was either electrified or not. This explains the large size and cost of early computers as the size of memory required was directly proportional to the volume of data to be processed. Virtually all modern computer systems utilize hexadecimal (or base 16) arithmetic operations made up of the digits 0 through 9 and the letters A through F.

In either system, symbolic logic sequences are constructed which represent the program being run. Such sequences consist of logical "elements", logical "statements", and logical "operators". A logical "element" is a proposition which may take one of two values, true (1) or false (0). A logical "statement" is a combination of two or more logical statements connected by a logical "operator". The logical operators are "OR" (logical inclusive — either or both), AND (logical connective — both) and NOT (logical inequation).[11]

11 Rosenberg, *op. cit.*, at 290.

The CPU of a computer is constructed to simulate the rules of symbolic logic. An open switch, in which there is no current flow, corresponds to a false value, while a closed switch, in which current is flowing through the circuit, corresponds to a true value. There are also three types of circuits or gates, an "AND" circuit, an "OR" circuit, and an inverter or "NOT" circuit corresponding to the three types of logical operators. These three types of circuits are also used in various combinations such as NAND (NOT-AND) and NOR (NOT-OR) to effect logical operations.[12]

(iv) *Control*

The control elements of a computer interpret and implement the instructions of a user, such as instructions to perform direct arithmetic operations, to transmit instructions to and from memory and to control the input and output devices. The control elements of a computer perform in two cycles, namely instruction cycles (whereby the computer determines the next step in the program) and execution cycles (whereby the computer executes that step).

(v) *Output*

The most common output element of a computer is a printer, which produces a durable record or "hard" copy, usually on paper media, of information generated by a computer. Printers vary widely in terms of reliability, speed, print quality, and, of course, cost. "Impact" printers usually produce the highest quality of print, with the most expensive models capable of printing a few thousand lines per minute. The next most commonly used form of printer is the dot matrix printer in which each character printed is composed of a matrix of dots burned into the paper, and which produce print quality generally inferior to that of impact printers. Other types of printers used include xerographic printers (especially for producing computer graphics), photographic printers (such as typesetting machines) and flat bed printer/plotters used for producing high quality graphic output.[13]

It is neither desirable nor necessary to have a hard copy of all information generated by the computer. For an examination of transitory information, the most commonly used output device is a Cathode Ray Tube ("CRT") which displays information by controlled electron beams in a manner similar to a television screen. Each screen of information is retained for the time necessary to complete the output on the screen

12 See Bender, *op. cit.*, at 16.

13 See Hoffman, "Computer Input and Output Devices", in Bigelow, *supra*, footnote 1, at 23-27.

after which it is erased to be superseded by another screen. CRT's vary in quality and sophistication from home television sets to high quality graphic output screens.

A CRT may also form part of a video terminal which includes a keyboard as an input device for sending information to the computer. Such terminals may be connected to the central processing unit of a computer by telecommunicating facilities or by a local area network ("LAN").

In the preceding discussion, it has been assumed that the various elements of a computer are made up of separate devices performing discrete functions. In actual fact, however, such elements may be combined in a device which performs more than one function. This is especially true in the case of memory, logic and control elements which are usually combined as part of the central processing unit.

(c) The Historical Evolution of Computers

Computers had their origins in mechanical tabulating machines which were manufactured in the late 19th century and first used to tabulate the results of the 1890 United States Census. The machines were manufactured by the Tabulating Machine Company (a predecessor of International Business Machines Corporation) and were capable of reading data contained on punched cards which were then processed using sorters and printers. The first electrical computers were used in the fields of science and engineering where mathematical expressions were modelled by the electrical circuits of the early analog computers. Of course, if the mathematical model was changed, it was necessary to rewire the entire computer, thus severely limiting its flexibility. The first digital computers had main memories comprised of magnetic cores (from which the term core memory was derived), and were able to perform their operations by either magnetizing or not magnetizing each core.

The evolution of computers has been marked by major reductions in size, energy and cost required to obtain the same computational power. Computers have evolved through four chronological generations, which have used increasingly sophisticated types of electronic circuitry. The first generation of computers, using vacuum tube technology, began with a system constructed by John Atanasoff in 1939. The second generation of computers, based on transistors, began with the Univac Solid State 80 in 1958 and continued with products developed by other manufacturers until approximately 1965. The third generation of machines was based on integrated circuitry and began with IBM's System 360 in 1964. The fourth generation of computers consists of machines using Large-Scale (LS) and Very Large Scale (VLS) integrated circuitry, in which large numbers of circuits are encoded on a single semiconductor chip.

The "fifth generation" of computers are those which are capable of "symbolic inference . . . and the symbolic representation of the knowledge to be used in making inference."[14] These so-called "artificial intelligence" computers are based on the principle that a computer can be programmed to make inferences in a manner approximating human reasoning. The development of computers which are capable of performing such knowledge-based inferences requires a substantial increase in computing capabilities from those of ordinary computers.[15]

(d) An Overview of Computer Terminology

(i) *Computer Software and Programs*

Although the term "software" is most often used to refer to computer programs and accompanying documentation, it may also be used to refer to computer data bases or indeed any "information" embodied in computer media or concerned with the operation of a computer system. However, the legal concepts applicable to these different types of computer software vary considerably. Accordingly, the term software will be used in this text in the narrower sense to refer only to computer programs and documentation.

A computer program may be defined as any set of instructions used to operate a computer. In other words, a program represents a sequence of operations to be performed by a computer without human intervention.[16]

The process of producing a program takes place in several stages. The first stage involves problem definition and flow charting. At this stage of development, a schematic representation of the logic or algorithm underlying a program is produced. A flow chart graphically illustrates the step-by-step procedures to be followed in solving a problem under any set of circumstances. In conjunction with the "flow-charting" of a program, the programmer must determine the computer hardware on which the program is to operate, any auxiliary programs with which the program is to be used, the specific programming language to be used, and the nature and form of the input and output. Such information may be compiled as a set of technical specifications which describe in detail the manner in which the program is to operate and the tasks which it is to perform.

The next stage in the creation of a program is that of "coding", that is, the actual writing of instructions in a standard programming language such as FORTRAN or COBOL. These are examples of high order or higher

---

14 E. Feigenbaum & P. McCorduck, *The Fifth Generation* (Signet, 1984) at 312.

15 For a discussion of artificial intelligence, see A.J. Cordell, "Artificial Intelligence" in *The Uneasy Eighties, The Transition to an Information Society* (Science Council of Canada, Background Study 53, 1985).

16 See Halton, "The Anatomy of Computing" in *The Information Technology Revolution*, T. Forester, ed. (MIT Press, 1985) at 6.

level programming languages which utilize commands and short form expressions in a symbolic language similar to English. A program written in a high level programming language is commonly referred to as source code and is comprehensible to any person familiar with that language.

The program in source code form is input into the computer where a special program known as a compiler or an interpreter is used to convert the program into "machine-readable" or "object code" form. The two types of programs differ in that an interpreter transforms each line of source code (usually corresponding to a separate instruction) into a line of machine code, while a compiler transforms the entire set of source code into an object code program based on the over-all logic structure of the program. Consequently, compilers have much faster execution times than an equivalent interpreter for the same language.

Whether an interpreter or compiler is used, the program in source code form is converted into a machine-readable (or object code) program which may be represented or written in either a binary or hexadecimal notation. The binary system uses only the two digits "1" and "0" corresponding to the two states of "electronically on" or "electronically off" which can be understood directly by the computer. The hexadecimal notation is a base 16 system in which the various electrical states are represented by any combination of digits from "0" to "9" and letters from "A" to "F".

The process of converting a program from source code to object code form involves two kinds of transformation. The first is that there is a conversion of the program from a form in which it was written by a human programmer to a series of electrical states so that it can be executed directly by the computer. In addition, there is a transformation of the expression of the program, in that the digits or symbols by which the object code may be expressed or represented are different from the kind of programming language in which a source code is written. Thus, even where a program in object code form is "read" out of a computer to be produced in printed form, the expression of that program is different from that of the source code.[17]

The final phase in the development of computer programs involves the preparation of program documentation. The term documentation refers to any writings which describe the operation and/or applications of a program. The complexity and scope of the documentation that accompanies a program will vary depending on the needs of the intended user. Technical documentation, containing detailed descriptions of the internal operations of a program, is generally directed towards computer programmers and other specialists who are responsible for correcting errors

17 For a discussion of the distinction between source code and object code programs, see *Apple Computer Inc. v. Mackintosh Computers Ltd.* (1986), 10 C.P.R. (3d) 1 (Fed. T.D.).

or "de-bugging" programs and for developing enhancements and new features. As noted above, such technical documentation or "design specifications" may be prepared, at least in draft form, prior to the actual writing of the programs. Less technical operational documentation, on the other hand, may consist of a user's manual describing the nature and form of the input and output of a program, together with illustrative examples of its applications. As corrections or enhancements are made to any computer program, it is usually necessary to develop updated versions of the accompanying program documentation.

(ii) *Application versus Operating System Programs*

Computer programs fall into two broad categories according to their functions. An application program is designed to perform a specific task, usually user oriented. Examples include word processing, spread sheet analysis, game playing or the control of the operation of an industrial process. An operating system program, on the other hand, controls the basic operation of the computer, and manages its internal functions. For example, an operating system program may be used to control access to information contained in auxiliary storage devices or facilitate the use of application programs.

Both categories of programs may be embodied on different kinds of machine-readable media in the form of source and/or object code. The most common types of such media are punch cards, magnetic tapes, magnetic discs, floppy discs, magnetic drums or magnetic bubbles. Both types of programs may also be embodied as a permanent part of the electronic circuitry of a computer. Such programs are usually engraved in circuits which form part of the ROM. They can also be incorporated in a PROM (programmable read-only memory) or an EPROM (erasable programmable read-only memory).

(iii) *Hardware versus Software*

There is a clear conceptual distinction between the hardware and the software components of a computer system. Thus, while hardware consists of the electronic circuitry and other physical apparatus of a computer, software consists of the programs or instructions used to operate a computer, as well as the documentation which relates to the operation of a computer.

As a practical matter, however, the distinction between hardware and software has become blurred insofar as computer programs embedded in ROM form an integral part of the computer hardware. Various control and/or operating system programs in particular are quite commonly built into the computer since the information making up such programs must be accessed every time the machine is used, regardless of the application program being used. Computer programs which are incorporated as part of the electronic circuitry of a computer are commonly referred to as firmware.

Even for programs which are not a part of the permanent hardware of a computer, the hardware/software distinction may be difficult to make, in that all programs and other information must be stored on some form of physical device such as tapes, discs or cartridges. For purposes of understanding the legal incidents of computer technology, it is important to recognize the distinction between the representations of information which are embodied in various physical devices and the tangible devices themselves.

(iv) *Computer Data Bases*

In the broadest sense, a computer data base is any collection or aggregation of information which is represented, stored or expressed in machine-readable form. The term data base is generally used in a narrower sense, however, to refer to an organized collection or compilation of information, such as a customer list arranged in a certain way, a collection of cases that may be searched by a legal research system, an organized compilation of articles written in a newspaper, or a directory of information such as airline schedules or accommodations available in any given city.

Whatever the information that they contain, all data bases consist of highly ordered collections of data. Each unique item of data (for example, a customer name) has its own record, and may have a number of different fields associated with it (for example, customer address, special shipping instructions, etc.). The unique aspect of modern data bases is that the records may be sorted, with relative ease, on one or more key fields. Data bases may vary in size from the financial information of a small firm to the taxation data base maintained by Revenue Canada.

The information in a computer data base may be collected by a corporation to facilitate the conduct of its internal business. Alternatively, a data base may be established specifically to provide information to third parties for a fee. In the latter case, users may wish to obtain access to the data base by remote communications facilities. The user in such event will be concerned with ensuring the compatibility of the various hardware and software components making up the network of computer/communications facilities whereby such access is to be obtained. Specifically, the user will be concerned with determining what communications standards must be complied with and what passwords must be used to obtain access to remote data bases. A primary objective of communications and computer companies in the late 1980's is the implementation of a common international standard, that is, the Integrated Services Digital Network, to allow equipment to be plugged into a common jack to send voice, data or images anywhere in the world.[18]

(v) *Computer Systems*

Computer systems may be divided into the three categories of mainframes, minicomputers and microcomputers (or personal computers). A mainframe computer has virtually unlimited capacities in terms of volume of work, types of applications and number of users it can handle. However, it is the most expensive form of computer and usage is usually restricted to governmental agencies and large corporations, often with multiple user access requirements.

Minicomputers offer a number of the benefits of mainframes but on a smaller scale. They are commonly used by medium sized businesses, often with multiple user requirements. Marketing of minicomputers is usually carried out as a turnkey operation by independent dealers, who offer their own packaged and/or customized software in conjunction with a given piece of computer hardware.

The microcomputer (also commonly referred to as the personal or business computer) was made possible by the use of small semiconductor chips which can be mass-produced. A wide variety of application programs have been developed for microcomputers, providing such functions as financial analysis, word processing, education or training and game playing. The power and flexibility of such computers also span a very wide range, with the most powerful microcomputers providing computing capabilities similar to older generations of minicomputers or even mainframes. They are used primarily in a single user situation, although the development of Local Area Network (LAN's) makes possible the linking together of a number of microcomputers controlled by a minicomputer or another microcomputer. The microcomputer market is highly competitive, and rapid advances in production technology and the proliferation of clones have lead to very sharp decreases in the price of computing power.

(vi) *Computer Services*

In addition to hardware and software, a large part of the computer industry in Canada is devoted to the provision of computer services. Computer services may include software development, ongoing support services, in-house programming and complete data processing services such as those provided by a computer service bureau. In the latter case, the service bureau generally provides all facilities necessary for processing

18 See "Scientists try to find a link for phones", *The Globe and Mail*, June 24, 1986, p. B2, col. 4.

a customer's data, including the complete hardware configuration, all necessary operating system and application programs, trained personnel to run data on the computer, and input and output preparation services.

Processing services by a service bureau may be provided over-the-counter (in which event the customer provides the data in a form suitable for processing while the service bureau provides the finished product on tapes, discs or paper) or by remote on-line access (in which event the customer has direct telecommunications access to the computer facility). In either case, the service bureau will be concerned with preventing unauthorized access to its computer facility, while the customer will be concerned with maintaining the security of its data and its unauthorized or inadvertent disclosure or destruction.

(vii) *Data Transmission Facilities*

In the early days of computer communications, the only method for the transmission of data was by telephone networks. Telephone systems were designed, however, specifically for voice transmission rather than the transmission of data in digital form. Accordingly, a device called a modem (modulator/demodulator) was required to transform information in digital form into analog signals and back again to digital form at the receiving end. At present, however, nearly one-half of telephone communications facilities use digital rather than analog signals.[19] The use of such digital facilities together with computerized switching systems have increased the message handling capacity of telephone networks as well as the quality of data transmission.

Historically, telephone messages were transmitted along copper cables. In recent times copper cable transmission systems have been supplanted in large part (except for short distances) by microwave and satellite transmission systems. A microwave transmission system consists of two points "linked" together by a number of transmitters and receivers to form a transmission route. Such chains of transceivers are necessitated by the relatively short range of microwave signals. Furthermore, the terrain over which the microwaves travel can often cause fading or "hits" on transmitted microwaves. While this has relatively few consequences in terms of voice communication, the absence of even one or two bits of information can often prove critical in altering the meaning of transmitted data. Satellite transmission overcomes some of the limitations encountered in microwave transmission facilities in that the cost of such transmission is distance-insensitive, it is less susceptible to interference, and it provides

19 See Branscomb, "Moving Data Between Computers", in Bigelow, *supra*, footnote 1, at 27-34.

the capability of transmitting messages to a number of earth receiving stations simultaneously.

A recent form of data transmission (voice and digital) consists of optical fibre transmission systems. In these systems the various signals are converted to pulses of laser light which are transmitted along glass fibres of microscopic diameter. Such fibres have a far greater data handling capacity than microwaves, satellites or conventional cables, while providing considerably less interference and signal distortion. This method of data transmission promises to herald new standards in the speed, quantity and quality of data transmission.

In the late 1970's, new electronic text transmission services were developed to provide home information services. The two principal types of text transmission systems are teletext (in which text is broadcast, often as part of the vertical blanking interval (VBI) of an ordinary television signal) and videotext (an interactive communication system in which a decoder attached to a television set can be used to determine the content of the signal to be transmitted by telephone or cable). Videotext services may also be provided in conjunction with home computers in the place of decoders attached to television sets. Canada, through its Telidon technology, was a world leader in the development of interactive text transmissions systems. However, such systems have failed to gain wide acceptance in consumer and business markets.[20]

20 "The fading of a high-tech dream", *The Toronto Star*, June 15, 1986, p. F2, col. 1-4.

2

The Constitutional Parameters for the Regulation of Computer Technology in Canada

1. INTRODUCTION

Because Canada is a federal state, legislative jurisdiction to regulate computer technology and to enact laws governing the rights of the creators and users of such technology is divided between the federal and provincial levels of government. Although there are few decided cases which have considered the scope of such jurisdiction in the context of computer technology, the underlying principles have been considered in other areas.

One of the major constitutional issues involves the jurisdiction of each level of government in Canada over undertakings used for both communications and data-processing or other computer functions. In this regard, computer facilities may be an integral part of a network used for the transmission of information, whether in the form of voice, text or data, which may extend across provincial and national boundaries.

As a practical matter, no major jurisdictional disputes have arisen to date with respect to the regulation of either the physical undertaking of computer facilities or the contents of computer data banks *per se*. However, such disputes may arise in the future if the Canadian Parliament extends the powers which it has exercised to date in the area of communications.

A second major constitutional issue involves the powers of each level of government in Canada with respect to the regulation of the rights of creators, distributors and users of computer data and programs. Such rights arising from intellectual property laws involve matters within both federal and provincial jurisdictions. For example, insofar as the rights of program owners are based on contractual or common law actions for breach of confidence or protection of trade secrets, they involve matters of "property and civil rights in the Province."[1] On the other hand, such rights may be based on patent, copyright, industrial design or trade mark law, all of which are within federal jurisdiction. An increasingly important issue with respect

---

1 Constitution Act, 1867, subs. 92(13).

to the protection of information in computerized form is the application of criminal law to the protection of the rights of the owners of computer programs as well as computer data base compilations.

The third major area of concern with respect to the constitutional parameters for the regulation of computer technology involves the Canadian Charter of Rights and Freedoms. Although there are few provisions of the Charter which appear to be directly applicable to computer technology, certain provisions will be of increasing importance in determining the manner in which the owners of computer programs and data may be able to enforce their rights.

2. THE REGULATION OF COMPUTER/COMMUNICATIONS FACILITIES

The regulation of communications in Canada is a matter of both federal and provincial responsibility. This follows from subsection 92(10) of the Constitution Act, 1867, which provides that the legislature of each province has exclusive jurisdiction over "local works and undertakings" other than those coming within one of three classes. One of the classes which is excepted from provincial jurisdiction under paragraph 92(10)(*a*) consists of "Lines of Steam or other Ships, Railways, Canals, Telegraphs and *other Works and Undertakings connecting the Province with any other or others of the Provinces, or extending beyond the Limits of the Province.*"[2] Other works and undertakings which are excluded from provincial jurisdiction are those declared by Parliament to be for the general advantage of Canada or for the advantage of two or more provinces.[3] By virtue of subsection 91(29) of the Constitution Act, works and undertakings which are excepted from provincial regulatory authority come under federal jurisdiction.

In view of the nature of the undertakings which are specifically listed in paragraph 92(10)(*a*) (that is, ships, railways, canals and telegraphs) it would appear that the term "other works and undertakings" must be interpreted restrictively to apply only to the works and undertakings involved in transportation or communication.[4] A distinction has also been made between "works" and "undertakings", in that a work is a physical thing whereas an undertaking is "an arrangement under which . . . physical things are used."[5]

2 Constitution Act, 1867, para. 92(10)(*a*) (emphasis added).
3 *Ibid.*, para. 92(10)(*c*).
4 See Hogg, *Constitutional Law of Canada*, 2nd ed. (Carswell, 1985) at 486.
5 *Re Regulation and Control of Radio Communication in Canada*, [1932] A.C. 304 at 315 (P.C.).

In the case of *Re Alberta Government Telephones and C.R.T.C.*,[6] the test that was applied in determining whether the federal Parliament has jurisdiction over a communications undertaking is whether the undertaking engages in a "significant amount of continuous and regular extra-provincial activity." Reed J. of the Federal Trial Court held that "this test does not demand that the extra-provincial portion of an undertaking's enterprise must dominate but it does mean that occasional and irregular extra-provincial activity will not lead to a characterization of the enterprise as falling within federal jurisdiction."[7]

The issue in dispute in the *Alberta Government Telephones* case was whether AGT was a local work and undertaking and accordingly excluded from regulation by the CRTC under the federal Railway Act. In her judgment, Madam Justice Reed found that AGT's undertaking could not be characterized as local since it was operated as an integral part of a national telecommunications system.

Although AGT did not have any facilities outside Alberta, the fact that it offered its customers inter-provincial and international telecommunications services through the Trans-Canada Telephone System (now Telecom Canada) was sufficient to make it subject to federal jurisdiction. Although AGT could retain control over its own facilities as a legal proposition, as a practical reality it could not separate itself from the joint Telecom Canada enterprise without destroying its telecommunications system in its present form. Reed J. relied on the decision of the Supreme Court of Canada in *Capital Cities Communications Inc. v. C.R.T.C.*,[8] in which it was held that inter-provincial undertakings are those "which reach out beyond the Province in which their physical apparatus is located."[9]

The Federal Court of Appeal upheld the finding of Reed J. that AGT was an inter-provincial work or undertaking. However, the Court rejected her finding that AGT was entitled to exercise provincial Crown immunity as an agent of the provincial government. By operating a federal undertaking, AGT had stepped outside the authority conferred on it by the Alberta Legislature and accordingly was not entitled to Crown immunity.

In general, federal jurisdiction will extend to all aspects of a business carried on by an organization which engages in inter-provincial activity. However, different considerations may apply if there is a fundamental separation between the different activities in which the organization is engaged. In the case of *C.P.R. v. A.G.B.C.*,[10] it was held that in operating

6 (1984), 15 D.L.R. (4th) 515 at 532 (Fed. T.D.); reversed on other grounds (1985), 24 D.L.R. (4th) 608 (*sub nom. Re CNCP Telecommunications and Alta. Govt. Telephones*) (Fed. C.A.).

7 *Ibid.*, at 532 (Fed. T.D.).

8 [1978] 2 S.C.R. 141 (S.C.C.).

9 *Ibid.*, at 159.

10 [1950] A.C. 122 (P.C.).

a hotel the Canadian Pacific Railway Company was engaged in a separate undertaking from that of operating an inter-provincial railway. It was noted that the two parts of the enterprise were carried on as separate businesses. On the other hand, when two enterprises are separately owned, they may nevertheless be considered part of a single inter-provincial undertaking. In *R. in Right of Ontario v. Board of Transport Commissioners*,[11] the use by a provincial railway commuter service of a local part of an inter-provincial railway line made the commuter service part of the inter-provincial system.

The scope of the matters coming within federal jurisdiction under paragraph 92(10)(*a*) of the Constitution Act, 1867 has been considered in several cases relating to broadcasting. In *Capital Cities Communications v. C.R.T.C.*[12] federal jurisdiction was found to extend over the entire field of radio broadcasting.[13] Similarly, in the case *Re C.F.R.B. and A.G. Can.*,[14] federal jurisdiction was held to extend over "the whole of the undertaking of broadcasting."[15] Such jurisdiction included the regulation of not only the technical aspects of transmitting and receiving radio signals but also the "intellectual content of radio communication."[16] The Court thus upheld the validity of a provision of the federal Broadcasting Act which prohibited the broadcasting of partisan political messages on the day before an election.

Federal jurisdiction over broadcasting has also been found to extend to the activities of cable operators that receive and re-transmit off-air signals. In the case of *Capital Cities Communications v. C.R.T.C.*,[17] the Supreme Court of Canada found that a cable distribution system which receives broadcast signals and transmits them to the homes of cable subscribers was part of a communications undertaking within the competence of the federal Parliament. The Court noted that it would be incongruous "to deny the continuation of regulatory authority because the signals are intercepted and sent on to ultimate viewers through a different technology."[18] In a decision handed down the same day, *Public Service Board v. Dionne*,[19] the Court denied provincial regulatory authority with respect to the activities of a cable television system within the province. The Court noted that the cable undertaking was part of "what is

---

11 [1968] S.C.R. 118 (S.C.C.).
12 [1978] 2 S.C.R. 141 (S.C.C.).
13 *Ibid.*, at 162.
14 [1973] 3 O.R. 819 (Ont. C.A.); leave to appeal to Supreme Court of Canada, refused, November 13, 1973.
15 *Ibid.*, at 822.
16 *Ibid.*, at 824.
17 *Op. cit.*, footnote 12.
18 *Ibid.*, at 162.
19 [1978] 2 S.C.R. 191 (S.C.C.).

functionally an inter-related system."[20] Accordingly, jurisdiction could not be divided between the federal Parliament and the provinces.

Notwithstanding the wide jurisdiction that the federal Parliament may exercise over broadcasting and cable undertakings, the program content of such undertakings may nevertheless be subject to limited provincial regulation. In *A.G. Quebec v. Kellogg's Co. of Canada*,[21] the Supreme Court of Canada upheld a Quebec law that prohibited the use of cartoons in advertising intended for children. Since the main purpose of the law was to control commercial activity in the province, there was no objection that it incidentally affected what could be shown on television.

The foregoing cases provide some indication of the extent to which the operations of companies offering communications and/or computer services may come within the jurisdiction of the federal Parliament. Such jurisdiction will likely be found to exist where data processing and other computer services are offered by a business or enterprise as part of a communications network having a regular and continuous inter-provincial aspect. Enterprises offering such services may be subject to federal jurisdiction notwithstanding that they are separately owned. However, in the case of a company having computer facilities only in one province which uses the communications facilities of another company, there would have to be an integration of management or operational dependence or control to constitute what would otherwise be a "local" system as part of an inter-provincial undertaking.

Among the factors that may be considered in determining whether computer facilities with added data communications links are subject to federal jurisdiction are the following: the degree of integration of data communications and data processing facilities, the situs of ownership and control, the mode of electronic transmission, the degree to which remote links are provided by a telecommunication service and whether the service is federally or provincially regulated, whether the remote facilities are co-used for other purposes, whether the computer/communications undertaking can be separated functionally or operationally, whether the links extend over provincial boundaries, the frequency with which inter-provincial links are used, and the identification of who manages, markets and controls the undertaking.[22]

In applying these factors, as long as computer facilities are used solely for the internal purposes of a local business which is itself within provincial jurisdiction, it would not be considered as part of a federal undertaking.

20 *Ibid.*, at 197.

21 [1978] 2 S.C.R. 211 (S.C.C.).

22 See "Constitutiuonal Considerations" from *Branching Out: Report of the Canadian Computer/Communications Task Force* (Ottawa: Department of Communications, May, 1972) at 31.

Similarly, if a computer service bureau offers services on an "over-the-counter" basis, without providing remote on-line access, it would generally be considered a local work or undertaking. This would also be the case if the service bureau provides remote on-line access that is restricted to a particular province or that involves communications across national or provincial boundaries only on an irregular basis. It should not matter in this regard whether the computer service bureau offers its services to customers by means of leased private lines, by dedicated lines linked to connection centres in various cities or by use of the regular lines of a telephone system.

At the other extreme, in the case of a communications carrier such as Bell Canada which is subject to federal jurisdiction, that jurisdiction would extend to "all matters which are a vital part of the operation of an inter-provincial undertaking as a going concern."[23] Accordingly, to the extent that such an enterprise offers data processing or other computer services to its customers, that part of its undertaking would be subject to federal jurisdiction. Similarly, if a provincial telephone company such as AGT becomes involved in the designing, managing and marketing of integrated computer communications facilities in more than one province, the entire undertaking of such enterprise may be subject to federal jurisdiction.[24] Federal jurisdiction would also extend to the data processing operations of other businesses such as banks which are subject to federal control.

To the extent that the federal Parliament may exercise jurisdiction over telecommunications carriers or private companies which offer a combination of computer and communications services on an inter-provincial basis, such jurisdiction would appear to extend not only to the technical and physical aspects of such enterprises but also to the contents of computer data bases which they establish, maintain and market. In this regard, in the same way as the jurisdiction of Parliament over broadcasting undertakings extends to program content, so too its jurisdiction over communciations/computer facilities should extend to the information content which is processed or communicated by such facilities.

To date, federal jurisdiction has not been exercised to regulate the activities of private companies offering computer services. In a decision announced in July, 1984,[25] the Canadian Radio-Television and Telecommunications Commission decided that "enhanced services" provided by parties other than common carriers would not be regulated and would be subject to a competitive market. An enhanced service was defined as "any offering over the telecommunications network which is more than

23 *Comm. du salaire minimum v. Bell Telephone Co. of Can.*, [1966] S.C.R. 767 at 772 (S.C.C.).

24 See *Branching Out, op. cit.*, at 33.

25 The "Enhanced Services Decision", C.R.T.C. Telecom Decision 84-18, July 12, 1984.

a basic service." A basic service, in turn, was defined as "one that is limited to the offering of transmission capacity for the movement of information," that is, that involves a transmission capability that is virtually transparent in terms of time and substance.[26] An enhanced service would thus include a basic service combined with a computer processing application that acts on the content, code, protocol or other aspects of the subscribers' information.[27]

The Commission imposed a number of restrictions on the right of carriers to provide enhanced services. These restrictions include a prohibition against electronic publishing involving editorial control over content, a prohibition against the creation and distribution by carriers of their own data bases, a provision whereby carriers which provide enhanced services are deemed to have acquired the underlying basic services at tariff rates, and a prohibition against carriers funnelling charges for terminal equipment with other components of an enhanced service. However, no restrictions were imposed on the provision of enhanced services by non-common carriers.

3. THE REGULATION OF THE RIGHTS ATTACHED TO COMPUTER PROGRAMS AND DATA

The second major constitutional issue that arises in the area of computer technology involves the respective jurisdictions of the federal Parliament and provincial legislatures to enact legislation for the protection of the rights of creators and distributors of computer programs, data and other technology.

As a general rule, the jurisdiction of the provinces over "property and civil rights in the Province" extends to all matters comprising "primarily proprietary, contractual or tortious rights [which] exist when the legal rule stipulates that in certain circumstances one person is entitled to something from another."[28] Such matters will generally include all causes of action involving claims for breach of contract, claims for breach of a confidential or fiduciary relationship and claims for tortious interference with private rights.

The federal Parliament also has a limited jurisdiction to create civil remedies insofar as they are incidental to legislation enacted under one of the federal heads of power set out in section 91 of the Constitution Act, 1867. For example, Parliament may create civil remedies as incidental

26 See C. Bredt, "Emerging Issues: The Convergence of Computers and Communications", Paper presented at Canadian Bar Association, Program on Basic Computer Law, Toronto, November 23, 1985, at 9-10.

27 *Ibid.*, at 10.

28 Hogg, *Constitutional Law of Canada*, 2nd ed. (Carswell, 1985) at 455.

to its power to legislate with respect to patents of invention and discovery,[29] copyrights[30] and the criminal law.[31] Such legislation may also be justified based on the federal trade and commerce power under subsection 91(2), which has been found to be the source of constitutional power for federal legislation relating to trade marks.[32]

An example of valid federal legislation affecting property rights is section 21 of the Copyright Act. Section 21 provides that all infringing copies of any work in which copyright subsists and plates used in the production of such copies are deemed to be the property of the copyright owner who may take proceedings for the recovery of possession or in respect of their conversion. In the case of *Bishop v. Tele Metropole Inc.*[33] the Court rejected the contention that section 21 was *ultra vires* Parliament as involving the alteration of substantive property rights. The Court found that in the exercise of its exclusive jurisdiction over copyright, Parliament could incidentally affect property rights which in other circumstances would fall within provincial jurisdiction. Although copyright represents an intangible property right, the protection of such rights may make it necessary for Parliament to limit or affect incidentally rights in tangible property such as books or records.

Federal jurisdiction with respect to the creation and enforcement of civil remedies relating to certain trade practices has been found to be limited in the context of trade marks legislation. Section 7 of the Trade Marks Act[34] sets forth the following prohibitions:

> 7. No person shall
> (*a*) make a false or misleading statement tending to discredit the business, wares or services of a competitor;
> (*b*) direct public attention to his wares, services or business in such a way as to cause or be likely to cause confusion in Canada, at the time he commenced so to direct attention to them, between his wares, services or business and the wares, services or business of another;
> (*c*) pass off other wares or services as and for those ordered or requested;
> (*d*) make use, in association with wares or services, of any description that is false in a material respect and likely to mislead the public as to
> (*i*) the character, quality, quantity or composition,
> (ii) the geographical origin, or
> (iii) the mode of the manufacture, production or performance of such

29 Constitution Act, 1867, subs. 91(22).

30 *Ibid.*, subs. 91(23).

31 *Ibid.*, subs. 91(27).

32 See *Ref. re Dom. Trade and Industry Comm. Act*, [1936] S.C.R. 379 (S.C.C.), varied [1937] A.C. 405 (P.C.); *Motel 6 Inc. v. No. 6 Motel Ltd.*, [1982] 1 F.C. 638 (Fed. T.D.).

33 (1985), 4 C.P.R. (3d) 349 (Fed. T.D.).

34 R.S.C. 1970, c. T-10.

wares or services; or

(*e*) do any other act or adopt any other business practice contrary to honest industrial or commercial usage in Canada.

In the case of *MacDonald v. Vapour Canada Ltd.*,[35] the Supreme Court of Canada found that paragraph 7(*e*) of the Trade Marks Act was *ultra vires* Parliament. The Court noted that the prohibition set out in paragraph 7(*e*) was closely related to the tort of conversion involving persons in business.[36] The Court held that subsection 91(2) of British North America Act (now the Constitution Act, 1867) did not authorize "federal legislation that merely creates a statutory tort, enforceable by private action, and applicable, as here, to the entire range of business relationships in any activity, whether the activity be itself within or beyond federal legislative authority."[37] The Court noted that other provisions of section 7 could be upheld insofar as they rounded out "regulatory schemes prescribed by Parliament in the exercise of its legislative power in relation to patents, copyrights, trade marks and trade names."[38] However, after such other provisions were taken into account, there was no subject matter left to paragraph 7(*e*) in relation to such matters. Accordingly, paragraph 7(*e*) could not be relied on to restrain an alleged breach of contract and breach of confidence by a former employee when the issue of patent infringement which had been considered by the lower courts was no longer before the Supreme Court.

The validity of paragraph 7(*b*) of the Trade Marks Act was considered in the case of *Motel 6 Inc. v. No. 6 Motel Ltd.*[39] The Federal Court found that the section could not be upheld as being "necessarily incidental" to a general regulatory scheme to govern trade marks. In view of the essentially private nature of the method of enforcement, paragraph 7(*b*) was an intervention into matters of a local or private nature within the province. The court referred to the decision of the Federal Court in *S.C. Johnson & Son, Ltd. v. Marketing International Ltd.*,[40] in which paragraphs 7(*b*) and (*c*) were described, by way of *obiter dicta*, as a statutory embodiment of the common law action of passing-off which could not be validated as part of the regulatory scheme governing trade marks. Similarly, in the case of *Seiko Time Canada Ltd. v. Consumers Distributing Ltd.*,[41] the Ontario Supreme Court found that the whole of section 7 was

---

35 (1976), 22 C.P.R. (2d) 1 (S.C.C.).

36 *Ibid.*, at 22.

37 *Ibid.*, at 27-28.

38 *Ibid.*, at 34.

39 *Supra*, footnote 32.

40 (1977), 32 C.P.R. (2d) 15 at 31 (Fed. T.D.); reversed [1979] 1 F.C. 65 (Fed. C.A.); affirmed [1980] 1 S.C.R. 99 (S.C.C.).

41 (1980), 29 O.R. (2d) 21 (Ont. H.C.); affirmed (1981), 34 O.R. (2d) 481 (Ont. C.A.); varied [1984] 1 S.C.R. 583 (S.C.C.).

an invasion of provincial power. However, this issue was not expressly considered by the Supreme Court of Canada.

Notwithstanding the foregoing decisions, it would appear to be premature to conclude that section 7 of the Trade Marks Act is unconstitutional in its entirety.[42] In *Pilote Ready Mix Inc. v. Racois Construction Inc.*[43] the Federal Court of Appeal upheld paragraph 31.1(1)(*a*) of the Combines Investigation Act (now the Competition Act, 1986) which grants a civil right of action to a person who suffers loss or damage as a result of an offence in relation to competition. The Court in that case noted that paragraph 31.1(1)(*a*) had a rational functional connection with overall federal plans in relation to competition which came within the federal trade and commerce power.

Until the question of the constitutionality of section 7 is ultimately resolved by the Supreme Court of Canada or until the provision is amended to come more clearly within federal jurisdiction, there would be some risk in relying on the section as a basis for the protection of industrial property rights. Accordingly, apart from actions brought under other provisions of intellectual property legislation enacted by Parliament, owners and creators of computer programs and data would be advised to rely on contractual and other civil rights of action brought in provincial courts for the protection of their rights.

4. IMPLICATIONS OF THE CANADIAN CHARTER OF RIGHTS AND FREEDOMS

In any review of Canadian laws relating to the regulation of computer technology, consideration must be given to the provisions of the Canadian Charter of Rights and Freedoms. Particular reference should be made to section 2 of the Charter which gives to everyone certain fundamental freedoms, including freedom of thought, belief, opinion and expression, section 7 which confers on everyone "the right to life, liberty and security of the person and the right not to be deprived thereof except in accordance with the principles of fundamental justice," and section 8 which grants

42 See Bell and Probert, "The Constitutionality of Canadian Trade Mark Law" (1985), 4 C.P.R. (3d) 305. In *Riello Can. Inc. v. Lambert* (1986), 9 C.P.R. (3d) 324 (Fed. T.D.), paragraph 7(a) of the Trade Marks Act was found to be "a legitimate exercise of the federal legislative authority" when invoked in an action for damages in respect of false statements made as to infringement of the defendant's patents. By contrast, in *Asbjorn Horgard A/S v. Gibbs/Nortac Indust. Ltd.* (1986), 9 C.P.R. (3d) 341 (Fed. T.D.), and *Aca Joe Int. v. 147255 Can. Inc.* (1986), 10 C.P.R. (3d) 301 (Fed. T.D.), it was held, respectively, that para. 7(a) and paras. 7(b) and 7(c), were constitutionally invalid and that the Federal Court had no jurisdiction in respect of any claims brought under such provisions.

43 (1985), 8 C.P.R. (3d) 145 (Fed. C.A.).

to everyone the right "to be secure against unreasonable search or seizure."

Several of these provisions have been considered in cases involving copyright and other intellectual property rights. For example, in the case of *R. v. James Lorimer Co.*,[44] it was held that the defendant's right of freedom of expression did not include the right to reproduce and distribute materials published by the federal government. It was noted that the works in question involved the expression of the Crown in right of Canada rather than that of the defendant itself.

In the case of *Cat Productions Ltd. v. Macedo*,[45] the defendants alleged that the plaintiff was estopped from bringing an action for copyright infringement on the grounds that he had deprived the defendants of certain rights under sections 7 and 8 of the Charter by procuring the Metropolitan Toronto Police to seize the defendants' goods. The Court found that the allegations complained of did not raise a triable issue since the party who allegedly infringed the defendant's rights, namely, the Metro Toronto Police, were not a party to the action. The defendants were found to be invoking the Charter as an aid to bringing an extraneous matter which had nothing to do with the real merits of the controversy between the parties.

In *Smith Kline & French Laboratories v. A.G. of Canada*,[46] the plaintiff argued on the basis of section 7 and other provisions of the Charter of Rights and Freedoms and the Bill of Rights that subsection 41(4) of the Patent Act was *ultra vires* Parliament. Subsection 41(4) provides for the grant of a compulsory licence for any patented invention which is intended to be used for medicine. In considering the possible application of section 7 of the Charter, Strayer J. took the view that the concepts of "life, liberty and security of a person" have to do with the bodily well-being of a natural person and refer to freedom from arbitrary arrest and detention. Such concepts were found not to apply to describe any rights of a corporation or any purely economic interests of a natural person. Strayer J. found further that property rights were not implicitly protected by section 7, particularly in light of the fact that an amendment to include "property" in the protection of section 7 was withdrawn during consideration of the Charter by the Joint Parliamentary Committee on the Constitution. He distinguished in this regard certain statements by Wilson J. (Dickson C.J. and Lamer J. concurring) in *Singh v. Minister of Employment & Immigration*,[47] which indicated that the concepts set forth in section 7 were capable of a broad meaning.

The Charter of Rights and Freedoms does not confer an express right of privacy and it has been held that such right does not fall within the

---

44 (1984), 77 C.P.R. (2d) 262 (Fed. C.A.).

45 (1985), 5 C.P.R. (3d) 71 (Fed. T.D.).

46 (1985), 7 C.P.R. (3d) 145 (Fed. T.D.).

47 [1985] 1 S.C.R. 177 at 205 (S.C.C.).

phrase "security of the person" set forth in section 7 of the Charter.[48] The right to be secure against "unreasonable search" may in some circumstances be invoked to restrict the manner in which governments may collect information about private citizens to be held in governmental data banks and the uses to be made of such information, and to limit the cross-linking of data banks. However, it appears unlikely that such provision would have any application with respect to the regulation of data bases held by private companies.[49]

48 *Charbonneau v. College of Physicians and Surgeons of Ont.* (1985), 52 O.R. (2d) 552 (Ont. H.C.).

49 See Tarnopolsky and Beaudoin, The *Canadian Charter of Rights and Freedoms* (Carswell, 1982) at 44-48.

Part II

INTELLECTUAL PROPERTY LAWS AND COMPUTER TECHNOLOGY IN CANADA

3

An Introduction to Intellectual Property Laws Relating to Computer Technology

The following chapters will review various Canadian laws for the protection of intellectual property as they apply to computer technology in Canada. The primary focus will be a review of three major areas of law, namely, copyright, patents and trade secrets as they apply to computer programs and data.

The essential elements of these different forms of protection are well established. In the case of copyright law, the basic conditions of protection are first, that the subject matter come within one of the categories of "works" protected under the Copyright Act; second, that the work be "original" in that it be created through the expenditure of skill, effort and judgment of the author rather than being derived from another work; and third, that it be fixed in some material form. Copyright law is intended to protect only the expression of ideas rather than the ideas themselves: in other words, another person cannot be prevented from creating a similar or even identical work as long as it is created through his independent efforts. Finally, copyright protection in any work arises from the time of creation; although registration of copyright creates certain presumptions that may facilitate the exercise of the rights of the owner, it is not a condition of protection.

By contrast to copyright, patent protection arises only after an application is filed and letters patent are granted. A patent may be granted only for an invention (which may include a process) that meets certain strict statutory tests, including novelty, inventiveness and utility. The grant of a patent gives the owner a true monopoly so that subject to certain conditions, any other person may be prevented from exploiting the work during the life of the patent.

Finally, trade secrecy law represents a means of protecting information of virtually any kind, including a process, formula, plan, method or idea. The two conditions for the protection of information under trade secrecy law are first, that the information be truly confidential rather than a matter of public or common knowledge, and second, that any person to whom the information is disclosed receive it in circumstances of confidentiality. Trade secrecy protection may arise based on the nature of the relationship between the holder of information and the recipient in the absence of an express contract. However, the terms of a contract may expand the scope of protection available for confidential information and indeed provide protection for information which is otherwise not confidential.

It will be seen from this brief overview that certain conditions must be met both by statute and at common law in order to rely on each of these forms of protection. However, as will be indicated by the detailed analysis in the following chapters, the scope and application of these areas of law to computer technology remain a matter of some uncertainty. The relationship between various forms of statutory protection is also uncertain and recent case law suggests that reliance on one form of protection such as the Patent Act may preclude reliance on other statutory protection such as the Copyright Act.[1]

The three areas mentioned do not necessarily represent all possible forms of legal protection that may be asserted by the creators or owners of computer technology and particularly computerized information. As discussed in greater detail in Part III, the fraud, theft and other provisions of the Criminal Code may apply to the misappropriation of information, whether in computerized form or otherwise, as well as the unauthorized distribution of counterfeit or pirated copies of computer programs. Trade mark law may also be relied upon to enable the owners of either computer programs or equipment to assert exclusive rights to certain names or marks for the sale or licensing of such technology to end-users. A common law action for passing off may also be brought where unauthorized copies of computer programs, or "clones" of microcomputers or other hardware, are marketed under the guise of legitimate copies and with the purported authority of the rights owners.[2]

Apart from the specific kinds of legal protection mentioned above, a review of recent case law in Canada and other countries suggests that there is a growing tendency by the courts to rely on a broad concept of "property" rights as a means of protecting computer technology, particularly programs and data. Such cases appear to suggest that information may be protected whether or not the conditions prescribed by the "traditional" forms of protection such as trade secrecy or copyright law are met. Moreover, there are suggestions that the rights of the "owners" of information may not be subject to the restrictions normally associated

---

1 *Rucker Co. v. Gavel's Vulcanizing Co.* (1985), 6 C.I.P.R. 137 (Fed. T.D.).

2 For an overview of the general principles of an action for passing off see *Seiko Time Can. Ltd. v. Consumers Distributing Co.* (1980), 29 O.R. (2d) 21 (Ont. H.C.); affirmed (1981), 34 O.R. (2d) 481 (Ont. C.A.); varied [1984] 1 S.C.R. 583 (S.C.C.).

with such areas of law.

The uncertainty that can arise from a reliance on such broad "property" concepts in this area of law is demonstrated by the decision of the Ontario Supreme Court in the case of *Canavest House Ltd. v. Lett.*[3] In that case, the plaintiff sought to restrain his former employee from infringing copyright in computer programs developed while the defendant was in the plaintiff's employ and from competing in relation to contracts or emerging business opportunities. Callon J. found that "ownership" of the programs was vested in the plaintiff and on this basis the defendant could be restrained from providing copies of the programs or of the notes or materials leading to their creation to any other person. Although it is uncertain whether it was a separate ground for his decision, Callon J. also indicated that by virtue of the plaintiff's ownership of *copyright* in the programs, the defendant could be restrained from copying, reproducing or *using* the software.

While suggesting that the plaintiff's rights in the *Canavest* case were based on its ownership of the programs, there was no indication of whether there was a distinction between the ownership of intellectual property rights in the programs and that of the media in which such programs were contained. Moreover, it was not clear whether the court considered that there may be a distinct "property" interest in the information making up programs that is not dependent on copyright or other "traditional" forms of intellectual property protection.

The confusion is exacerbated by the suggestions in the *Canavest* case that by virtue either of the general ownership rights or the more specific right of copyright in computer programs, the defendant could be restrained either from providing copies of the programs to any other person or from using such programs. As will be discussed in greater detail below, copyright law as such does not prevent either the disclosure of any work to third parties or the use of legitimately made copies of such works. However, in appropriate circumstances, such acts may be restrained on other grounds, such as if they represent a breach of a fiduciary or contractual relationship. The facts of the *Canavest* case would appear to have justified an order restraining the acts complained of on these grounds without reliance on either an ambiguous concept of "ownership" of programs or the misapplication of copyright law to restrain acts which are truly not an infringement of copyright.

The foregoing analysis demonstrates the general confusion that has arisen in the courts and indeed within the computer industry, as to the meaning of concepts of "property" and "ownership" as they are applied to information. The increasing tendency of the courts to apply general "property" concepts for the protection of information without regard to

---

3 (1984), 2 C.P.R. (3d) 386 (Ont. H.C.).

the principles of intellectual property law would appear to represent a departure from traditional legal thinking. Such protection has been generally obtained within the ambit of trade secrecy, patent, copyright and industrial design law, even though the scope and application of these laws have been subject to some uncertainty. The detailed analysis of Canadian cases that have considered the copyright status of computer programs will demonstrate the confusion that has arisen with respect to this particular issue.

The tendency of the courts to apply more generic "property" concepts as a basis for the legal protection of information raises a number of complex issues. For example, what are the rights and duties of a third party who acquires knowledge of information "belonging" to another person, without necessarily acquiring possession of any tangible media in which such information is embodied? Should such a person be deemed to be in "possession" of property belonging to another person. Can this person be a party to an offence if it is to be treated as "stolen" property? Can such a person be prevented from *using* such information when he is not subject to any contractual or other duties with respect to the manner in which such information was obtained?

It would be hoped that if the Canadian Parliament eventually comes to terms with the application of copyright law to computer technology, a firm statutory foundation for the protection of computer programs and data will be established. At such time, Canadian courts may be better able to deal with computer technology based on the application of well-established legal principles rather than based on general property concepts.[4]

4 For a further discussion of the applicability of concepts of "property" to information, and the various forms of legal protection available for information, see G. Hammond, "The Misappropriation of Commercial Information in the Computer Age" (1986), 64 Can. Bar Rev. 342. See also chapter 7, *infra*, for a discussion of whether information can be classified as "property" for purposes of criminal law.

4

Computers and the Copyright Law of Canada

1. INTRODUCTION

Copyright law has been defined as being concerned "with the negative right of preventing the copying of physical material existing in the field of literature and the arts."[1] Copyright law has evolved, however, both in terms of the works to which copyright extends, and the acts which are deemed to constitute infringement. The present Canadian Copyright Act,[2] which was enacted in 1921 and came into force in 1924, protects literary, artistic, dramatic and musical works as well as sound recordings and cinematograph works. In addition, the sole rights granted to authors include not only that of reproduction but also such rights as public performance, publication, the making of a record or other contrivance for the mechanical performance of a work, and the communication of a work by "radiocommunication".

In the same way as the 1921 Copyright Act extended both the rights of copyright owners and the works in which copyright subsists, the determination of the proper scope of such rights and works represents the most contentious issue in copyright revision today. In this regard, at the time of writing, the copyright status of computer programs under the 1921 Canadian Copyright Act remains a matter of some uncertainty. Although a number of judgments have been rendered in which programs in various forms of expressions were found to be protected by copyright, many of these judgments involved interlocutory injunctions and did not include a detailed analysis of the applicable legal principles. Several judgments which contained such detailed analysis relied in large part on decisions of the courts of other jurisdictions, the laws of which are now in many cases very different from the 1921 Canadian Act. In particular, Canadian courts have placed considerable reliance on a judgment of the Australian Federal Court of Appeal which was subsequently reversed by the Australian High Court.[3]

1 *Skone James on Copyright*, 12th ed. (1980) at 3.

2 R.S.C. 1970, c. C-30.

3 *Apple Computer Inc. v. Computer Edge Pty. Ltd.* (1984), 50 A.L.R. 581 (Aust. Fed. Ct.); reversed (1984), 53 A.L.R. 225 (Aust. Fed. C.A.); reversed (1986), 65 A.L.R. 33 (Aust. H.C.).

The present chapter will review the major principles of the 1921 Canadian Copyright Act, including the works to which the Act extends and the rights of copyright owners. This is followed by a review of Canadian jurisprudence dealing with copyright protection of computer programs, and the corresponding jurisprudence of several other jurisdictions. The application of the Act to other computer-related subject-matter such as designs for computer hardware, semiconductor chips, computer data bases and computer-created works is also considered. Other issues which are examined include the ownership of copyright in works created by employees and commissioned works, the principles relating to registration of copyright and of assignments of copyright, and the remedies available for copyright infringement. Finally, recent proposals for the revision of Canadian copyright law as it relates to computer-related subject matter are considered.

2. GENERAL PRINCIPLES OF COPYRIGHT LAW

In order to be protected by copyright, it must be shown that the subject-matter for copyright protection is an "original" work, that the work has been fixed or expressed in some material form, that the author is a qualified person for purposes of the Copyright Act, and that the work comes within one of the recognized categories of copyright protection. Each of these conditions is reviewed below.

(a) The Requirement of "Originality"

The requirement of "originality", as a condition of copyright protection, was described as follows by Peterson J. in the often cited case of *University of London Press Ltd. v. University Tutorial Press, Ltd.*[4]

> The word original does not . . . mean that the work must be the expression of original or inventive thought. Copyright Acts are not concerned with the originality of ideas, but with the expression of thought, and in the case of a 'literary work' with the expression of thought in print or writing. . . . But the Act does not require that the expression must be in an original or novel form, but that the work must not be copied from another work — that it should originate from the author.

It was noted in the foregoing case that a work was entitled to copyright protection notwithstanding that it drew upon a stock of common knowledge and that only a short time was expended in its creation. Peterson J. noted that the rough practical test to be applied was that "what is worth copying is *prima facie* worth protecting."[5]

---

4 [1916] 2 Ch. 601 at 608-609.

5 *Ibid.*, at 610.

In *Canadian Admiral Corp. v. Rediffusion Inc.*,[6] Cameron J. held that

> [f]or a work to be original it must originate from the author; it must be the product of his labour and skill and it must be the expression of his thoughts.

Copyright law does not prevent a person from creating a work that is the same or similar to that of another, provided that it is created through the expenditure of his own skill, labour and judgment. In other words, copyright law does not give any monopoly to the creator of a particular form of expression. According to *Skone James on Copyright*,[7] copyright

> is thus to be distinguished from the rights conferred by patent, trade mark and design legislation, which give to the registered proprietor an exclusive right to the registered material, even as against a person who has reproduced such material innocently and from an independent source.

(b) The Author of a Work as a "Qualified Person"

In order to be protected by copyright, it must be shown that the author of the work is a person to whom the Copyright Act extends. Section 4 of the Act provides that the author at the time of the making of the work must be a British subject, a citizen of any country which has adhered to the Berne Convention, or a resident of "Her Majesty's Realms and Territories". In the case of a published work, it must be shown that the work was first published within Her Majesty's Realms and Territories or in such foreign country.

Subsection 4(2) provides that the Copyright Act may also be extended to any country which had not adhered to the Berne Convention but which has undertaken to provide to Canadian citizens the benefit of copyright on substantially the same basis as to its own citizens. In accordance with subsection 4(2), copyright protection in Canada has been granted to citizens of the United States, which is not a party to the Berne Convention, although it has adhered to the Universal Copyright Convention of which Canada is also a member.

Subsection 4(2) reflects the so-called "national treatment" principle. This principle requires each state which has adhered to the Berne Convention and to the Universal Copyright Convention to grant to the works of nationals of other Convention countries substantially the same protection as is granted to the works of nationals of the legislating state. However, the national treatment principle would appear not to be applicable to works such as semiconductor chips which are not protected subject-matter for purposes of the Conventions.

---

6 [1954] Ex C.R. 382 at 398.

7 *Supra*, footnote 1, at 3.

(c) The Requirement of "Fixation"

A further requirement for a work to be protected by copyright law is that it be expressed in some tangible form. In *Jefferys v. Boosey*,[8] it was stated that copyright extends "not to ideas, but to the order of words, and this order has a marked identity and a permanent endurance." In *Canadian Admiral Corp. v. Rediffusion Inc.*,[9] it was held that "for copyright to subsist in a 'work' it must be expressed in some material form, capable of identification and having a more or less permanent endurance." This test meant that a live broadcast of a sporting event was not protected by copyright although a fixation of the broadcast would be protected.

The requirement of fixation would appear to mean that protection would not be granted to computer programs as they are used in the active memory of a computer, or to computer programs or data which are merely displayed on a CRT. Programs or data in such form are expressed only as electronic impulses rather than in any media having a permanent endurance.

(d) The Proper Subject-Matter of Copyright Protection

Section 45 of the Copyright Act provides that no person is entitled to copyright except in accordance with the provisions of the Act. In *Canadian Admiral v. Rediffusion Inc.*,[10] Cameron J. noted that since copyright is purely statutory,

> [i]t follows, therefore, that no matter how "piratical" the taking by one person of the work of another may appear to be, such taking cannot be an infringement of the rights of the latter unless copyright exists in that "work" under the provisions of s. 3.

Accordingly, to rely on the Copyright Act, it must be shown that the subject-matter for which protection is claimed is a "work" within the meaning of the Act.

Subsection 4(1) of the Copyright Act provides that "copyright shall subsist in Canada . . . in every original literary, dramatic, musical and artistic work. . . ." Subsection 4(3) of the Act further provides that copyright shall subsist in "records, perforated rolls, and other contrivances by means of which sounds may be mechanically reproduced, in like manner as if such contrivances were musical, literary or dramatic works." The Act also grants protection to cinematograph productions which are protected either as photographs (that is, artistic works) or as dramatic works if the author

---

8 (1854), 4 H.L.C. 815 (H.L.).

9 *Supra*, footnote 6, at 394.

10 *Ibid.*, at 390.

has given such works an original character.[11]

The term "every original literary, dramatic, musical and artistic work" is defined in section 2 of the Act to include "every original production in the literary, scientific or artistic domain, whatever may be the mode or form of its expression. . . ." The definition goes on to set out certain examples of works coming within that expression such as books, pamphlets and other writings, lectures, dramatic or dramatical-musical works.

In spite of the broad definition set out in section 2, it appears that any work must come within one of the established categories to be protected. In *Cuisenaire v. South West Imports Ltd.*,[12] Noel J. held that it could not have been the intention of Parliament in enacting this definition to protect all original productions in the scientific field for a period of life of the author plus 50 years, particularly in view of the limited protection granted to inventions under the Patent Act. Noel J. indicated that although the definition may have extended the meaning to be given to the works included within each of the four categories, it was still necessary to show that a work came within one of such categories to be protected.

The following sections will review in further detail the nature of the various categories of copyright works. This will be followed by a review of the scope of the exclusive rights granted to copyright owners. On the basis of this review, recent jurisprudence dealing with the application of copyright law to computer programs and related subject-matter will be considered.

3. THE CATEGORIES OF COPYRIGHT-PROTECTED WORKS

(a) Literary Works

The concept of a "literary work" is not defined in the Copyright Act, although section 2 states that a "'literary work' includes maps, charts, plans, tables and compilations." According to *Skone James on Copyright*,[13] in order to be protected as a literary work, "so long as something in 'writing' exists, it is sufficient, and it is not necessary that what is written should express a meaning in ordinary language." In *University of London Press Ltd. v. University Tutorial Press Ltd.*,[14] Petersen J. stated that "in my view the words 'literary work' cover work which is expressed in print or writing, irrespective of the question whether the quality or style is high." This test was cited with approval by the English House of Lords in the case of *Ladbroke*

11 Copyright Act, ss. 2, 3(1)(*e*).

12 (1967), 54 C.P.R. 1 (Ex. Ct.); affirmed (1968), 57 C.P.R. 76 (S.C.C.).

13 *Supra*, footnote 1, at 61. Unlike the Canadian Copyright Act, the United Kingdom Act contains a definition of the word "writing": see p. 66, *infra*.

14 *Supra*, footnote 4, at 608.

(Football) Ltd. v. William Hill (Football) Ltd.,[15] by the Canadian Federal Court in *Bulman Group Ltd. v. "One Write" Accounting Systems Ltd.*[16] and in the decision of the Australian High Court in *Apple Computer Inc. v. Computer Edge Pty. Ltd.*[17]

Some cases have suggested that, in addition to any requirement that a literary work be in writing, the work must be "intended to afford either information and instructions, or pleasure, in the form of literary enjoyment."[18] This test was recently applied by the English Court of Appeal in the case of *Exxon Corp. v. Exxon Insurance Consultants International Ltd.*[19] However, it would not appear necessary that a literary work be capable of conveying information which can be understood by an ordinary human reader. In *D.P. Anderson & Co. v. Lieber Code Co.*,[20] a code of made-up words which was meaningless of itself was found entitled to copyright protection. It was noted, however, that a meaning could be attached to the code by the sender of the message as well as a recipient.

In *Bulman Group Ltd. v. "One Write" Accounting Systems Ltd.*,[21] Collier J. held that the imparting of intelligible information was not a fundamental requirement for a compilation to qualify as a literary work. However, in the case of a compilation of a commercial type, it was necessary to show "a literary sense of functionally assisting, guiding or pointing the way to some end."[22] In other words, it may be necessary to show that a literary work serves some communicative purpose even if it is not expressed in ordinary language.

In general, a literary work must be distinguished from a device or apparatus which is used for a utilitarian purpose. In the case of *Hollinrake v. Truswell*,[23] the Court refused to grant copyright protection to a cardboard sleeve pattern containing scales and descriptive words. Lindlay L.J. noted that what was sought to be protected was a measuring instrument whose utility lay in the device itself rather than in the information which it conveyed. Davie L.J. also noted that a literary work did not include a mechanical contrivance, appliance or tool which was intended for practical use rather than for the purpose of giving information.

In *Cuisenaire v. South West Import Ltd.*,[24] the plaintiff claimed that the manufacture of coloured rods which were used to teach arithmetic

---

15 [1964] 1 W.L.R. 273 at 285 (H.L.).
16 (1982), 16 B.L.R. 16 at 22 (Fed. T.D.).
17 *Supra*, footnote 3.
18 *Hollinrake v. Truswell*, [1894] 3 Ch. 420 at 428 (C.A.).
19 [1982] Ch. 119 at 143-144 (C.A.).
20 [1917] 2 K.B. 469.
21 *Supra*, footnote 16.
22 *Ibid.*, at 23.
23 *Supra*, footnote 18.
24 *Supra*, footnote 12.

to school children was an infringement of copyright in a written text in which the system of using such rods was described. Noel J. of the Exchequer Court found that the mere fact that an object was of a functional or utilitarian nature and could form the subject of a patent of invention did not by itself invalidate its protection under the Copyright Act, provided that the object came within one of the categories of copyright-protected subject-matter. In finding that the coloured rods could not be considered as reproductions of the written text in which the rods were described, Noel J. relied on the following extract from the decision of Pape J. in the Australian case of *Cuisenaire v. Reed*,[25] which dealt with a similar fact situation:

> Where, as here, you have a literary copyright in certain tables or compilations, there is in my view no infringement of the copyright in those tables or compilations unless that which is produced is itself something in the nature of a table or compilation which, whether it be in two dimensions or three dimensions, and whatever its material form, reproduces those tables.

The Supreme Court of Canada upheld the finding of Noel J. that coloured rods were not "things in which copyright can be had."[26] The Court cited with approval the finding of Pape J. in the Australian *Cuisenaire* case that the construction of a model in accordance with the instructions in a handbook could not constitute an infringement of the plaintiff's literary copyright.

The foregoing cases indicate that there may be some difficulties in characterizing computer programs, at least in their machine-readable object code form, as coming within the category of literary works for purposes of the 1921 Copyright Act. First, in the absence of a definition of "writing" in the Canadian Copyright Act, it is questionable whether programs expressed in machine-readable form would be works expressed in print or writing for purposes of the Act. Second, insofar as there is a legal requirement that literary works serve a communicative function, programs expressed in object code form may not meet that criterion. There is also some question as to whether programs in their electrical code form are "in the nature" of literary works which reproduce the underlying source code programs so as to be "things in which copyright can be had" for purposes of the Copyright Act. In spite of these difficulties, as will be reviewed in detail below, a number of Canadian cases have found that programs, whatever their form of expression, qualify for protection under the Copyright Act either as literary works or as reproductions or translations of such works.

---

25 [1963] V.R. 719 at 735-6 (Aust.).

26 (1968), 57 C.P.R. 76 at 81 (S.C.C.).

(b) Artistic Works

Section 2 of the Copyright Act defines an artistic work as including "works of painting, drawing, sculpture and artistic craftsmanship, and architectural works of art and engravings and photographs." Unlike the 1956 Copyright Act of the United Kingdom, the Canadian Act does not include maps, charts and plans as coming within the definition of artistic works. Instead, such works are protected in Canada as literary works.

In the case of *Cuisenaire v. South West Import Ltd.*,[27] Noel J. held that

> an artistic work, in my view, must to some degree at least, be a work that is intended to have an appeal to the aesthetic senses not just an incidental appeal, . . . but as an important or one of the important objects for which the work is brought into being.

Based on this test, Noel J. found that the coloured rods in question could not be protected as artistic works since any attraction that they had to children represented only "a very secondary purpose."[28] Moreover, such objects could not be considered as works of artistic craftsmanship first, because they were not artistic and second, because there was no particular skill or craftsmanship involved in their production.

The requirement that artistic works have an appeal to the aesthetic senses would appear to mean that computer programs would not be protected as artistic works under the Act. However, flow charts and diagrams prepared in the course of the development of programs may be considered as charts or plans coming within the definition of literary works for purposes of the Act. As noted below, the concept of an "artistic work" may include designs for computer equipment.

(c) Sound Recordings

Subsection 4(3) of the Copyright Act provides for copyright protection for "contrivances by means of which sounds may be mechanically reproduced." Such works are protected as if they were musical, literary or dramatic works. In *Compo Co. v. Blue Crest Music*,[29] the Supreme Court of Canada held that the copyright that subsists in a record under subsection 4(3) is separate from the copyright in the musical work embodied in the record.

Computer programs would not appear entitled to protection as sound recordings, whatever the mode of their expression, since they are not intended for the mechanical performance of sound. However, the concept of a sound recording is of some significance insofar as it indicates that

---

27 (1967), 54 C.P.R. 1 at 22 (Ex. Ct.)

28 *Ibid.*

29 [1980] 1 S.C.R. 357 at 374 (S.C.C.).

if computer programs in certain forms of expression, such as when embodied in ROM, are characterized as "contrivances" used to operate computers, special provisions may be required to protect such works under the Copyright Act in the same way as such provisions were enacted to extend the Act to sound recordings.

4. THE SCOPE OF THE RIGHTS GRANTED TO COPYRIGHT OWNERS

Subsection 3(1) of the Copyright Act provides as follows:

> For the purposes of this Act, 'copyright' means the sole right to produce or reproduce the work or any substantial part thereof in any material form whatever, to perform, or in the case of a lecture to deliver, the work or any substantial part thereof in public; if the work is unpublished, to publish the work or any substantial part thereof; and includes the sole right
> (*a*) to produce, reproduce, perform or publish any translation of the work;
>
> . . .
>
> (*d*) in the case of a literary, dramatic or musical work, to make any record, perforated roll, cinematograph film, or other contrivance by means of which the work may be mechanically performed or delivered;
>
> . . .
>
> (*f*) in the case of any literary, dramatic, musical or artistic work, to communicate such work by radiocommunication;
> and to authorize any such acts as aforesaid.

It will be noted that subsection 3(1) sets out certain acts which are deemed to come within the "meaning" of copyright, while certain other acts are deemed to be "included" within the rights of copyright owners. The principal rights which would appear to be relevant to the exploitation of computer-related subject-matter are first, the right to produce or reproduce the work or any substantial part in any material form; second, the right to publish the work; third, the right to produce or reproduce any translation of the work; fourth, the right to make any record or other contrivance for the mechanical performance or delivery of the work; and fifth, the right to communicate such work by radiocommunication. The nature and scope of these rights are reviewed below.

(a) The Right of Reproduction

In *Francis Day & Hunter Ltd. v. Bron*,[30] Lord Diplock set out the following elements as to what constitutes a reproduction of a work:

---

30 [1963] 2 All E.R. 16 at 27 (C.A.).

> [I]t is well established that to constitute infringement of copyright in any . . . work there must be present two elements: First, there must be sufficient objective similarity between the infringing work and the copyright work, or a substantial part thereof, for the former to be properly described, not necessarily as identical with, but as a reproduction or adaptation of the latter; secondly, the copyright work must be the source from which the infringing work is derived.

In order to constitute an infringement of copyright, it is not necessary that the reproduction be identical to the original. Instead, it is sufficient to show that a "substantial part" has been taken. In *Schauenberg Industries Ltd. v. Borowski,*[31] Craig J. adopted the following test set out in Fox, *The Canadian Law of Copyright and Industrial Design,*[32] as to what constitutes the taking of a substantial part of a work:

> It is always a question of fact whether the part taken is a substantial part or not. In determining what is a substantial part a variety of circumstances must be considered — the value of the part appropriated; its relative value to each work; the purpose it serves in each; and how far the pirated matter will tend to supersede the original or interfere with its sale.

The right of reproduction would appear not to include the making of a device or contrivance that implements the work. In *Boosey v. Whight,*[33] it was held that perforated rolls for the mechanical performance of musical works were not "copies" of those works. The Court noted that such sheets were produced as part of a mechanical contrivance for the production of musical notes rather than for the purpose of being read or understood.

Following the *Boosey v. Whight* decision, the Copyright Act of England was amended in 1911 to provide that copyright owners had the sole right to produce or reproduce the work in any material form. The Act also granted to copyright owners the sole right to make and to authorize the making of any record or other contrivance for the mechanical performance of the work. These provisions were also incorporated in the 1921 Copyright Act of Canada. Notwithstanding the general language set out in the amended U.K. Act, the right to reproduce a work in any material form was equated in the case of *Thompson v. Warner Bros. Pictures Ltd.*[34] with the "right of making copies". The right to make contrivances for the mechanical performance of a work was described as a new right which did not exist under the previous law.[35]

---

31 (1979), 50 C.P.R. (2d) 69 (Ont. H.C.).

32 2nd ed. (Carswell, 1967), at 341-2.

33 [1900] 1 Ch. 122 (C.A.).

34 [1929] 2 Ch. 308 at 333 (C.A.).

35 *Ibid.*, at 334. See also *Ladbroke (Football) Ltd. v. William Hill (Football) Ltd., supra,* footnote 15, at 276, in which Lord Reid stated that "broadly, reproduction means copying."

The "reproduction" right appears, however, to be broader than the right to make copies insofar at least as a "reproduction" need not be in the same dimensions as the original. In *King Features Syndicate Inc. v. O. & M. Kleeman Ltd.*,[36] Simmonds J. held that three-dimensional dolls or toys of "Popeye the Sailor" made from two-dimensional sketches were "reproductions" of those sketches on the following grounds:

> In my judgment, it would be contrary to the plain meaning and spirit of the Act if a copy of an artistic work were an infringement only if made in the same dimensions, so that, for example, copyright in a sketch would not be infringed if the exact design were reproduced either in a frieze made with some measure of relief or in the sculptured pediment.

The scope of the "reproduction" right was also considered in *Cuisenaire v. South West Import Ltd.*[37] As noted above, it was held in that case that coloured rods made in conformity with directions contained in a book could not be considered as reproductions of the tables or compilations in the literary work on which they were based. This conclusion was based on the finding of Pape J. in the Australian *Cuisenaire* case[38] that literary copyright in a table or compilation was not infringed "unless that which is reproduced is itself something in the nature of a table or compilation which, whether it be in two dimensions or three dimensions, and whatever its material form, reproduces those tables." Pape J. distinguished the *King Features Syndicate* case and other authorities dealing with the reproduction of two-dimensional drawings in three-dimensional objects on the grounds that in those cases, there was a clear visual resemblance between the alleged infringing article and the work in which copyright subsisted, sufficient to warrant the conclusion that one had been copied from the other.

The foregoing authorities suggest that in order to infringe the "reproduction" right in a literary work, the infringing work must be derived from the original and there must be an objective similarity between the two works. However, the scope of the reproduction right under the Canadian Copyright Act remains uncertain in three respects. First, it is not clear whether the similarity between a reproduction and an original must be capable of being detected as a matter of visual resemblance or otherwise be perceptible, or whether, as suggested in a recent Canadian case on computer programs, it is sufficient that there be a method by which the alleged infringing work can be compared to the original.[39] Second, it is uncertain whether there is a requirement, as suggested in the Australian

36 [1940] 2 All E.R. 355 at 358; affirmed [1941] A.C. 417 (H.L.).

37 (1967), 54 C.P.R. 1 (Ex. Ct.); affirmed (1968), 57 C.P.R. 76 (S.C.C.).

38 *Supra*, footnote 25, at 735-6.

39 *Apple Computer Inc. v. Mackintosh Computers Ltd.* (1986), 10 C.P.R. (3d) 1 (Fed. T.D.); see *infra*, at 54.

and Canadian *Cuisenaire* cases, that the reproduction must be "in the nature" of the original, such that a reproduction may not include the embodiment of a design or plan in a functional object or device. A related issue is whether apart from the circumstances in which the Industrial Design Act applies, the reproduction right in designs or plans for functional objects may be infringed by the indirect reproduction of the objects themselves.[40]

(b) The Right of Publication

Subsection 3(1) provides that one of the sole rights of a copyright owner is to publish the work or any substantial part. Although the provision states that such right exists only if the work "is unpublished", it would appear necessary to interpret the section to mean that the right continues subsequent to the first publication of a work.

Subsection 3(2) of the Act provides that "publication, in relation to any work, means the issue of copies of the work to the public. . . ." The word "copies" is not defined in the Act and, as noted above, it is uncertain whether a work may be "reproduced in any material form" without being "copied". There is also no definition in the Act of the expression "to the public", and it would be a question of fact in any case as to whether a work is distributed on a sufficiently wide scale so that it would be deemed to be distributed "to the public".

The "publication" right set forth in subsection 3(1) of the Act is separate from that of reproduction. However, as a practical matter, the "issue" of copies of a work to the public would require the "making" of copies, so that infringement of the publication right would also result ordinarily in an infringement of the reproduction right.

(c) Translation

Subsection 3(1)(*a*) of the Act provides that one of the sole rights of an author is to produce, reproduce, perform or publish any translation of the work. In the absence of a definition in the Copyright Act, reference must be made to the dictionary definitions of a translation. According to the *Oxford English Dictionary*[41] a translation is defined as follows:

> 1. The action or process of turning from one language into another; also, the product of this; a version in another language. . . . The expression or rendering of something in another medium or form.

Once a translation is prepared, the translation is itself recognized as

---

40 See *Bayliner Marine Corp. v. Doral Boats Ltd.* (1985), 5 C.P.R. (3d) 289 (Fed. T.D.); reversed on other grounds (1986), 67 N.R. 139 (Fed. C.A.); see also s. 7 of this chapter, *infra*, at 79.

41 3rd ed. (1977), Vol. 2, at 2347.

an original literary work. In *Byrne v. Statistical Co.*,[42] it was stated that a "translator of a literary work has for many years been held to be the author of his translations." In *Pasickniak v. Dojacek*,[43] the Court cited the following extract from Copinger on *The Law of Copyright*, 6th ed., at 53:

> Translations are also original literary works, and consequently entitled to protection; of course, without prejudice to the rights of the owner of the copyright (if any) in the work translated.

Apart from the computer copyright cases referred to below, there does not appear to be any case which has considered whether a translation of a literary work must itself meet all of the requirements of a literary work to be protected. However, the reference in the opening words of subsection 3(1) to "the work" would indicate that a translation must itself come within one of the recognized categories of copyright protection. Such a characterization is of some significance in view of the issue which has arisen in a number of copyright cases as to whether an object code program in ROM is a translation of the underlying written source code program.

(d) The Making of Contrivances

Paragraph 3(1)(*d*) of the Copyright Act provides that one of the rights of the author of a literary, dramatic or musical work is "to make any record, perforated roll, cinematograph film or other contrivance by means of which the work may be mechanically performed or delivered." The word "performance" is defined in section 2 of the Act to mean any "acoustic representation of a work or any visual representation of any dramatic action in a work, including a representation made by means of any mechanical instrument or by radiocommunication."

The reference to acoustic and visual representation would indicate that the rights granted to an author under paragraph 3(1)(*d*) relate only to the making of a contrivance for the performance of a work intended to be delivered as an acoustic or visual representation to a human listener or viewer. Accordingly, this section would not seem applicable to the transformation of a computer program into a form whereby it can be executed in connection with the operation of a computer.

(e) Communication of a Work by Radiocommunication

Paragraph 3(1)(*f*) of the Copyright Act grants to the owner of copyright in any literary, dramatic, musical or artistic work, the sole right to communicate such work by radiocommunication. The term "radiocom-

42 [1914] 1 K.B. 622 at 627.
43 [1928] 2 D.L.R. 545 at 550 (Man. C.A.).

munication" is defined in the Interpretation Act[44] to mean any "transmission, emission or reception of signs, signals, writing, images, sounds or intelligence of any nature by means of electromagnetic waves." However, a more restrictive interpretation has been given to such right for purposes of the Copyright Act. In *Composers, Authors and Publishers Association of Canada Ltd. v. CTV Television Network Ltd.*,[45] it was held that only public performances or representations by radio broadcasting came within the "radiocommunication" right set out in the Copyright Act.

The foregoing case suggests that the use of computer communications facilities to transmit a copyright-protected work would not come within the exclusive rights of the copyright owner unless the transmission is intended for reception directly by the general public. In other words, the private communication of a work from one computer facility to another by any form of telecommunications facilities would not be deemed to be a radiocommunication for purposes of the Copyright Act.

(f) Other Rights of Copyright Owners

Subsection 17(1) of the Copyright Act provides that copyright in any work is deemed to be infringed by any person who without the consent of the copyright owner does anything that only the owner of the copyright has the right to do under the Act. Subsection 17(4) states that copyright is also deemed to be infringed by any person who sells, lets for hire, offers for sale by way of trade, distributes for the purposes of trade, exhibits in public by way of trade, or imports for sale or hire into Canada any work that to his knowledge infringes copyright or would infringe copyright if it had been made within Canada. Accordingly, the unauthorized distribution of, or trade in, infringing copies of any work will itself be deemed to be an infringement of copyright, provided that the person carrying out such acts knows that the work infringes copyright.

Subsection 12(7) of the Copyright Act grants certain so-called "moral rights" in favour of the author of any work. These rights are to claim authorship of a work and to restrain any distortion, mutilation or other modification of the work prejudicial to the author's honour or reputation. These rights are vested in the author independently of his copyright and continue even after the assignment of copyright.

---

44 R.S.C. 1970, c. I-23, s. 28; see also Broadcasting Act, R.S.C. 1970, c. B-11, s. 2; and Radio Act, R.S.C. 1970, c. R-1, s. 2.

45 [1968] S.C.R. 676 (S.C.C.).

5. THE 1921 COPYRIGHT ACT AND COMPUTER PROGRAMS

(a) A Review of Canadian Jurisprudence

The foregoing review indicates that there may be some difficulty in applying the 1921 Copyright Act to computer programs in all possible forms of expression. Although Canadian courts have generally given a liberal interpretation to the Copyright Act to protect programs, most decisions have been at the level only of interlocutory injunctions and have not considered in detail the precise wording of relevant provisions of the Copyright Act or the case law in which such provisions have been considered. The absence of a clear rationale underlying many of the Canadian decisions underlines the need for a revision of the copyright law to deal expressly with this technology. The history of proposals for revision of the copyright law is reviewed in detail in the final section of this chapter.

Many of the Canadian cases have placed considerable reliance on court decisions rendered in other jurisdictions, such as the United States, the United Kingdom, South Africa and Australia. However, there are major differences between the laws of some of these countries, particularly the United States, and Canada. Other countries, such as the United Kingdom and Australia, have amended their copyright laws to deal expressly with computer programs and to overcome the difficulties with the earlier law. In the case of Australia, a decision of the Federal Court which was relied on in many Canadian cases was subsequently reversed by the Australian High Court.[46]

It is clear that various works produced in the preliminary stages of creating a program qualify as literary works. In this respect, flow-charts and logic-flow diagrams as well as written specifications for programs are works expressed in print or writing which are capable of being perceived and understood by trained human programmers, even if they do not express a meaning in ordinary language. In addition, the programs themselves when written in source code form with accompanying labels and comments would be literary works since they are expressed in print or writing and can be understood by any person skilled in the programming language which is used. Copyright protection also clearly extends to manuals, user guides and similar written material describing the operation of programs. That programs expressed in source code form and accompanying program materials are protected by copyright has been accepted by virtually all

---

46 *Apple Computer Inc. v. Computer Edge Pty. Ltd.* (1984), 50 A.L.R. 581 (Aust. Fed. C.); reversed (1984), 53 A.L.R. 225 (Aust. Fed. C.A.); reversed (1986), 65 A.L.R. 33 (Aust. H.C.).

courts which have considered the issue.[47]

Of greater difficulty is the question of whether the transposition of programs from source code to object code form by the use of compilers or interpreters represents one of the exclusive rights of the copyright owner in the underlying source code program. A further issue is whether programs as embodied in object code form on machine-readable media, such as tapes, discs or read-only memory, continue to constitute literary works for purposes of the Copyright Act. If they are not themselves literary works, are they nevertheless entitled to protection as "reproductions" or "translations" of the underlying source code programs? Conversely, does the unauthorized reproduction of programs in ROM or other machine-readable form represent the indirect copying or "reproduction in any material form" of the source code programs?

The following summary of a number of Canadian decisions will indicate that copyright appears to subsist in programs whatever their mode of expression. However, such decisions have generally failed to consider in detail the issues raised above with respect to the manner in which the Copyright Act applies to this technology.

In *Spacefile v. Smart Computing Systems Ltd.*,[48] Steele J. held that a computer package consisting of an instruction manual, a source code and programs was a proper subject-matter of copyright. It was noted that "the source code and the programmes are electronically recorded on discs but are capable of, and have been, printed from them." Because such works were "original ideas expressed in a particular form", they were entitled to copyright protection.[49]

Steele J. did not expressly consider whether the programs as electronically recorded were to be classified as literary works under the Copyright Act, whether they would have been protected if they had not been originally expressed in written or printed form or whether the reproduction of programs in machine-readable form was an infringement of copyright in the underlying source code programs.

The decision in the *Spacefile* case was followed by Montgomery J. in *F & I Retail Systems Ltd. v. Thermo-Guard Automotive Products Canada Ltd.*[50] In that case, the plaintiff sought to obtain an interlocutory injunction against the copying of its computer system screens with respect to its automobile dealer computer programs. Montgomery J. expressed

---

47 See, *e.g.*, *Apple Computer Inc. v. Mackintosh Computers Ltd.*, *supra*, footnote 39; and *Apple Computer Inc. v. Computer Edge Pty. Ltd.*, *ibid.*; in the latter case, all five judges of the Australian High Court were of the view that computer programs in source code form were protected by copyright as literary works, although a majority of the Court held that such copyright did not subsist in the programs in object code form embodied in ROM.

48 (1983), 75 C.P.R. (2d) 281 (Ont. H.C.).

49 *Ibid.*, at 282.

50 (1984), 1 C.P.R. (3d) 297 (Ont. H.C.).

agreement with the finding of Steele J. in the Spacefile case that copyright exists in computer programs, and that the source codes were original ideas "expressed in a particular form". An application for leave to appeal from this decision was rejected by Mr. Justice R.E. Holland on July 4, 1984.[51]

The copyright status of computer programs was considered by the Federal Court in *IBM Corp. v. Spirales Computers Inc.*[52] The plaintiff in that case sought an interlocutory injunction to restrain the defendants from infringing copyright in a computer program entitled "IBM Personal Computer Basis Input Output System — 1.0" ("BIOS"). The BIOS program was "burned into" or embedded in a chip mounted on the IBM Personal Computer. Reed J. found on the evidence that IBM had published a listing of the source code and object code together with other technical specifications in a reference manual for the Personal Computer. The issue, however, was not whether the defendant had copied such written work, but rather whether it had infringed the plaintiff's copyright through the import and sale in Canada of a computer containing a copy of the BIOS program recorded in a ROM.

Reed J. rejected the defendant's contention that a computer program was not a literary work because it was not in written or printed form. She noted that the program, at least in its source code form, met the requirement of being in writing. She found further that a program need not express a meaning in ordinary language in order to qualify as a literary work.

Madam Justice Reed also rejected the defendant's contention that a computer program is an idea and therefore not copyrightable because there might be a merger of the idea and its expression. Relying on the U.S. case of *Apple Computer Inc. v. Franklin Computer Corp.*,[53] she found that if an idea is capable of various modes of expression, a program embodying one such mode of expression is copyrightable. The defendants had not contradicted the plaintiff's evidence that other programs could be created to fulfill the same purpose as the plaintiff's BIOS program.

The defendant further contended that the plaintiff's computer program was not a literary work because it was merely a machine part and that its sole purpose was to make the computer function. Reed J. rejected this argument on the ground that the permanent mounting of a program embodied in a ROM was for convenience only and that its role could be compared to that of a cassette tape in a tape recorder. Specifically, a program embedded in a chip, like information recorded on a tape, operates in conjunction with the machine for which it is designed but is not part of the machine.

---

51 *Ibid.*

52 (1984), 80 C.P.R. (2d) 187 (Fed. T.D.).

53 714 F. 2d 1240 (1983).

Reed J. also rejected the defendant's contention that the program was not protected because it was designed to provide instructions to the computer rather than to communicate information or instructions to human beings. Reed J. found that this argument was answered by the decision of the Australian Federal Court of Appeal in *Apple Computer Inc. v. Computer Edge Pty. Ltd.*[54] in which Lockhart J. held that a computer program was not deprived of copyright protection merely because its function was to control the sequence of operations carried out by a computer. Reed J. noted that in the case at bar, the BIOS program had been published in both source code and object code form, and that this was not a case where there was no written version of the program in its object code manifestation.

The final argument of defendant's counsel was that copyright protection should extend only to the program as written in the technical manual, and not to the machine language embedded in the chip. It was argued that the manual itself had not been copied, and that the program was copied directly from the BIOS chip. Madam Justice Reed found that the method of reproduction was not material. The BIOS program embodied in ROM met the test set forth in *Canadian Admiral Corp. Ltd. v. Rediffusion Inc.*,[55] that it be expressed in some material form. The fact that it could not be read by the human eye was immaterial. Citing the Australian Federal Court of Appeal decision in the *Computer Edge* case, Reed J. held that the copying of the program embodied in a chip was a reproduction in material form of an "adaption" of the work.

The copyright status of computer programs was also considered by the Quebec Superior Court in the case of *Société d'Informatique R.D.G. Inc. v. Dynabec Ltée.*[56] In that case, the plaintiff sought to restrain its former employees from using or modifying certain computer packages in which the plaintiff claimed copyright. The plaintiff alleged that the defendants had converted such packages to themselves and to Dynabec, their new employer, and were using such packages to profit fraudulently from the plaintiff's investment.

Relying on expert evidence, Hannon J. held that the ensemble of each software package existed as a series of written instructions based on observation and analysis, which was expressed in literate terms destined for introduction in the computer. The manuals for the operation of each software package were also expressed in printed form. The function of translating the written instructions into machine language in the computer was described as being "a treatment, an abstract or a summary of the

---

54 The decision of the Federal Court of Appeal was subsequently reversed: see footnote 3, *supra*.

55 (1954), 20 C.P.R. 75 at 86 (Ex. Ct.).

56 (1984), 6 C.P.R. (3d) 299 (Que. S.C.); affirmed (1985), 6 C.P.R. (3d) 322 (Que. C.A.).

software package expressed in literate terms."[57] Accordingly, the software packages were the proper subject matter of copyright protection as "literary works of a certain mode or form of expression in the literary or scientific domain."[58]

In his decision, Mr. Justice Hannon cited the Australian Federal Court's decision in the *Computer Edge* case as authority for the proposition that the technological act of rendering literate software comprehensible to a computer was equivalent to preparing a translation of the work. The U.S. *Apple Computer* case was also cited to support the conclusion that a computer software package could properly be classified as a literary work. It was noted in this regard that the definition of a literary work to include any work "whatever may be the mode of expression" meant that tables and compilations could be in numerical form.

The decision of Hannon J. was upheld by the Quebec Court of Appeal which found that object code programs were translations in machine language, that is, in a form executable by the computer, of source programs which were written in a human comprehensible form. The Court noted further that a program in source code form was clearly a literary work and that it was conceivable that an object code program was simply a reproduction of the source code program in any material form whatever. Accordingly, the Court concluded that the trial judge had not erred in law in finding that the programs in question were entitled to protection under the Copyright Act.

Computer programs were also found to be protected by copyright in the case of *Canavest House Ltd. v. Lett.*[59] The plaintiff's business in that case consisted of providing computerized investment management products and services to institutional investors. The defendant had been employed by the plaintiff to write specialized programs for use in investment decision-making. Although the defendant admitted that he did not have ownership of the computer programs in question, he claimed the right to keep copies of such programs based on discussions at the time of his employment.

Callon J. found that a computer program was a "work" within the meaning of the Copyright Act. He noted that copyright may reside in "either a language, or form of language, where it is an original creation of an author . . . or, more traditionally, in the particular form of expression of an idea or concept."[60] Relying on this principle, Callon J. found that a program is either a novel language or an expression of thought, although in numeric form. Accordingly, an injunction was granted restraining the

57 *Ibid.*, 6 C.P.R. at 308.
58 *Ibid.*, 6 C.P.R. at 308.
59 (1984), 2 C.P.R. (3d) 386 (Ont. H.C.).
60 *Ibid.*, at 392.

defendant from infringing copyright in the programs which he developed while in the plaintiff's employ. At the same time, the defendant was found entitled to keep notes and other materials prepared in the course of developing the programs based on the provisions of subsections 17(2)(*a*) and (*b*) of the Copyright Act which permit, respectively, any fair dealing with any work for purposes of private study and research, and the use by an author of an artistic work of any plan, model or study made by him for purposes of the work.

In *Apple Computer Inc. v. Minitronics of Canada Ltd.*[61] the plaintiff sought an interim injunction to restrain infringement of copyright in computer programs. The plaintiff also sought to restrain the infringement of copyright in the designs of cases for its computer hardware as well as its trade marks.

With respect to the alleged infringement of copyright in the computer programs, Walsh J. noted that evidence had been adduced to show that two EPROM's removed from one of the defendant's computer contained 8,188 out of 8,192 bytes of information identical in content and location to those contained in the ROM's in the plaintiff's Apple IIe computer. In another EPROM, 8,176 out of 8,192 bytes of information were identical to those in the plaintiff's Apple IIe ROM. The information included the name of the plaintiff's programmer which had been inadvertently inserted in the original program. Based on this evidence, Walsh J. found that the defendants had copied either the chips in the plaintiff's computer, or programming information obtained from the plaintiff's copyright-protected technical manuals. His Lordship noted that the legal issues raised by the defendants as to whether the information embodied on the microprocessor chip was protected by the copyright protecting the program itself (that is, in source code form) were of a "serious nature" and that they appeared to have at least an arguable defence. However, based on the heavy losses that the plaintiffs would suffer if the defendants continued to sell their computers, and in view of the defendants' "flagrant copying" and the possibility that they would go into bankruptcy so as to prevent the plaintiff from recovering damages, Walsh J. granted an interim order restraining the defendants from selling computers pending a hearing for an interlocutory injunction.

The copyright status of computer programs was also considered by the Federal Court in the case of *Apple Computer Inc. and Apple Canada Inc. v. Mackintosh Computers Ltd.* Two judgments were rendered in that case, the first involving an interlocutory injunction, and the second after a full trial of the issues. In the first judgment (hereinafter referred to as

61 (1985), 7 C.P.R. (3d) 104 (Fed. T.D.); affirmed but "without necessarily subscribing to all of the principles of law which the learned motions judge enunciated." (1985), 8 C.P.R. (3d) 431 (Fed. C.A.).

Apple Computer Inc. v. Mackintosh (No. 1)),[62] Cullen J. rejected an application for an interlocutory injunction to restrain the alleged infringement of copyright in the plaintiff's Applesoft and Autostart ROM operating system programs. Although the decision was based primarily on a finding that the balance of convenience was against the plaintiff, Cullen J. also considered the merits of the plaintiff's legal argument. He noted that, although the plaintiff had established a *prima facie* case that its programs were copyrightable, its case was not a strong one. Cullen J. cited evidence by an expert witness who stated that while a computer program written on paper was an expression of the writer's instructions, a computer ROM chip was an "implementation of those instructions in a useful device."[63] It was noted that copyright is a statutory right, and that legislation in each country was different in specifics. In particular, the Australian cases which relied on United States law were described as "suspect" since the American law had been amended to bring it into the computer age. His Lordship concluded that the defendants had established to his satisfaction that they had an arguable case on all of the issues which they were entitled to put at trial.

The full trial in the *Apple Computer v. Mackintosh* case took place before Madam Justice Reed who rendered her decision on April 29, 1986 (which judgment is hereinafter referred to as *Apple Computer v. Mackintosh Computers Ltd.*).[64] Madam Justice Reed noted that the sole legal issue in dispute was whether a computer program when embodied in a computer was a subject-matter in which copyright exists. The plaintiffs asserted copyright in the original assembly (or source code) versions of the programs and claimed that the hexadecimal (or object code) versions of such programs in machine-readable form were translations of the original works. Evidence adduced by the plaintiff indicated that out of 12,288 bytes encoded on one of the chips sold by the defendants, 12,261 were absolutely identical in content and location to the plaintiff's corresponding chips; in the case of the other chip sold by the defendant, 12,277 out of 12,288 bytes were identical to those of the plaintiff.

Madam Justice Reed reviewed in detail the nature of computer programs and how they are created. She noted that when a program is embodied in ROM, a pattern corresponding to the binary pattern of the written program is etched into the memory cells of the chip by a photolithographic process. Thus, the chip is structured so that it can replicate the program in its electrical code form as required whenever an electrical current is passed through the appropriate address of the ROM.

Madam Justice Reed rejected the defendant's contention that a

62 (1985), 3 C.I.P.R. 133 (Fed. T.D.).

63 *Ibid.*, at 138.

64 (1986), 10 C.P.R. (3d) 1 (Fed. T.D.).

program should be characterized as merely specifications for a machine part. Instead, a program represents instructions to the computer for moving information between registers and performing operations; in other words, an instruction manual addressed to the CPU of a computer rather than to another human being. The program retains this character as an instruction manual in the computer, since it can be read out of ROM as and when required by the use of a "disassembler" program. In this regard, the hexadecimal code version of the program can easily be displayed on the screen of the monitor or produced in a printout and can also be converted to the original assembly language (source code) version.

While stating that the programs could be read out of ROM and are occasionally so read by human beings for diagnostic purposes, Reed J. conceded that this was not their primary purpose. Instead, programs are designed primarily to be "communications" to a computer and are used to cause a computer to perform certain operations. It was recognized that no other form of written text operates in this way and that it stretches ordinary conceptual notions of the nature of a written text to conceive of it in operating in this fashion. This uniqueness was the essence of the dispute as to whether copyright exists.

Madam Justice Reed noted that there were three aspects of the case that were not in issue; first, that copyright subsisted in the written assembly language versions of the programs; second, that there was a substantial copying of such versions (assuming that they could be copied through the copying of the ROM chips); and third, that sufficient evidence was not adduced to rebut the presumption of ownership of copyright of the written program by the plaintiff arising from the production of a certificate of registration of copyright. The sole issue, instead, was whether a computer program which originates in a written text continues to be covered by that copyright when it is converted into its electrical code version or when it is embodied in a device designed to replicate that code. The defendant made four arguments for contending that copyright protection does not extend in this fashion.

The first argument was that the conversion of the program from assembly code to hexadecimal code was not a translation of the program but a different literary work. Reed J. found, however, that the conversion of a work into a code, or of a work written in one code to another code, constitutes a translation for purposes of the Act. The hexadecimal form of a program was described as being not a different literary work but rather a translation of the original.

The second argument was that copyright does not extend to a computer program embodied in a ROM chip because the program in such form exhibits a merger of the idea and its expression. Reed J. found that this argument would mean that the assembly code version of the program itself would not be copyrightable. She questioned the scope and even the existence of any legal rule that copyright ceases to subsist when there is a merger of the idea with its expression. Moreover, the evidence indicated that there was a multitude of forms of expression in which any given program could be written.

Reed J. referred to the Canadian and Australian *Cuisenaire* cases in which the physical objects in question were found not to be a type of work covered by the Act. She found, however, that these cases were not applicable, since the computer program when written was clearly a literary work, and its embodiment in a silicon chip retained the form of expression of the original work. The program in its source code version could be retrieved (read) by a process of translation or translations from the ROM chip.[65]

The third issue considered by Reed J. was whether the provisions of section 3 of the Act, and particularly the opening words to the effect that "'copyright' means the sole right to produce or reproduce the work or any substantial part thereof in any material form whatever", would encompass the embodiment of a program in a ROM chip. Reed J. held that these words were purposely drafted to encompass new technologies developed after the Act was enacted. Moreover, the 1911 amendments of the English Act, which were continued in the 1921 Canadian Act, did away with any requirement that in order to be covered by copyright, the copy or reproduction of the work had to be in a human-readable form. Such amendments also did away with any rule denying copyright protection to a work merely because the copy or reproduction could be characterized as being part of a machine. The opening words of section 3 covered the plaintiff's program as embodied in the ROM chip, since such embodiment was the production or reproduction of the work in a material form in the same way as a record or cassette tape. She noted that the requirement of readability or appearance to the eye means no more than that there be a method by which a work in which copyright is claimed and an alleged infringing work can be visually compared for the purpose of determining whether copying has occurred. Since the programs in this case could be read out of ROM and so compared, this requirement was met.

Reed J. also rejected the argument that the programs were not entitled to copyright protection because they were not intended for communication to human beings. It was noted that although a program could be used for such purpose, that was not its primary purpose. She indicated, however, that based on the opening words of section 3, there was no requirement that it be used for such purpose. The various paragraphs of subsection 3(1) (for example, paragraph 1(*d*) referring to contrivances for the mechanical performance or delivery of the work, that is, by acoustic or visual representation) should not be interpreted as limiting the opening words of section 3.

---

65 *Ibid.*, at 23-24.

Reed J. also rejected various public policy arguments that were made for the contention that a program recorded in ROM should not be protected by copyright. In particular, she rejected the arguments that the extension of copyright to such works constituted a granting of a monopoly on an item of commerce, that such extension would result in an overlap of copyright and patent law, that the interpretation of the Act to computer programs would usurp the function of Parliament in view of various proposals made to amend the Copyright Act, and finally that the defendants should not be penalized for acting as they did when there was so much debate and uncertainty as to whether the legislation extended to computer programs.

Based on the conclusion that copyright subsisted in the programs encoded in the ROM chips, Reed J. found that in addition to the corporate defendants which copied the programs, certain of the individual defendants were also liable for copyright infringement on the grounds that they had "authorized" the infringing activity. Certain other defendants were also found liable for such infringement on the grounds they "offered for sale" the infringing articles for purposes of subsection 17(4) of the Copyright Act.

In most of the cases discussed above, the main issue was whether as a matter of law, copyright in a computer program in its written form extends to its expression in a machine-readable form embodied in ROM. In general, there was no factual dispute as to whether the plaintiff's program was copied. By contrast, in *Gemologists International Inc. v. Gem Scan International Inc.*,[66] the issue was whether copyright in computer programs was infringed in circumstances where the literal code itself was not directly copied.

The applicant in that case sought an injunction to stop the use of pirated computer programs and solicitation of its customer lists. Montgomery J. found that the respondents conspired to commence business in competition with the applicant while still in its employ. As part of that conspiracy, the respondents purchased computer hardware capable of converting the applicant's software package for use on a different make of computer. After receiving advice that they could not use the applicant's programs, they hired computer programmers to write a program.

Based on the evidence, Montgomery J. found that the respondents infringed the applicant's copyright in its computer programs and program manuals by "copying the overall logical structure of the programs and substantially copying the program sequence of menus and menu options and by use of the applicant's flow chart."[67] He found further that the respondents' computer program prepared was created in large measure

66 (1986), 7 C.I.P.R. 225 (Ont. H.C.).
67 *Ibid.*, at 228.

by using and modifying the applicant's programs. Accordingly, Montgomery J. granted an injunction which provided, *inter alia*, that the respondents were restrained from copying, using or modifying the applicant's computer appraisal system and the appraisal system developed by the respondents, including program disks, source code and documentation. The respondents were also required to deliver up originals and copies of all such materials, including any modifications and changes made to such appraisal systems.

The foregoing review indicates that the weight of Canadian decisions is clearly in favour of the protection of computer programs, whatever their form of expression. However, notwithstanding the liberal interpretation given to the provisions of the 1921 Act, it cannot be said with certainty that all questions that arise in applying that Act to computer programs have been resolved. The uncertainty that remains will be illustrated by a review of the copyright laws of some other jurisdictions.

(b) Protection of Computer Programs Under Copyright Laws of Other Jurisdictions

(i) *United States*

Under the 1976 Copyright Act, a work is entitled to copyright protection if it is an original work of authorship and if it is fixed in a tangible medium of expression.[68]

The U.S. Act sets out seven categories of works of authorship including literary works, which are defined as follows:

> "Literary works" are works, other than audio visual works, expressed in words, numbers or other verbal or numerical symbols or indicia, regardless of the nature of the material objects, such as books, periodicals, manuscripts, phono records, film, tapes, disc, or cards, in which they are embodied.[69]

Section 101 also sets out when a work is deemed to be "fixed" in a tangible medium of expression, that is, "when its embodiment in a copy . . . is sufficiently permanent or stable to permit it to be perceived, reproduced, or otherwise communicated for a period of more than transitory duration." The word "copies" is also defined in section 101 to mean "material objects . . . in which a work is fixed by any method now known or later developed and from which the work can be perceived, reproduced, or otherwise communicated, either directly or with the aid of a machine or device".

In 1980, the U.S. Copyright Act was amended to include a definition of a computer program as "a set of statements or instructions to be used

68 U.S.C. Title 17, s. 102(a).

69 *Ibid.*, s. 101.

directly or indirectly in a computer in order to bring about a certain result."[70] Under the same Act, section 117 of the Copyright Act was amended to provide that it would not be an infringement of copyright in a computer program for the owner of a copy of the program to make or authorize the making of another copy or adaptation when necessary for its utilization or for archival purposes.

That computer programs in any form of expression are protected under the U.S. Copyright Act is indicated by two decisions of the U.S. Court of Appeals for the 3rd Circuit. In *Williams Electronics Inc. v. Artic International Inc.*,[71] it was held that a computer program embodied in a ROM chip used to operate a video game was entitled to copyright protection. The Court noted that the words of the 1976 Copyright Law precluded any requirement that a work be intelligible to human beings.

In *Apple Computer Inc. v. Franklin Computer Corp.*,[72] the Court of Appeals for the 3rd Circuit confirmed that operating systems programs expressed in object code form and embodied in ROM devices were entitled to copyright protection. The court noted that the 1976 Act removed any requirement which may have existed under the 1909 Act that a work be capable of communicating information to individuals.[73] Moreover, the definition of computer programs to mean statements or instructions to be used *directly* or *indirectly* in a computer indicated that instructions in object code form were protected, since only instructions in such form could be used "directly" by the computer. The court also referred to the definition of literary works to include works expressed in "numbers . . . or numerical symbols or indicia" as indicating an intention to expand the common usage of that expression.

The court also held that the statutory requirement of fixation was clearly met with respect to the embodiment of the computer program on a ROM, as distinguished from its expression in a traditional writing. The court noted in this regard that there was no distinction between application and operating system programs. Moreover, the fact that a program is used ultimately in the implementation of a process in no way affected its copyrightability. Finally, the Court of Appeals found that by providing for copyright protection of a computer program, protection was not thereby granted to an idea itself. The Court noted that as long as the same idea could be expressed in a plurality of different manners, and as long as other methods of expressing the idea were not foreclosed as a practical matter, there was no merger of the expression and the idea. In the case at bar, there was no such merger since other programs could be written to perform

70 Act of December 12, 1980, Pub. Law No. 96 — 517, s. 10, 94 Stat 3015, 3028.

71 685 F. 2d 870 (1982).

72 714 F. 2d 1240 (1983).

73 See *White-Smith Music Publishing Co. v. Apollo Co.*, 209 U.S. 1 (1908).

the same functions as the Apple operating system program.

The *Apple v. Franklin* decision has been generally regarded as an authoritative finding that computer programs in machine-readable form are entitled to copyright protection in the United States. However, the judgment is clearly based on the interpretation of the 1976 statute, as amended in 1980 to deal expressly with computer programs, which the court found reflected a clear legislative intent to cover programs in any form of expression. Accordingly, it is questionable whether the decision should be regarded as persuasive authority for the interpretation of the 1921 Canadian Act provisions, which are substantially different from those of the United States law.

Subsequent to the decision in *Apple v. Franklin*, various other United States cases considered what is commonly referred to as certain "second generation" issues relating to copyright and computer programs. These issues involve whether copyright protection extends not only to the literal computer code but also the structure, sequence and organization of a program. The leading case dealing with this question is *Whelan Associates Inc. v. Jaslow Dental Laboratory Inc.*[74] In that case, the Court of Appeals for the Third Circuit upheld the finding of the District Court[75] that the U.S. Copyright Act covers the non-literal elements of a computer program.

The evidence in that case showed that the defendant's computer program was written in a different computer language and was not a direct transliteration of the plaintiff's program. However, its structure and overall organization were found to be substantially similar based on the similarity of the file structures and screen outputs and the fact that various sub-routines performed almost identically in the two programs.

The Court of Appeals found that the application of copyright to the structure of a computer program did not mean that copyright protection was being extended to the idea of the program rather than its expression. In the case at bar, the idea underlying the plaintiff's computer program was the management of a dental laboratory, but the structure of the program was not essential to that task since other programs on the market performed the same functions but with different structures and designs. The expression of the idea in the plaintiff's computer program was the manner in which the program operates, controls or regulates the computer in receiving, assembling, calculating, retaining, collating and producing useful information. The court referred to the definitions of a "compilation" and a "derivative work" in the U.S. Copyright Act to support its conclusion that the sequencing and ordering of materials could be copyrighted and could be considered as part of the expression rather than the idea of a work.

In considering the particular elements of the structure of a program

74 797 F. 2d 1222, 55 U.S.L.W. 2094 (1986).
75 609 F. Supp. 1307 (E.D. Pa., 1985).

which were found entitled to copyright protection, the court first noted that complex and detailed file structures of a program which require certain information and order that information in a certain fashion are sufficiently informative to deserve copyright protection. With respect to the screen outputs, the court noted that these could be protected as audio-visual works rather than literary works. Although screen outputs were admissible as indirect evidence of copyright infringement in the underlying program, their probative value was limited since different programs could create the same screen output. Finally, with respect to various sub-routines in the two programs in question, the court noted that it was necessary to do a qualitative rather than quantitative analysis to determine whether the most significant steps of the programs were similar. The court accepted expert evidence which indicated that the two programs were doing the same thing with the same fields in the same files and accomplishing roughly the same results. In other words, the same file structures and program steps were followed and the functions performed by the programs in question were substantially the same.

(ii) *United Kingdom*

Prior to 1985, no express references were made in the Copyright Act of the United Kingdom to computer programs. However, certain provisions of the 1956 U.K. Act could be relied upon to extend protection to computer programs. For example, unlike the Canadian Act, the U.K. Act defines the word "writing" to include "any form of notation, whether by hand or by printing, typewriting or any similar process."[76] The 1956 Copyright Act also provides that in addition to the reproduction of a work in any material form, the making of an adaptation of the work (which is defined to include a translation) is a restricted act for copyright purposes.

In *Sega Enterprises Ltd. v. Richards*,[77] the Court considered whether a computer program used in a video game was entitled to copyright protection. The Court found that the assembly code program of the video game was a literary work under the 1956 Copyright Act and the machine code program derived from it by the operation of the assembler was to be regarded as "either a reproduction or an adaptation of the assembly code program, and accordingly . . . copyright does subsist in the program."[78] A similar conclusion was reached in the case of *Thrustcode Ltd. v. W.W. Computing Ltd.*[79] Both of these cases involved applications for interlocutory injunctions and did not contain a detailed analysis of the applicable legal principles.

---

76 Copyright Act, 1956 (4 & 5 Eliz. 2), c. 74, subs. 48(1).
77 [1983] F.S.R. 73.
78 *Ibid.*, at 75.
79 [1983] F.S.R. 502.

The Copyright (Computer Software) Amendment Act 1985 came into force on September 16, 1985. Section 1 of that Act provides that "the Copyright Act, 1956 shall apply in relation to a computer program (including one made before the commencement of this Act) as it applies in relation to a literary work." Section 1 also provides that in applying the Act to a computer program, a version of the program in which it is converted into or out of computer language or code, or into a different computer language or code, is an adaptation of the program. Section 2 states that references in the Copyright Act, 1956 to the reduction of a work to a material form, or the reproduction of any work in a material form should include references to the storage of that work in a computer. Accordingly, it appears that a computer program in machine-readable form would be subject to the same protection as a literary work in the United Kingdom and that its conversion from one to another would come within the rights of a copyright owner.

(iii) *South Africa*

The decision of the South Africa Supreme Court in the case of *Northern Office Microcomputers (Pty.) Ltd. v. Rosenstein*[80] has been cited in a number of Canadian cases. The court in that case considered whether programs were literary works within the meaning of sections 1 and 2 of the South Africa Act. Section 2(2) of that Act provides as follows:

> A literary, musical or artistic work shall not be eligible for copyright unless
> (*a*) sufficient effort or skill has been expended on making the work to give it a new and original character; and
> (*b*) the work has been written down, recorded or otherwise reduced to material form.

Based on the foregoing definition, the Court held that it was not necessary for a work to be "written down" but it was sufficient that it be "reduced to material form" to be eligible for copyright protection. It was noted that a computer printout which was in the form of a series of markings such as apertures of various geometrical shape and which constituted a code which was decipherable to those familiar with it was a work "reduced to material form" within the meaning of Section 2(2)(*b*) of the Act. Moreover, the instructions recorded upon a floppy disc also represented a work which was reduced to material form.

The *Rosenstein* case was based in large part on the provisions of section 2 of the South Africa Act which made it sufficient for a literary work to be "reduced to material form" and it is unclear whether the Court would have reached the same conclusion if the Copyright Act did not contain

80 [1982] F.S.R. 124.

such a provision precluding the necessity for a literary work to be expressed in writing.

(iv) *Australia*

The copyright status of computer programs in Australia was considered in detail in the case of *Apple Computer Inc. v. Computer Edge Pty. Ltd.*[81] The issue in that case was whether the defendant infringed copyright in the plaintiff's Applesoft and Autostart operating system programs by manufacturing "Wombat" computers which embodied certain ROM's and an EPROM containing copies of the plaintiff's programs.

The 1968 Copyright Act of Australia which was considered in that case granted to copyright owners the rights "to reproduce the work in a material form" and "to make an adaptation of the work." The word "adaptation" was defined to include, in relation to a literary work, the making of a translation of the work.

At trial, Beaumont J. held that a computer program, even in source code form, was not a literary work because it did not meet the test set forth in the English case of *Hollinrake v. Truswell*[82] that it afford "either information and instructions or pleasure in the form of literary enjoyment." He found further that a computer program was intended to assist in the functioning of a mechanical device rather than to serve a literary function.

On appeal, the Full Federal Court of Australia held that computer programs embodied in ROM were entitled to copyright protection. Fox J. noted that the object code programs could properly be described as adaptations since they were "a straightforward electronic translation" of the protected source code programs. Accordingly, the copying of an object code program embodied in a ROM was a reproduction of an adaptation of a literary work, whether or not the adaptation itself was a literary work.

Lockhart J. also found that an object code program was essentially a "mechanical translation of the source program into another language" and accordingly was an adaptation of the written program notwithstanding that it was not perceptible. Lockhart J. held further that the reproduction of an adaptation was itself an act comprised in copyright, and it was unnecessary to decide whether object code programs were themselves original literary works, although he expressed "considerable reservations" about this point. He also found it unnecessary to consider whether copies of programs expressed in ROM were reproductions of the source programs.

In his dissenting judgment, Sheppard J. stated that programs in object code form were not literary works. As fixed in ROM, they could not be seen and only the machine could "understand" the object code. Based

---

81 *Supra*, footnote 43.

82 [1894] 3 Ch. 420 at 428 (C.A.).

on dictionary meanings of a "translation", Sheppard J. suggested that it would be an abuse of language to describe the conversion of source code into object code as a translation. Moreover, based on the context in which the word "translation" is used, an adaptation of a literary work, like a literary work itself, must be capable of being seen or heard.

The decision of the Full Federal Court in the *Apple Computer v. Computer Edge* case was subsequently reversed by the Australian High Court.[83] The judgment was rendered subsequent to the Australian Copyright Act being amended to deal expressly with computer programs, but was based on the law as it was prior to such amendments, which was substantially the same as the 1921 Copyright Act of Canada.

The five judges of the High Court were of the unanimous view that a set of instructions expressed in a high level computer language or a low level assembly language (that is, source code form) qualified for protection as a literary work. However, in separate judgments, three judges of the High Court held that copyright did not subsist in programs expressed as sequences of electrical impulses stored permanently in a ROM containing connected electrical circuits. In a separate minority judgment, two judges expressed agreement with the judgments of Fox and Lockhart J.J. of the Full Federal Court.

In his reasons for judgment, Gibbs C.J. noted that the sequence of electrical impulses making up the object program could not be perceived although it could be described either in binary notation or in hexadecimal notation. Such notation, in turn, could be either displayed on the VDU of a computer or printed out. However, in the case at bar, the appellant had apparently not made use of any writing or other visual display of Apple's programs in either source code or in binary or hexadecimal notation.

Gibbs C.J. noted that the making of ROM's and an EPROM reproducing Apple's programs would have constituted an infringement of its copyright in any one of three cases:

(1) Apple's source programs were original literary works, its object programs were adaptations of those source programs and the Wombat object programs embodied in the ROM's and an EPROM were reproductions of those adaptations; or
(2) Apple's source programs were original literary works and the Wombat object programs were either adaptations or reproductions of those source programs; or
(3) Apple's object programs embodied in its ROM's were themselves original literary works and the Wombat object programs were reproductions of those object programs.

83 (1986), 65 A.L.R. 33.

Gibbs C.J. noted that the decision of the Full Federal Court was in favour of the first proposition. In considering whether this finding or any of the other propositions was correct, Gibbs C.J. reviewed the authorities dealing with the meaning of a literary work. He found that such a work must be expressed in print and writing and must afford either information and instruction or pleasure in the form of literary enjoyment. He concluded that insofar as the object code programs existed either as a sequence of electrical impulses, or in the pattern of circuits that when activated generated those impulses, they were not literary works:

> Although the electrical impulses could be represented by words or figures, the impulses themselves did not represent or produce any words or figures. They were not visible or otherwise perceptible, and they were not, and were not intended to be, capable by themselves of conveying a meaning which could be understood by human beings. Obviously, the patterns of the circuits in the ROMS also did not represent or reproduce any words or figures, and were incapable of conveying any meaning.
>
> . . .
>
> It seems to me a complete distortion of meaning to describe electrical impulses in a silicon chip, which cannot be perceived by the senses and are not intended to convey any message to a human being and which do not represent words, letters, figures or symbols as a literary work; still less can a pattern of circuits be so described.[84]

Gibbs C.J. also rejected the view that object code programs embodied in ROM's were adaptations or translations of the source programs. Relying on the definition of a "translation" set forth in the *Shorter Oxford English Dictionary*,[85] he noted that the source programs were not turned into another language, even into another computer language, but rather were turned into electrical impulses. He also indicated that the secondary definition of a translation as the "expression or rendering of something in another medium or form" was a "transferred and figurative use" of the word which did not come within the meaning of an adaptation for the purposes of the Act. In any event, the object programs in the ROM's were not a translation of source programs since they did not express or render the source programs, but were only a means of making the instructions written in the source programs effective. He noted further that an adaptation within the meaning of the Act must itself be a work and that the object programs in ROM's were not works.

Gibbs C.J. held further that the object programs embodied in the Wombat ROM's were not reproductions of Apple's source programs since it was not possible to show any objective similarity between the two works.

84 *Ibid.*, at 39-40.

85 *Supra*, footnote 41.

Although the Wombat ROM's might well be reproductions of the Apple ROM's, they were not reproductions of the underlying source programs. The Wombat ROM's embodied the idea and logical structure of Apple's source programs, but, in the absence of any resemblance, did not *reproduce* the expression of the idea and the logical thought of the source program.[86]

Gibbs C.J. concluded that he was not persuaded by decisions in other jurisdictions which appeared to come to a contrary conclusion that a sequence of electrical impulses in a silicon chip, not capable itself of communicating anything directly to a human recipient, and designed only to operate a computer, was either a literary work or a translation of a literary work within the Copyright Act. He noted further that

> [i]t may be regretted that the respondents have no remedy in copyright against the appellants who pirated their programs. However although it would be no doubt right to give the *Copyright Act* a liberal interpretation, it would not be justifiable to depart altogether from its language and principles in an attempt to protect the products of scientific and technological developments which were not contemplated, or only incompletely understood, when the statute was enacted.[87]

In a concurring judgment, Brennan J. also expressed the view that the object code programs were not literary works. He indicated that such a work must be capable of being perceived by the senses and the electrical charges which constituted the object programs did not meet this test:

> [They] cannot be seen or touched or heard or, if they can, they do not communicate the letters of the original literary work, the source program. Nor, for that matter, do the electrical charges communicate the letters or figures by which an object program may be represented. The object programs are not literary works. Apple's right to relief, if any, must depend on its ownership of the copyright in the Applesoft and Autostart source programs.[88]

Brennan J. also found that the process of turning the source programs into object programs could not be characterized as a translation. He concluded that reliance on the secondary meaning of that word (that is, the expression of a work in another medium) was precluded by the necessity for an adaptation itself to be a work. Since a sequence of electrical charges constituting an object program was neither a literary work nor any other kind of "work" within the meaning of the Copyright Act, it was not an "adaptation" of the source program.

Brennan J. also found that the Wombat ROM's were not reproductions of the underlying source programs on the following grounds:

---

86 *Op. cit.*, footnote 83, at 42.
87 *Ibid.*, at 42.
88 *Ibid.*, at 54.

> Reproduction in any material form is a notion wider than the notion of copying which defined the scope of copyright protection before the 1911 Act, but the notion of reproduction still connotes a resemblance between the work in which copyright subsists and the work which is copied from it. The material form to which the respective works are reduced need not be the same, but if the forms are so dissimilar as to deny resemblance between the works, one cannot be said to be a reproduction of the other.[89]

Brennan J. rejected the view that it was sufficient to look only at the collocation of ideas expressed in a written compilation, rather than the manner in which those ideas are expressed. Such a view would be tantamount to conferring by copyright the protection given by the grant of a patent. He went on to state

> The exclusive right to reproduce a work in which copyright subsists is not calculated to prohibit the use of ideas expressed in the work, though the form in which those ideas may be expressed without the licence of the owner of the copyright is limited. But copying in the sense of producing something which embodies the same ideas is not, without more, an infringement; the form of the thing produced must be such as to resemble the work in which the copyright subsists. Here there is no resemblance between the Wombat ROM's and EPROM and the written compilation of the Applesoft and Autostart source programs. It is immaterial that the former were derived from the latter when there is no resemblance between them, for both resemblance and derivation are essential to reproduction.[90]

In a separate concurring judgment, Deane J. held that the embodiment of a program in object code form on a ROM was not a reproduction in material form of either the underlying source program, or of the written expression of the program expressed in object code. The written object code denoted the series of electrical impulses produced by following the directions of the source code while the charged silicon chip reflected the application of that series of electrical impulses. However, the notation of the electrical impulses in object code (as distinct from the electrical impulses themselves) played no part in the production of the Wombat ROM's and it would be a distortion to regard the latter as a reproduction or adaptation of the former.

Deane J. also rejected the argument that the actual series of electrical charges constituting each operational program was itself an original literary work when embodied in a ROM on the following grounds:

> It is not written. It is not in a comprehensible language. It cannot be read. It cannot even be seen. Nor is it designed or produced to be read or seen. It

89 *Ibid.*, at 57.
90 *Ibid.*, at 58.

is, and was designed and produced to be, an attribute of a functioning part of an operating machine.[91]

In a single dissenting judgment, Mason and Wilson J.J. expressed agreement with the majority of the Full Federal Court that the programs embodied in ROM's were entitled to copyright protection. Their Lordships found that the object code was a translation of the source code, and therefore an adaptation, notwithstanding that it took a form which could only be read and understood by a machine. Their Lordships also found there was no need for a translation to convey a clear idea of the literary work to human beings. In any event, an object program could be reduced to writing understood by those who were skilled in the art.

Their Lordships noted that *prima facie*, an adaptation of a literary work must itself satisfy the description of a literary work, and that such a work "ordinarily and traditionally" would take a written form, but this was not a requirement of the Act:

> There seems to be no reason to doubt that a literary work is made and entitled to copyright protection from the time it is first recorded on tape, if that be the first material form that the work takes. In our opinion, an object code, although brought into existence by mechanical means, takes on the same literary character as is possessed by the source code from which it is derived. This conclusion seems necessarily to follow, if the protection secured by the Act of source programs as original literary works is to be effective. If there is no copyright in the object programs which are a natural and necessary derivative of the source programs, then there is no point in protecting the source programs.[92]

Their Lordships found further that once it was determined that the object programs contained in the Apple ROM's were adaptations of original literary works, it was unnecessary to decide whether there was any reproduction in a material form of the original literary works, that is, the source programs. It was sufficient that Apple's object programs were entitled to protection as adaptations of the original literary works.

Following the decision of the Full Federal Court but before that of the High Court in the *Apple Computer v. Computer Edge* case, the Australian Parliament proceeded with amendments to the copyright law to deal with computer programs. The new Act, which received Royal Assent on June 15, 1984, extended the definition of a "literary work" to include a computer program. The new Act protects such a work if it is stored in any form from which it can be reproduced. A computer program is defined in the Act to mean

> an expression, in any language, code or notation, of a set of instructions

---

91 *Ibid.*, at 63.

92 *Ibid.*, at 48.

(whether with or without related information) intended, either directly or after either or both of the following:
(*a*) conversion to another language, code or notation;
(*b*) reproduction in a different material form,
to cause a device having digital information processing capabilities to perform a particular function.[93]

The new Act also expanded the definition of an "adaptation" to mean, in relation to a computer program, a version of the work (whether or not in the language, code or notation in which the work was originally expressed) not being a reproduction of the work. This definition would appear to be wide enough to include object code versions of programs originally written in source code form.[94]

(v) *Other Jurisdictions*

In addition to the foregoing jurisdictions, a number of other countries have enacted revisions to their copyright laws to deal expressly with computer programs. Such countries include France, Germany, Japan, Hungary, India and Taiwan. Other countries which are currently considering amendments to the copyright law include Denmark, Israel, Norway, Singapore and Sweden.[95] In still other countries, such as Italy and the Netherlands, the courts have upheld copyright protection notwithstanding the absence of legislation dealing expressly with such protection. Copyright protection would also appear to be available in the Philippines, Hong Kong, Singapore and New Zealand.[96]

(c) Copyright and Computer Programs: Conclusions

The foregoing review indicates that there is a growing tendency at the international level to grant copyright protection to computer programs, in whatever form of expression. In most jurisdictions, however, the extension of the copyright laws to computer programs has been based on legislative changes made to deal expressly with such technology. In the case of Canada, on the other hand, the 1921 Copyright Act has generally been given a liberal interpretation so as to encompass computer programs.

93 Copyright Amendment Act (Australia), 1984, s. 3(b).

94 See Millard, *Legal Protection of Computer Programs and Data* (Toronto: Carswell, 1985) at 73-74; Davidson, Gregarous and Bahrick, "International Software Protection" (1985), 4 Computer Law Reporter 322 (Part II) at 323-325.

95 See Susan Nycum, "Current Activity in the Legal Protection of Computer Programs", Paper presented at the Pacific Rim Symposium on Technology and the Law, Singapore, October, 1985. See also Davidson, Gregarous & Bahrick, "International Software Protection" (1985), 4 Computer Law Reporter 167 (Part I) and 322 (Part II).

96 *Ibid.*, at 13-15.

However, since most of the Canadian cases have been decided only at the level of interlocutory injunctions and, since to date, no appellate decision has been rendered after a full trial of all of the issues, the legal status of computer programs in Canada remains a matter of some uncertainty. Such uncertainty is highlighted by the decision of the Australian High Court in the *Apple Computer v. Computer Edge* case which, in the context of a copyright statute substantially the same as the 1921 Copyright Act, came to the opposite conclusion from that reached in the majority of the Canadian cases as to the application of copyright to computer programs in their object code form.

The Canadian computer copyright cases appear to assume that a program in its object code form embodied in a ROM can be properly characterized as a literary work. According to Madam Justice Reed in the *Apple Computer v. Mackintosh (No. 2)* case, if there is any requirement that such a work be expressed in print and writing, it is deemed to be met by the fact that the work is *capable* of being expressed in such form, that is, that it can be replicated in its object code form or disassembled into its original source code form to be read on the screen of a monitor or to be used to produce a printout. Similarly, any requirement that a literary work be intended for human communication is deemed to be met by the fact that it is *capable* of being used for such purpose even if that is not its primary purpose. By contrast, the majority of the Australian High Court in the *Apple Computer v. Computer Edge* case found that a program in ROM was not a literary work because it was not expressed in print or writing or any other human perceptible form, and it was not intended for human communication.

The Canadian copyright cases also appear to assume that an object code program in its machine-readable form can properly be described as a translation of the underlying source code program. Accordingly, the reproduction of the object code program is assumed to be an infringement of copyright in the source code program based on the provisions of paragraph 3(1)(*a*) of the Copyright Act which gives to the owner of copyright the sole right "to produce or reproduce any translation of a work." The Canadian cases have generally not considered whether, as a matter of law, a translation of a literary work must itself be a literary work. Such cases have also generally not included a detailed analysis of the meaning of the concept of a "translation" for purposes of the Copyright Act. Such an analysis lead the Australian High Court in the *Computer Edge* case to conclude that the process of a transformation of a program from source code to object code machine-readable form could not be characterized as a translation even within the secondary meaning that may be given to that term.

The Canadian computer copyright cases raise difficult issues as to the meaning of the concept of reproduction of a work in any material form. Although the precise rationale of the Canadian cases is somewhat unclear, they would appear to assume that even if a program embodied in a ROM is not itself a literary work, it is nevertheless a reproduction in a material form of the underlying written source code program. They also appear to assume that the reproduction is itself protected by copyright whether or not it is a "work" for purposes of the Copyright Act.

The reliance placed on the opening words of section 3 of the Copyright Act, particularly in the *Apple Computer v. Mackintosh (No. 2)* case, would appear to be somewhat misplaced. In this regard, such opening words are of assistance only if it is first shown that what is reproduced is in fact a "work", that is, that the program as embodied in a ROM is a literary work for purposes of the Act, or otherwise a translation of such a work for purposes of paragraph 3(1)(*a*) of the Act. If that initial criterion is satisfied, then the process whereby the program embodied in a chip is replicated in another chip would be a reproduction of an "original" work for purposes of the Act. Alternatively, it would have to be shown that the copying of the program embodied in a chip represents an indirect copying of the original assembly or source code program.

The finding that an object code program embodied in ROM is a reproduction of the underlying source code program would appear not to be supported by some of the earlier authorities dealing with the concept of a reproduction in a material form, as reviewed by the Australian High Court in the *Apple Computer v. Computer Edge* case.[97] In this regard, the requirement that there be an objective similarity, capable of being detected as a matter of visual resemblance, between the original and the copy would not be met. Even in the case of a written notation or representation of the object program, there would not be such similarity detectible as a matter of visual resemblance with the underlying source program, although expert evidence could be adduced to show that the former is derived from the latter. Although it has been held in a series of English cases that copyright in an artistic work may be infringed by the indirect reproduction of an object based on that work, in all such cases there was a visual resemblance with the underlying work that could be detected by a non-expert.[98]

The authorities which have dealt with the concept of a reproduction in any material form also indicate that a sound recording is not a reproduction of the underlying musical work, and that the making of such a recording is deemed to be a "copyright" use of the musical work only because of the express wording to such effect in paragraph 3(1)(*d*).[99]

---

97 See *supra*, at 53.

98 See *Dorling v. Honnor Marine Ltd.*, [1965] Ch. 1 (C.A.); *L.B. (Plastics) Ltd. v. Swish Products Ltd.*, [1979] R.P.C. 611 (H.L.); *British Leyland Motor Corp. v. Armstrong Patents Co.*, [1986] 2 W.L.R. 400 (H.L.).

99 See, *e.g.*, *Thompson v. Warner Bros. Pictures Ltd.*, [1929] 2 Ch. 308 (C.A.).

Moreover, a sound recording is protected not as a reproduction of a musical work but as a separate category of copyright protected material under subsection 4(3) of the Act. In other words, the making of a sound recording does not appear to come within the inclusive language of the reproduction of a work in any material form, but instead represents a separate right for which express provision has been made in the Copyright Act. Accordingly, there would appear to be some difficulty in relying on the analogy between programs embodied in ROM chips and musical works embodied in cassettes as a basis for contending that the former should be protected in the same manner as the latter.

In the absence of an authoritative appellate decision in which all issues relating to the application of copyright to computer programs in their machine-readable form are fully considered, it would be premature to conclude that the 1921 Copyright Act of Canada applies without qualification to programs in all forms of expression.

6. COPYRIGHT PROTECTION FOR VIDEO GAMES: LITERARY OR AUDIO-VISUAL WORKS

Several Canadian cases have considered whether copyright protection may be obtained for coin-operated video games. The nature of such games and their method of operation were described as follows, in the case of *Nintendo of America Inc. v. Coinex Video Games Inc.*:[100]

> Before a coin is inserted in the machine, the games operate in a repeated 'attract mode', playing continuously and automatically in repeating sequences which display the name "Donkey Kong" or "Donkey Kong Junior." as the case may be. The attract mode also displays the plaintiff's copyright notice, a sample of the game in the form of a series of images accompanied by musical and sound effects, previous game scores and instructions as to how the game is to be played. Upon the deposit of a coin and the depressing of the start button, the game shifts into the "play mode". In this mode, the figure (subject to certain controls which are exercised by the game player) and other objects and obstacles move on the screen in a predetermined sequence according to the stored information in the electronic circuitry to interact with the figure controlled by the player in a pre-set manner, also according to the stored information in the electronic circuitry.

Video games are thus different from other audio-visual works since the player exercises some control over the manner in which the work is displayed in the "play mode". Accordingly, the question arises whether there is a fixed audio-visual work, or whether a new work is created every time the game is played. A further question is whether the game displayed in the "attract" and "play" modes is protected by copyright as an audio-

100 (1982), 69 C.P.R. (2d) 122 at 125 (Fed. C.A.).

visual work separate from any copyright that may subsist in the computer program used to operate the game.

In the *Nintendo* case,[101] the Federal Court of Appeal granted an "Anton Piller" order[102] compelling the preservation of all documents, games, apparatus, plans, drawings, circuitry and component parts directly related to the plaintiff's video game. The plaintiff's affidavit evidence in that case included an admission by the defendant that he had copied the plaintiff's game boards and installed them in cabinets enabling them to be operated. The court noted that the unique video images and sounds of the game were generated by instructions in the form of digital memory chips. These were contained in the game board which were sold by the defendants apart from the cabinets themselves. The court found, based on this affidavit evidence, that the appellant had made out an extremely strong *prima facie* case of copyright infringement. The judgment suggests that copyright may subsist in the program used to operate a video game, although it is not clear whether a separate copyright subsists in the audio-visual displays of the game.

Anton Piller orders were also granted in respect of alleged copyright infringement in the cases of *Midway Manufacturing Co. v. Bernstein*[103] and *Bally Midway Manufacturing Co. v. Coinex Video Games Inc.*[104] However, there was no discussion in either of these cases as to whether a separate copyright subsists in a video game and the program used to operate the game. By contrast, in *Atari Inc. v. Video Amusements of Canada Ltd.*,[105] Jerome A.C.J. refused an interim injunction to restrain the defendant from infringing its copyright in video games. It was noted that there was a triable issue as to the applicability of copyright to computers and their programs.

In the United States, it would appear that copyright protection extends both to the images and sounds that make a video game an audio-visual work as well as the program used in the operation of the game. In *Midway Manufacturing Co. v. Dirkshneider*[106] it was held that video games were copyrightable as audio-visual works under section 102(*a*)(6) of the Copyright Act. The Court held that such works, as fixed in printed circuit boards, were "tangible objects from which the audio-visual works may be perceived for a period of time more than transitory."[107] Accordingly, the statutory requirement for fixation was met.

In *Midway Manufacturing Co. v. Strohon*,[108] the Court held that there

101 *Ibid.*
102 See *Anton Piller K.G. v. Mfg. Processes Ltd.*, [1976] 1 All E.R. 779 (C.A.).
103 [1983] 1 F.C. 510 (Fed. T.D.).
104 (1983), 71 C.P.R. (2d) 105 (Fed. T.D.).
105 Summarized at (1982), 19 A.C.W.S. (2d) 4 (Fed. T.D.).
106 543 F. Supp. 466 (D.Neb., 1981).
107 *Ibid.*, at 480.
108 564 F. Supp. 741 (N.D. Ill., 1983).

was no infringement of copyright in the plaintiff's video game as an audio-visual work since the characters in the defendant's game had been changed. However, the Court held that the program used to operate the game was entitled to a separate copyright protection. The Court noted in particular that the Copyright Act of 1976 was intended to protect object code as well as source code and the physical medium in which the object code was stored did not affect such protection.

7. COPYRIGHT PROTECTION FOR DESIGNS OF COMPUTER HARDWARE

The question of the copyright status of designs used in the manufacture of computer hardware raises complex issues as to the interpretation of the Copyright Act and its relationship to the Industrial Design Act.[109] One such question is assuming that the designs are protected by copyright as either literary works (that is, charts or plans) or artistic works (that is, drawings), whether the reproduction of physical objects reflecting such designs infringes copyright in the designs themselves? A further question is whether such designs are excluded from copyright protection by virtue of subsection 46(1) of the Copyright Act, which provides as follows:

> This Act does not apply to designs capable of being registered under the *Industrial Design Act*, except designs that, though capable of being so registered, are not used or intended to be used as models or patterns to be multiplied by any industrial process.

The question of whether copyright in designs may be infringed by the reproduction of a physical object based on such designs was considered in the case of *Bayliner Marine Corp. v. Doral Boats Ltd.*[110] In that case, Walsh J. found that the defendant infringed copyright in the plaintiff's designs for the hull and deck portions of its boats by the reproduction of the boat parts themselves. In reaching this conclusion, Walsh J. relied on the following extract from Fox, *Canadian Law of Copyright and Industrial Design*:[111]

> There will, of course, be infringement whether the defendant copied the work directly from the work protected by copyright, or whether he reproduced it from memory, or derived it indirectly from an intermediate copy, and it makes no difference that the intermediate copy is unprotected. While in order to succeed a plaintiff must show that the copyright work is the source from which the infringing work is derived, it need not be the direct source.

---

109 R.S.C. 1970, C. I-8.

110 (1985), 5 C.P.R. (3d) 289 (Fed. T.D.); reversed on other grounds (1986), 67 N.R. 139 (Fed. C.A.).

111 2nd ed. (Carswell, 1967) at 329.

Reference was also made to the *King Features Syndicate* case,[112] in which it was held that the right to reproduce sketches or drawings includes their reproduction in three-dimensional form, such as by dolls or toys. The test of copying adopted in that case was that set forth in *Mast v. Francis*[113] in which it was held that "a copy is that which comes so near the original as to give to every person seeing it the idea created by the original."

Walsh J. noted that the Canadian *Cuisenaire* case[114] stood for the proposition that copyright in a work in two dimensions could be infringed by the production and sale of an article in three dimensions provided that there was a clear visual resemblance between the infringing work and the work in which copyright was alleged to exist. Reference was also made to the case of *L.B. (Plastics) Ltd. v. Swish Products Ltd.*,[115] in which it was held that the correct approach was to compare object to drawing and to look at both in establishing the scope of the copyright and determining whether there was infringement.

Based on the jurisprudence, Walsh J. concluded that it could be seen from looking at the defendant's boats that they were derived from the plaintiff's designs. Accordingly, the copying of the intermediate objects, that is, the boats made from the designs, constituted infringement of copyright in the designs themselves.

On appeal, the Federal Court of Appeal found on other grounds as noted below that the designs in question were not entitled to copyright protection. Accordingly, it did not find it necessary to consider whether copyright in a plan could be infringed by the making of an object made according to the plan. The Court of Appeal referred to the decision of the House of Lords in the case of *British Leyland Motor Corp. v. Armstrong Patents Co.*,[116] in which it was held that copyright in plans for automobile mufflers was infringed by the making of copies of the mufflers themselves. However, in view of significant differences between the relevant Canadian and British legislation, the Court refused to express an opinion by way of *obiter dicta* on this issue.

In the case of *Rucker Co. v. Gavel's Vulcanizing Ltd.*,[117] Walsh J. found that drawings used for the construction of mechanical objects were not entitled to copyright protection when the objects themselves were covered by a patent. He indicated that to give copyright protection for the drawings would defeat the time limitation provided for in the Patent Act, and that it was not the intention of Parliament, nor from a practical point of view was it desirable, that the Patent Act, the Copyright Act or the Industrial

112 [1940] 2 All E.R. 355 at 358.
113 (1822), 5 B. & Ald. 737 at 743.
114 (1967), 54 C.P.R. 1 (Ex. Ct.); affirmed (1968), 57 C.P.R. 76 (S.C.C.).
115 [1979] R.P.C. 611 (H.L.).
116 [1986] 2 W.L.R. 400 (H.L.).
117 (1985), 6 C.I.P.R. 137 (Fed. T.D.).

Design Act be interpreted to give overlapping protection.[118]

The second issue that must be considered in determining whether the reproduction of articles is an infringement of copyright in the underlying designs is whether the designs are capable of being registered under the Industrial Design Act, in which event they will be excluded from copyright protection. However, such exclusion will not apply if the designs are not used or intended to be used as models or patterns to be multiplied by any industrial process.[119]

The Industrial Design Act does not state expressly what designs are capable of registration under the Act. Section 6 provides that the Minister shall register a design for which an application for registration is deposited under the Act if he finds that it is not identical with or does not so closely resemble any other design already registered so as to be confounded therewith. The registration of a design gives the registrant the exclusive right to the design for a term of five years, which term is renewable for a further five year period.[120] Section 14 of the Act sets out two conditions of registration: first, that the design must be registered within one (1) year from its publication in Canada, and second, that the article to which the design applies must be marked with the name of the proprietor together with the letters "Rd".

To be registered under the Act, a design must meet the requirement of originality. In the case of *Clatworthy & Son Ltd. v. Dale Display Fixtures Ltd.*,[121] it was held that

> there must be some substantial difference between the new design and what had theretofore existed. A slight change of outline or configuration, or an unsubstantial variation, is not sufficient to enable the author to obtain registration.

In the case of *Bata Industries Ltd. v. Warrington Inc.*,[122] Reed J. held that the jurisprudence demanded "a higher degree of originality than is required with regard to copyright." It was necessary to show "a spark of inspiration on the part of the designer either in creating an entirely new design or in hitting upon a new use for an old one."[123] It was not sufficient that the design was merely a trade variant of what had gone before.

While earlier authorities took the position that the only designs that could be registered under the Act were those to be applied to the surface of an article, subsequent cases indicate that a design relating to the shape or configuration of an article may be registered provided that it is to be

118 *Ibid.*, at 155.
119 Copyright Act, subs. 46(1).
120 S. 10, Industrial Design Act, *supra*, footnote 109.
121 [1929] 3 D.L.R. 11 at 14 (S.C.C.).
122 (1985), 5 C.P.R. (3d) 339 at 347 (Fed. T.D.).
123 *Ibid.*, at 347.

applied to the ornamenting of an article.[124] However, such designs will be registrable only if they give no protection against the function served by the objects. In *Mainetti S.P.A. v. E.R.A. Display Co.*,[125] Walsh J. held that

> [i]f the primary purpose served by the article in question is one of function, the decorative ornamentation being merely incidental, the article should be patented under the Patent Act, if it is patentable, rather than seeking an industrial design registration for it.

Walsh J. held further in the *Mainetti* case that in order for a design not to be rejected for registration on the ground of functionality, the designer must depart from a commonplace form based on requirements of utility alone and must achieve a distinctive appearance. It was necessary to show an artistic design which might appeal to a purchaser apart from the useful function which the design may serve. Because the more significant design features in the *Mainetti* case were hidden and were primarily functional, Walsh J. concluded that they should not have been subject to industrial design registration.

It will be seen that in order for a design to be capable of registration under the Industrial Design Act, it must create a visual appeal to a potential purchaser. In addition, a design is not registrable if it is wholly functional, that is, if its only object is to perform a functional purpose. Accordingly, designs for computer hardware which provide a visual appeal may be capable of registration under the Industrial Design Act and as such, may be excluded from protection under the Copyright Act, provided that they are not used or intended to be used as patterns or models to be multiplied by an industrial process for purposes of subsection 46(1) of the Copyright Act, in which event the exclusion from copyright protection does not apply.

Subsection 46(2) of the Copyright Act provides for the adoption of rules under the Industrial Design Act for determining the conditions under which a design shall be deemed to be used for the purposes set out in subsection 46(1). In accordance with this provision, Rule 11(1) of the Industrial Design Rules,[126] enacted pursuant to the Industrial Design Act, provides as follows:

> 11(1) A design shall be deemed to be used as a model or pattern to be multiplied by any industrial process within the meaning of section 46 of the Copyright Act,
>
> (a) where the design is reproduced or is intended to be reproduced in more than 50 single articles, unless all the articles in which the design is

124 *Cimon Ltd. v. Bench Made Furniture Corp.*, [1965] 1 Ex C.R. 811 (Ex. Ct.).
125 (1984), 80 C.P.R. (2d) 206 at 224-225 (Fed. T.D.).
126 C.R.C. 1978, c. 964.

reproduced or is intended to be reproduced together form only a single set as defined in subsection 2; and

(b) where the design is to be applied to
   (i) printed paper hangings,
   (ii) carpets, floor cloths, or oil cloths manufactured or sold in lengths or pieces,
   (iii) textile piece goods, or textile goods manufactured or sold in lengths or pieces, and
   (iv) lace, not made by hand.

A number of cases have considered whether the word "and", as linking sub-paragraphs (*a*) and (*b*) of Rule 11(1), should be read conjunctively or disjunctively. If read disjunctively, it would mean that any design which is capable of being registered under the Industrial Design Act, and which is reproduced or intended to be reproduced in more than 50 articles, would be deemed to be used as a model or pattern to be multiplied by an industrial process and would be excluded from copyright protection. This would mean that engineering drawings and designs for computer hardware may be excluded from copyright protection since such designs are generally reproduced in more than 50 articles. If read conjunctively, however, a design would be excluded from copyright protection only if it is intended to be reproduced in more than 50 articles *and* if it is to be applied to specific articles such as paper hangings, carpets, textile piece goods and lace.

In several cases before the Trial Division of the Federal Court, it was held that the word "and" as used in Rule 11(1) should be read conjunctively.[127] One of these decisions involved the case and keyboard for computer hardware which were found entitled to copyright protection.[128] The correctness of these decisions was questioned, however, in *Jeffrey Rogers Knitwear Productions Ltd. v. R.D. International Style Collections Ltd.*,[129] in which Dube J. noted that a conjunctive interpretation of the word "and" in Rule 11(1) would make it impossible for any design to qualify for protection under the Industrial Design Act and would thus serve as a bar to any registrations under the Act. Dube J. also referred to other authorities which supported a disjunctive interpretation being given to Rule 11(1) of the Industrial Design Rules.[130] In view of these authorities, Dube

127 See *Royal Doulton Tableware Ltd. v. Cassidy's Ltd.* (1984), 1 C.P.R. (3d) 214 (Fed. T.D.); *Interlego AG. v. Irwin Toy Ltd.* (1985), 3 C.P.R. (3d) 476 (Fed. T.D.); *Bayliner Marine Corp. v. Doral Boats Ltd.* (1985), 5 C.P.R. (3d) 289 (Fed. T.D.); *Apple Computer Inc. v. Minitronics of Can. Ltd.* (1985), 8 7 C.P.R. (3d) 104 (Fed. T.D.); affirmed (1985), C.P.R. (3d) 431 (Fed. C.A.).

128 *Apple Computer Inc. v. Minitronics of Can. Ltd., ibid.*

129 (1985), 6 C.P.R. (3d) 409 (Fed. T.D.).

130 *Con Planck Ltd. v. Kolynos Inc.*, [1925] 2 K.B. 804; *King Features Syndicate Inc. v. O. & M. Kleeman Ltd.*, [1941] A.C. 17 (H.L.); *Eldon Industries Inc. v. Reliable Toy Co. Ltd.* (1964), 44 C.P.R. 239 (Ont. H.C.); affirmed [1966] 1 O.R. 409 (Ont. C.A.).

J. would have reached a different conclusion as to the interpretation of Rule 11(1) from that reached in other judgments of the Federal Court; however, he felt compelled to follow the recent decisions of the Federal Court, particularly in the circumstance of that case where there was a "blatant violation of the plaintiffs' design by way of knock-off by the defendant."[131]

The proper interpretation to be given to the Industrial Design Rules would appear to have been resolved by the judgment of the Federal Court of Appeal in the case of *Bayliner Marine Corp. v. Doral Boats Ltd.*[132] The Court of Appeal reversed the finding of Walsh J. and held that the word "and" at the end of paragraph 11(1)(*a*) was intended to link joint and several notions, not only joint notions. Mahoney J. on behalf of the Court adopted the reasoning of Dube J. in the *Jeffrey Rodgers* case as to the interpretation to be given to Rule 11(1). He found further that plans for boat parts were "designs capable of being registered under the Industrial Design Act" since the details of the shape of the design were essentially ornamental, and accordingly they qualified as designs for purposes of the Act. He found it unnecessary to enter into any inquiry as to whether the designs met the necessary test of novelty. He also indicated that it was irrelevant that the plaintiffs had failed to register the designs within the time prescribed by the Act.

The appellate decision in the *Bayliner* case provides some indication of the relationship between the Industrial Design Act and the Copyright Act. In particular, the decision indicates that where designs for three-dimensional objects are intended primarily to serve an ornamental rather than functional purpose, they will be regarded as designs capable of registration under the Industrial Design Act. Moreover, where such designs are intended to be reproduced in more than 50 countries, they will be excluded from copyright protection.

In the case of designs or plans which serve only a functional or utilitarian purpose, however, considerable uncertainty remains as to the scope of protection that may be available under the Copyright Act. In particular, it is uncertain whether the copying of a three dimensional object should be considered a reproduction of the two dimensional drawing, design or plan used to make such article.

It would now appear to be settled law in England that the reproduction of a functional object will be deemed to be an indirect infringement of copyright in the drawing on which the object is based.[133] However, the House of Lords indicated in the *British Leyland* case that, although it regarded itself as bound by earlier decisions, Parliament did not intend

131 *Op. cit.*, footnote 129, at 414.

132 (1986), 10 C.P.R. (3d) 289 (Fed. C.A.).

133 *British Leyland Motor Corp. v. Armstrong Patents Co.*, [1986] 2 W.L.R. 400 (H.L.).

for the protection afforded by copyright to be exploited so as to prevent the reproduction of a functional object depicted in a drawing. Accordingly, the plaintiff was unable to assert its copyright so as to derogate from the right of a car owner to repair his car which entitled him to obtain replacement parts from competing manufacturers.

In view of the refusal of the Federal Court of Appeal in the *Bayliner* case to apply the finding of the *British Leyland* case, it is uncertain whether the concept of "indirect reproduction" is a part of existing Canadian law. In the *Cuisenaire* cases in Australia and Canada, it was suggested that the object which is reproduced must be "in the nature" of the underlying work, that is, that it must itself be a literary or other work protected by copyright in order for the copyright in the underlying work to be infringed. Such a narrower interpretation would mean that the reproduction of a three-dimensional utilitarian object which is not itself either a literary or artistic work would not be a reproduction of the underlying work since it is not in the nature of that work. In any event, the uncertainty and confusion in this area of the law demonstrates the need for legislation to clarify the circumstances in which the reproduction of three-dimensional objects may constitute an infringement of copyright in underlying plans or designs used to create such works.

8. COPYRIGHT AND OTHER FORMS OF LEGAL PROTECTION FOR SEMICONDUCTOR CHIPS

In 1984, the Semiconductor Chip Protection Act[134] was enacted in the United States. The present section will review the nature of semiconductor chip products, the legislative history and scope of the U.S. legislation and the possible application of the 1921 Canadian Copyright Act to the design, layout or architecture of semiconductor chip products.

(a) Nature of Semiconductor Chips

Semiconductor chips are an integral part of the architecture of many modern computers and are used to operate a wide variety of business, consumer and industrial products. Such chips are manufactured by a complex process that takes place in several stages. First, the physical and electrical characteristics of a chip are defined according to the functions to be served by the chip and any object in which it is to be incorporated. Second, a circuit design engineer prepares a schematic outline or series of drawings representing the method of connection of as many as 250,000 or more transistors to implement appropriate electronic functions. Third, based on the paper document, a layout design engineer decides on a three-

134 U.S.C. Title 17, Chap. 9.

dimensional layout pattern to be imprinted onto a silicon or other semiconductor substance to form a chip. The fourth stage is the manufacturing process, whereby composite drawings of the various layers of a chip are used to generate a series of "masks" or stencils. These masks are used to etch or deposit material on the silicon wafer so as to create the final chip product. The chip itself is thus configured to take on the patterns comprising the various mask works. As a result of this manufacturing process (known as photolithography or masking), several hundred chips may be created which are then sawed apart and wired into ceramic or plastic packages for use in electronic equipment.[135]

Several reasons have been advanced as to why the layout or architecture of semiconductor chips should be protected independently of any copyright protection that may subsist in the programs embodied in the chips. These reasons were summarized as follows in testimony by J. Baumgarten, General Counsel, U.S. Copyright Office, before a Subcommittee of the House Committee on the Judiciary which considered the adoption of the Semiconductor Chip Protection Act:

> a) the chips for which they seek protection include types that do not embody computer programs at the time they are exposed to duplication (for example, 'blank' chips sold for customer programming, or not designed to hold program material); b) unauthorized duplication of even 'programmed' chips may avoid duplication of that part of the chip carrying the program; and c) the copyright owner in the program may not be the same as the claimant of proprietary rights in the chip pattern design.[136]

In recommending the adoption of legislation to provide for the protection of semiconductor chips, the U.S. Senate noted that the semiconductor industry was an essential component of the U.S. economy. The U.S. chip market in 1983 was estimated at $7 billion, with the worldwide market being $16 billion. It was noted that the development of a new family of semiconductor chips could cost up to $100 million, while a pirate firm could make a perfect copy of a semiconductor chip for as little as $50,000 to $100,000 and an entire family of chips for less than $1 million.[137]

---

135 See Report on Semiconductor Chip Protection Act of 1984, Committee on the Judiciary House of Representatives, 98th Congress, 2d Session, Report 98-781, May 15, 1984 (reproduced in *Nimmer on Copyright*, Vol. 4 (Matthew Bender, 1986), Appendix 30).

136 Hearings on H.R. 1007 before the Subcommittee on Courts, Civil Liberties and the Administration of Justice of the House Committee on the Judiciary, 1979 CCH Copyright Law Reports 20,029 at 10,049, cited in Roberts, "Protection of Semiconductor Chip Design under Canadian Copyright Law: Will Canada Follow the Lead of the United States" (1985), 23 U.W.O.L. Rev. 101 at 105.

137 S. Rep. No. 98-425, 98th Congress, 2d Sess. (1984), at 4-6; cited in Roberts, *ibid.*, at pp. 105-106.

(b) The U.S. Semiconductor Chip Protection Act

The United States Semiconductor Chip Protection Act was enacted on the grounds that existing intellectual property laws did not offer adequate protection against the duplication of computer chips. In its Report to the House of Representatives, the Committee of the Judiciary[138] noted that patent law did not protect the particular layouts and design work performed by chip manufacturers in adapting the electronic circuits for a particular industrial purpose. Moreover, copyright law could not be relied upon since it did not protect utilitarian objects or useful articles. The Committee recommended that the Copyright Act not be extended to deal with mask works used in the creation of semiconductor chips since this involved a manufacturing process. The Committee noted further that copyright law did not prevent the unauthorized duplication of a finished article represented by a technical drawing, and protected only the artistic features of a useful article if it was capable of an independent existence as a work of art. The Committee also referred to international considerations which militated against granting copyright status to semiconductor chip products. Specifically, the extension of the Copyright Act in this manner would require the United States to grant the same protection to such products of foreign nationals based on the "national treatment" principle, but there was no assurance that other countries would protect the works of U.S. nationals. For the foregoing reasons, the Committee recommended the adoption of separate legislation to protect semiconductor chip products.

The U.S. Semiconductor Chip Protection Act[139] defines a mask work as follows:

> [T]he 2-dimensional and 3-dimensional features of shapes, pattern and configuration of the surface of the layers of a semiconductor chip product regardless of whether such features have an intrinsic utilitarian function that is not only to portray the appearance of the product or to convey information.[140]

A "semiconductor chip product" is defined as

> the final or intermediate form of any product
> (A) having two or more layers of metallic, insulating, or semiconductor material, deposited or . . . removed from, a piece of semiconductor material in accordance with a predetermined pattern; and
> (B) that is intended to perform electronic circuitry functions.[141]

Subsection 902(a)(1) of the Chip Protection Act provides that "an

138 Report on Semiconductor Chip Protection Act of 1984, 98th Congress, 2d Session, Report 98-781, *op. cit.* footnote 135.

139 *Supra*, footnote 134.

140 *Ibid.*, subs. 901(2).

141 *Ibid.*, subs. 901(1).

original mask work fixed in a semiconductor chip product" is eligible for protection provided that the owner of the work is a national or domiciliary of the United States, or if the owner is a national, domiciliary or sovereign authority of another country that is party to a treaty affording protection to mask works to which the United States is also a party, or if the work is first commercially exploited in the United States, or if the work originates in a country to which protection under the Act has been extended by Presidential proclamation under subsection 902(a)(2).

Protection is not available under the Act to a mask work that is not "original" (that is, an independent creation of the author who did not copy it from another source) or that consists of designs that are staple, commonplace or familiar in the semiconductor industry or that are variations of such designs not combined in an original way.[142] Similarly, protection for a mask work does not extend to any idea, procedure, process, system, method of operation, concept, principle or discovery.[143]

The exclusive rights given to the owner of a mask work are as follows:

(1) to reproduce the mask work by optical, electronic, or any other means;
(2) to import or distribute a semiconductor chip product which the mask work is embodied; and
(3) to induce or knowingly to cause another person to do any of the acts described in paragraphs (1) and (2).[144]

However, the owner of a mask work cannot prevent the reproduction of the work solely for the purpose of teaching, analyzing or evaluating the concepts or techniques embodied in the work or the circuitry or organization of the components used in the mask work.[145] In other words, a competitor can use the results of reverse engineering to study the design of a mask work with a view to designing and distributing semiconductor chip products embodying his own original mask work. The owner of the mask work also cannot restrict the further sale or other disposition of a chip product after its first sale.[146]

Protection under the Act lasts only for a period of ten years from the date of registration of a mask work or the date on which it is first commercially exploited, whichever first occurs.[147] If a work is not registered within two years from the date of first commercial exploitation, protection under the Act will cease.[148] The Act provides for the affixing of a notice to a mask work or a semiconductor chip product; such notice is not a

142 *Ibid.*, subs. 902(b).
143 *Ibid.*, subs. 902(c).
144 *Ibid.*, s. 905.
145 *Ibid.*, subs. 906(a).
146 *Ibid.*, subs. 906(b).
147 *Ibid.*, s. 904.
148 *Ibid.*, subs. 908(a).

condition of protection but affords *prima facie* evidence of notice of ownership.[149]

The Act provides for various remedies in respect of the infringement of the exclusive rights of the owner of a mass work, including an injunction, actual damages which may be supplemented by an account of profits or, if the owner elects instead of actual damages and an account for profits, statutory damages of up to $250,000.[150] In addition, a court may order the destruction or other disposition of any infringing semiconductor chip products and any masks or other articles by means of which such products may be reproduced.[151] The Act contains a provision preserving any rights or remedies to which a person is entitled under the Copyright Act or the Patent Act.[152]

Subsection 914(a) of the Act sets out a procedure whereby interim protection may be granted to works of nationals of countries which do not otherwise qualify for protection under subsection 902(a) of the Act. Specifically, subsection 914(a) provides that the Secretary of Commerce may extend protection to nationals of other countries on three conditions: first, that the foreign nation is making progress (either by treaty negotiations or legislative enactment) toward a regime of mask work protection similar to that under the Act; second, that the nationals and persons controlled by the nationals of such other country are not engaging and have not in the recent past engaged in chip piracy or the sale of products containing infringing semiconductor components; and third, that entry of the Secretary's order would promote the purposes of the Act of achieving international comity toward mask work protection.

The Secretary of State is authorized to determine whether any country qualifies for an order for interim protection for the works of its nationals and the length of any such order. After the expiration of three years, the Secretary's power to provide for interim protection and all orders thereunder will expire. After such time, applications for registrations will not be accepted from foreign nationals or domiciliaries, other than those of countries which are parties to a treaty providing for the protection of semiconductor chips or which come within the scope of a Presidential Proclamation declaring that the nation has enacted legislation offering protection to nationals of the United States on substantially the same basis as under the United States Act.[153] However, any registrations made in favour of foreign nationals under any interim orders will not be affected.

In accordance with section 914 of the U.S. Act, interim protection

149 *Ibid.*, s. 909.
150 *Ibid.*, s. 911.
151 *Ibid.*, subs. 911(f).
152 *Ibid.*, subs. 912(a).
153 *Ibid.*, para. 902(a)(2).

was extended to Canada for a one-year period commencing on June 27, 1985. This interim protection was subsequently extended to November 8, 1987, which is the date on which all orders made under subsection 914(a) are to expire. After such date, no registrations will be accepted for semiconductor chip products of Canadian nationals unless a treaty has been adopted or legislation providing for the protection of semiconductor chips has been enacted by Canada. Alternatively, it would have to be shown that the provisions of the existing Copyright Act extend to semiconductor chips and that such protection would be extended to United States nationals based on the national treatment principle of the Universal Copyright Convention.[154]

(c) The 1921 Canadian Copyright Act and Semiconductor Chip Products

There remains to consider whether the 1921 Canadian Copyright Act may be interpreted as extending to the architecture or layout of semiconductor chips, regardless of the copyright status of programs that may be embodied in such chips. As noted above,[155] it would appear that at least in some circumstances, the reproduction of a three-dimensional object based on a two-dimensional design or plan may be an infringement of copyright in the design or plan itself. On this basis, it could be argued that the copying of a semiconductor chip represents the indirect reproduction of technical drawings from which the chip is manufactured.

For several reasons, it is questionable whether the foregoing proposition would be accepted under the 1921 Copyright Act. First, it appears unlikely that any requirement of visual resemblance between the drawings and a three-dimensional product would be met since any similarity could be detected only through microscopic examination of a semiconductor chip in relation to the underlying designs. Moreover, such similarity would be apparent only to a trained expert and not to an ordinary layman. The drawings used to manufacture a chip product would also not appear to qualify as artistic works since they would not have an aesthetic feature. Such drawings may qualify as charts or plans coming within the category of literary works. However, insofar as there may be a requirement that the object which is reproduced is "in the nature" of the underlying design, a chip product produced through a manufacturing process and used for

154 At the international level, the World Intellectual Property Organization (WIPO) has established a Committee of Experts on Intellectual Property in Respect of Integrated Circuits. The Committee presented a draft treaty at its Second Session at Geneva in June, 1986. For the full text of the draft treaty and commentary, see *Computer Software and Chips 1986, Protection and Marketing*, Vol. 1 (Practising Law Institute, 1986) at 299-361.

155 See pp. 79-85, *supra*.

a utilitarian process would not appear to meet that criterion in relation to the underlying design. For the foregoing reasons, it is questionable whether the 1921 Copyright Act would be interpreted as providing protection for semiconductor chips.

9. PROTECTION OF COMPUTER DATA BASES

A computer data base is any collection or compilation of information held in electronic or machine-readable form. Such a data base may be created for the internal purposes of a business or may be provided to third parties for a fee.

The information stored in a computer data base may consist of scientific, financial, legal or other data. In some cases, a data base may be an electronic "reproduction" of a compilation of information which already exists in printed form. In other cases, the computer may be used to collect or compile the information which is then held solely in computer-readable media.

Two questions arise as to the copyright status of information held in a computer data base. The first is whether the input into a computer of information which already exists in printed form represents one of the sole rights of the copyright owner. The second question is whether a compilation of information held in computer-readable media is entitled to copyright protection (a) where the information already exists in written form, or (b) where the information is created in the computer and no written version of such information exists.

With respect to the first question, in order for the input of information into a computer to be a copyright-protected act, it must be shown that such input represents either a reproduction in some material form or a translation of the underlying literary work. The Canadian cases dealing with copyright in computer programs support the proposition that the input of the printed work into the computer is a copyright-protected act. In this regard, the work embodied in computer media may be displayed on a computer VDU or printed out in the same manner in which it is input. Accordingly, the requirement of a visual resemblance between the original and the reproduction would be met, although a machine or device is required to display the material embodied in computer medium.

Similar questions arise as to whether a compilation of information originally created in a computer would be entitled to copyright protection. It may be argued that such a compilation is not a literary work because it is not expressed in print or writing although it is capable of being expressed in such form. If the compilation also exists in written form, a machine-readable version may be regarded as an "intermediate work", the reproduction of which would represent indirect copying of the underlying literary work. Accordingly, it may be of some significance if the compilation exists only in computerized form or whether a written or printed version also exists.

Assuming that copyright subsists in a computer data base, the author of such a work must be determined. In general, the person who compiles or edits the information contained in the data base would be considered the author. Where different parts of the work are contributed by different authors, copyright would subsist in the separate contributions as well as in the compilation as a whole. In this regard, a "collective work" is defined in section 2 of the Copyright Act to include "any work written in distinct parts by different authors, or in which works or parts of works of different authors are incorporated."

If copyright is found to subsist in a computer data base, the rights of the owner would include the reproduction of the work in any material form, its publication, the production or reproduction of any translation of the work, as well as the communication of the work by "radiocommunication". The latter expression has been interpreted restrictively to refer only to any radiocommunication intended for reception by the general public. Accordingly, the transmission of information by telecommunication facilities from one computer facility to another would not come within the rights of a copyright owner as long as the transmission is not "broadcast" to the general public. However, the improper use of a computer facility in order to obtain unauthorized access to information stored in a data base may be prohibited based on section 301.2 of the Criminal Code.[156] In addition, the mere display of information contained in a data base on a VDU would not be a "reproduction" of the work in any material form unless the information is "down-loaded" for storage in the permanent memory of another computer. In this connection, the display of the text of information on the VDU would appear to be equivalent to the reading of a book rather than a reproduction of the information contained in the book.

If information from a data base is reproduced in some form, it will be a question of fact as to whether the part that is taken is a "substantial" part of the data base. In this regard, the copying of a non-essential part of a data base would not be an infringing act in the same way as the taking of a small part of the information contained in a dictionary or encyclopedia would not represent a reproduction of a substantial part of the work. This determination must be made based on the quality and not just the quantity of the work that is taken. The copying of extracts of information from a data base may also be permitted based on the provisions of section 17(2)(a) of the Copyright Act which permit any "fair dealing" of a work for such purposes as private study, research or criticism.

A data base is also different from other copyright works insofar as

156 R.S.C. 1970, c. C-34, as amended.

it may be subject to constant revisions and updates. However, each version of the data base should be entitled to copyright protection, since it would be derived from the immediately preceding version. Each such version could also be considered as a copy of the earlier versions of the data base insofar as it incorporates a "substantial part" of such earlier works.

10. COPYRIGHT PROTECTION FOR WORKS CREATED BY THE USE OF A COMPUTER

Computers may be used in a variety of ways to create works which, if created directly by human effort, would be entitled to copyright protection. For example, graphic and other artistic works, musical compositions, economic and statistical analyses and other original compilations may be produced through the use of a computer. The question arises as to whether such works should be entitled to copyright protection and who is the author or creator of such works.

Under the 1921 Copyright Act, a work created through the use of a computer should be entitled to protection provided that it meets the other criteria for protection; that is, it is an original work, is fixed in a material form and comes within one of the recognized categories of copyright protection. It would have to be shown, however, that the "author" of the work is a qualified person within the meaning of section 4(1) of the Copyright Act. In other words, a human author must be shown to be responsible for creating the work.

Different views have been expressed as to whether ownership of copyright in computer-created works should be vested in the person who owns the programs used in the creation of the work, the person who is responsible for the use of the programs or some other person. The National Commission on New Technological Uses of Copyrighted Works (CONTU) in its final report[157] recommended that the author of a work created through the aid of a computer be the person who employs the computer. The Commission noted that the authorship of the program or of the input data was entirely separate from the authorship of the final work. It was recognized, however, that the creator of the final work may have to obtain permission from the persons owning copyright in the program or the input data. By contrast, in England, the Whitford Committee[158] recommended that the author of the output of a computer be the person or persons who devised the instructions and originated the data used to control and condition the computer to produce the particular result. It was recognized that in many cases that this would be a matter of joint authorship. This

157 Library of Congress, Washington, 1979.

158 Copyright and Designs Law — Report of the Committee to Consider the Law on Copyright and Designs, London HMSO, CMND 6732, 1977, para. 515.

recommendation was rejected, however, in a Green Paper prepared in response to the Whitford Report.[159] The Green Paper recommended, instead, that the person responsible for running the data through the computer in order to create the new work be considered the author of the work.[160]

In its report issued in 1985, the House of Commons Subcommittee on the Revision of Copyright,[161] recommended that copyright in a protected work created with the assistance of a computer be vested in the individual or entity principally responsible for the making of the work. The Committee recommended that copyright ownership not be vested in the creator of the original program since such an argument confused the creation of the work with the creator's materials. The Committee qualified its recommendation by noting that in the case of video games, which involve an interaction of an individual with a computer program, copyright should not vest in the player because the results follow directly from what is written in the program. Although the use of such games requires a certain intellectual effort on the player's part, the results are entirely predictable because the player only supplies directions to guide a predetermined process.

It is submitted that under the 1921 Copyright Act, the owner of copyright in any computer-created work which otherwise qualifies for copyright protection would be the person responsible for the operation or use of the computer. Such a person may require authorization from the owner of copyright in the program or input data used to operate a computer. However, ownership of copyright in a computer-created work must be distinguished from ownership of copyright in the program or data. Although practical difficulties may arise in a particular case as to who is actually responsible for the use of a computer, such difficulties may be best resolved by negotiation among the parties involved in the use of the computer and the development of the work.

11. COPYRIGHT IN WORKS CREATED BY EMPLOYEES AND COMMISSIONED WORKS

Subsection 12(1) of the Copyright Act sets out the general principle that the author of a work shall be the first owner of the copyright therein. Two qualifications to that general rule are set forth in subsections 12(2) and 12(3) of the Act.

---

159 Reform of the Law Relating to Copyright, Designs and Performers' Protection, London, HMSO, CMND 8302, 1981.

160 *Ibid.*, at 35, para. 7.

161 A Charter of Rights for Creators, 2nd Report of the Subcommittee on the Revision of Copyright (Supply and Services Canada, 1985).

Subsection 12(2) provides that where an original of an engraving, photograph or portrait is ordered by some person for valuable consideration, then in the absence of an agreement to the contrary, the person by whom the original is ordered is the first owner of copyright. In the absence of a corresponding provision applying to other works, ownership of copyright in such works will be governed by the general principle set out in subsection 12(1).

Subsection 12(3) of the Copyright Act provides as follows:

> Where the author was in the employment of some other person under a contract of service or apprenticeship and the work was made in the course of his employment by that person, the person by whom the author was employed shall, in the absence of any agreement to the contrary, be the first owner of the copyright ...

The effect of the foregoing provisions is that except for computer programs created by employees in the course of their employment, the person or persons who prepare such programs will be the first owner of copyright. Moreover, in accordance with subsection 12(4) of the Copyright Act, any agreement to provide for a contrary arrangement as between an employer or employee, or to provide for the assignment of copyright by the first owner to some other person, must be in writing.

Computer programs are often developed through the combined efforts of two or more persons. Such a program may be considered as a "work of joint authorship" within the meaning of section 2 of the Act; that is, a work "produced by the collaboration of two or more authors in which the contribution of one author is not distinct from the contribution of the other author or authors."

Several recent Canadian cases dealing with computer programs have involved disputes as to rights to computer programs developed by employees. For example, in the case of *Canavest House Ltd. v. Lett*,[162] there was evidence of oral discussions, a draft employment agreement, and memoranda setting out the rights of the defendant to computer programs which he developed while in the plaintiff's employment. Since there was no agreement providing that the employee would have ownership of copyright in the computer programs, it was held that such copyright belonged to the employer. However, the defendant was entitled to retain copies of research materials used by him in the course of designing the programs.

The question of the ownership of copyright in computer programs as between an employer and employee should be clearly addressed in any employment agreement. The question of such ownership must also be addressed in any software development agreement in which the software

162 (1984), 2 C.P.R. (3d) 386 (Ont. H.C.).

is commissioned by one party on behalf of another, or in any situation where computer programs are to be developed through the joint efforts of two or more persons. Any such agreement should also set out the rights of the parties to copies of notes, writings and other materials used in the preparation of computer programs. If applicable, the agreement should provide for the waiver by the creators of any "moral rights" to the programs under subsection 12(7) of the Copyright Act, that is, the right to claim authorship of the programs and to prevent any changes prejudicial to the creator's honour or reputation.

12. REGISTRATION OF COPYRIGHT AND ASSIGNMENTS OF COPYRIGHT

Section 37 of the Copyright Act provides for books to be kept at the Copyright Office in which the names or titles of works, the names and addresses of authors and other particulars prescribed under the Act may be entered. An application for registration is required to be made on the prescribed forms according to whether the work is published or unpublished, and must be accompanied by the prescribed fee. The work is identified solely by title and no copy of the work need be submitted with the application. Subsection 37(2) provides that particulars concerning a work may be caused to be entered by the author or publisher, the owner or other person interested in the copyright. By contrast, subsection 38(1) indicates that the persons by whom an application for the registration of copyright may be made are the author or his legal representative or any person purporting to be the agent of such author or legal representative.

Although registration of copyright in a work is not a pre-condition of copyright protection, a certificate of registration is evidence that copyright subsists in the work and that a person registered is the owner of copyright.[163] Accordingly, the registration of copyright may be of considerable advantage particularly in interlocutory injunctions since it may preclude the necessity of adducing extensive evidence to prove the existence of copyright or its ownership. Every register of copyright under the Act is evidence of the particulars entered therein and documents purporting to be copies of entries from the register certified by the Registrar of Copyrights and sealed with the seal of the Copyright Office are admissible in courts without further proof or production of the originals.[164] Registration is also deemed to give a defendant notice of the existence of copyright in a work so that he will be precluded from denying liability for damages on the grounds that he was not aware and had no reasonable

163 Copyright Act, subs. 36(2).
164 *Ibid.*, subs. 36(1).

grounds for suspecting that copyright subsisted in the work.[165]

It is uncertain how registration under the Copyright Act should be effected for a work created by an employee in the course of his employment. Although subsection 12(3) of the Act provides that the employer is deemed to be the first owner of copyright in such a work, the employee would appear to be the "author". Accordingly, the name of the employee who actually creates a work should be shown on the application form filed in the Copyright Office. This may be impracticable in the case of computer programs which are often prepared through the efforts of a number of employees over an extended period of time.

Subsection 40(1) of the Copyright Act provides for the registration of a grant of an interest in copyright either by assignment or licence. Under subsection 12(4) of the Act, the grant of any such interest must be in writing signed by the owner of the right in respect of which such grant is made or his agent. Registration is effected by producing to the Copyright Office the original assignment of the grant of an interest and a certified copy, together with the prescribed fee.[166] The failure to register a grant of an interest in copyright may result in the assignment or licence in question being adjudged void as against a subsequent assignee or licensee for valuable consideration unless such assignee or licensee has actual notice of the prior assignment or licence.[167]

The Universal Copyright Convention provides that if copyright registration is required in any Convention country, the formalities of registration are deemed to be satisfied in the case of a work published in another Convention country if all copies of the work bear a copyright notice with the year of publication and the name of the copyright owner. Such information must be printed in such a manner as to give reasonable notice of claims of copyright.[168] Because Canada and the United States are both parties to the Universal Copyright Convention, works published in Canada should be marked in accordance with the Convention, that is, with the symbol ©, the year of publication and the name of the copyright owner.

13. REMEDIES FOR COPYRIGHT INFRINGEMENT

Subsection 20(1) of the Copyright Act provides as follows:

> (1) Where copyright in any work has been infringed, the owner of the copyright is, except as otherwise provided by this Act, entitled to all such remedies by way of injunction, damages, accounts, and otherwise, as are or

---

165 *Ibid.*, s. 22. See *infra*, at 106-107.
166 *Ibid.*, subs. 40(1).
167 *Ibid.*, subs. 40(3).
168 Universal Copyright Convention, Article III.

may be conferred by law for the infringement of a right.

This section will review the various forms of interim relief, that is, an interlocutory injunction and an Anton Piller order, that may be sought in a copyright infringement action. This is followed by a review of various forms of permanent relief such as a final injunction, an assessment of damages, an accounting for profits and the recovery of possession of copies of the work. Some of the presumptions of copyright ownership and authorship as well as procedural provisions set out in the Act relating to the enforcement of the rights of copyright owners are also considered. Finally, the so-called "summary remedy" or quasi-criminal provisions of the Copyright Act are reviewed.

(a) Interim Relief

(i) *Interlocutory Injunction*

The most commonly sought remedy in a copyright infringement action is an interlocutory injunction to restrain the alleged infringing act until a full trial of the action. In the case of *Bulman Group Ltd. v. Alpha One-Write Systems B.C. Ltd.*,[169] Addy J. described the purpose of an interlocutory injunction as follows:

> . . . an interlocutory injunction is a practical device or procedure by means of which a Court may prevent what it perceives as a *prima facie* injustice causing continuing harm, from being prolonged until the issue is finally determined.

In *Yule Inc. v. Atlantic Pizza Delight Franchise (1968) Ltd.*,[170] Evans C.J.H.C., relying on the House of Lords decision in *American Cyanamid Co. v. Ethicon Ltd.*,[171] adopted the following test as to whether an application for an interlocutory injunction should be granted in an action involving a contractual dispute:

> It is sufficient if the applicant establishes that there is a substantial issue to be tried; that it has demonstrated a *prima facie* case, in the sense that the applicant has a legal right which it is attempting to protect pending trial; that the failure to grant the relief will result in a threatened harm to the applicant which may not be adequately compensable by way of damages and finally, that the preponderance of convenience must be in favour of the applicant if the application for injunctive relief is to succeed.

---

169 (1980), 54 C.P.R. (2d) 171 at 174 (Fed. T.D.); reversed on other grounds (1981), 54 C.P.R. (2d) 179 (Fed. C.A.).

170 (1977), 34 C.P.R. (2d) 284 at 286 (Ont. H.C.); affirmed (1977), 17 O.R. (2d) 505 (Ont. Div. Ct.).

171 [1975] A.C. 396 (H.L.).

Notwithstanding the tests adopted in the *American Cyanamid* and *Atlantic Pizza* cases, it would appear that at least in some circumstances involving copyright infringement, a plaintiff must establish more than that there is a "serious" or "substantial" issue to be tried. In *Syntex Inc. v. Apotex Inc.*,[172] it was held that the judge should consider more carefully the merits of the underlying claim, where the granting of an injunction would effectively dispose of the action, and there would be nothing left for the unsuccessful party to proceed to trial. In *Canavest House Ltd. v. Lett*,[173] which involved an application to restrain copyright infringement by a former employee, it was held that, because the remedy sought went straight to the defendant's means of livelihood, the plaintiff had to establish a strong *prima facie* case.

Some authorities have indicated that different tests are also to be applied in copyright infringement actions with respect to proof of "irreparable injury" and the balance of convenience. In *Jeffrey Rogers Knitwear Productions Ltd. v. R.D. International Style Collections Ltd.*,[174] Dube J. noted that

> in matters of clear violations of copyrights registered under the *Copyright Act* . . . the jurisprudence of the Federal Court has been to grant interlocutory injunctions without concerning itself with irreparable injury or the balance of convenience.

In *Universal City Studio Inc. v. Zellers Inc.*,[175] Walsh J. made the following comments:

> I do not believe that the possibility of collecting damages is in all cases a sufficient answer to an action brought against an infringer when plaintiffs have a very strong *prima facie* case. The protection of industrial property rights from counterfeiting is an increasingly important question. In principle, these rights should be protected whether or not breach of them causes serious damages.

In *Duomo Inc. v. Giftcraft Ltd.*,[176] Strayer J. found that a strong *prima facie* case had been made out and it was not necessary to consider further the balance of convenience. Moreover, it was no answer to the plaintiff's claim for an interlocutory injunction that the plaintiff's loss would be compensated by damages on the following grounds:

> It should be for the owner of the copyright to decide whether he wishes to give a licence for particular uses of it and the court should not automatically endorse what is in effect a compulsory licence by requiring him to accept

---

172 (1984), 1 C.P.R. (3d) 145 (Fed. C.A.).
173 (1984), 2 C.P.R. (3d) 386 (Ont. H.C.).
174 (1984), 6 C.P.R. (3d) 409 at 410-411 (Fed. T.D.).
175 (1983), 73 C.P.R. (2d) 1 at 10 (Fed. T.D.).
176 (1984), 3 C.I.P.R. 70 (Fed. T.D.).

payment of money by an infringer in lieu of protecting his exclusive rights.[177]

In the case of *IBM v. Spirales Computer Inc.*,[178] Reed J. noted that different formulations had been set forth in various cases as to the degree of injury or damage required to be proven in order for a plaintiff to be granted an interlocutory injunction. Some cases formulated a test of "irreparable damage", others expressed the test as "whether or not damages would be an adequate remedy" and still others referred to whether there was threatened harm to the applicant which might not be "adequately compensated" by way of damages. In the case at bar, Reed J. noted that in view of the relative size of the defendant and the plaintiff, it was hard to conclude that the plaintiff would suffer "irreparable harm" from the actions of the defendant alone. However, it might nevertheless suffer "substantial harm" not only because of the defendant's activities but also because the withholding of an injunction would encourage others to import and sell computers infringing the plaintiff's copyright. This was described as "a flood gates effect" or "death by 1000 cuts". Reed J. concluded that in cases where the copying is "blatant", it was appropriate to consider a less stringent test of potential damages without requiring proof of irreparable harm or proof that damages would not be an adequate remedy.

The reasoning of Reed J. in the *Spirales Computer* case was also adopted by Walsh J. in the case of *Apple Computer Inc. v. Minitronics of Canada Ltd.*[179] It was noted in that case that the plaintiff had established a clear case of copyright infringement as well as of trade mark infringement and the only valid defences would be based on legal issues, some of which were of a serious nature so as to give the defendants an arguable defence. In an action for copyright infringement, Walsh J. held that it was not necessary to prove "irreparable harm" as would be necessary in an action for patent infringement.

In the case of *Apple Computer Inc. v. Mackintosh Computers Ltd. (No. 1)*,[180] Cullen J. refused to grant an interlocutory injunction to restrain the defendants from importing, selling or distributing copies of computer programs for which the plaintiff claimed copyright. Cullen J. found that although the plaintiffs had established a *prima facie* case, it was not a strong one and the defendants had an arguable case on all of the issues. The plaintiffs had also failed to establish that they were losing a substantial sum of money and customers because of the defendant's actions. Cullen J. noted that the affidavit evidence adduced by the plaintiffs did not contain any indication of irreparable harm and the information provided was based on speculation and conjecture without reference to direct sources.

177 *Ibid.*, at 74-75.
178 [1985] 1 F.C. 190 (Fed. T.D.).
179 (1985), 7 C.P.R. (3d) 104 (Fed. T.D.); affirmed (1985), 8 C.P.R. (3d) 431 (Fed. C.A.).
180 (1985), 3 C.I.P.R. 133 (Fed. T.D.).

An application for an interlocutory injunction was also refused in the case of *Digital Equipment Corp. v. C. Itoh & Co. (Canada),*[181] which involved an alleged infringement of copyright in certain terminals, menus and owner's manuals applicable to computers. Although the plaintiff had established that there was a very serious issue to be tried, the defendant had shown an arguable case. McNair J. found that the plaintiff had failed to establish, on the balance of probability, that it would suffer irreparable harm or damage if the defendant was not enjoined from selling competitive terminals.

The foregoing authorities indicate that somewhat different tests have been applied by the courts in determining whether an interlocutory injunction should be granted in copyright infringement actions. In some cases, it has been held sufficient to demonstrate that there is a "serious" or "substantial" issue to be tried without the need for the plaintiff to establish a strong *prima facie* case. However, if the granting or refusing of an injunction would effectively dispose of the action, or if the application for an injunction is made in the context of an employer-employee relationship, the establishment of a strong *prima facie* case would appear necessary.

Different tests have also been applied as to the degree of loss or injury that must be shown by the plaintiff as well as whether an injunction should be granted on the balance of convenience. Some authorities indicate that where the plaintiff has made out a strong *prima facie* case, it need not show irreparable loss or injury or deal with the issue of balance of convenience. Other authorities, such as the *Spirales Computer* case, indicate that a less stringent test of "substantial harm", rather than irreparable injury which cannot be adequately compensated by damages, should be adopted to take into account the effect that the failure to grant an injunction may have in encouraging other infringers.

It would appear that in cases of blatant or flagrant copying, the courts will be less stringent in requiring proof of irreparable injury suffered by the plaintiff which cannot be adequately compensated by damages. However, if the defendant succeeds in showing that it has an arguable defence in law, the courts will be more reluctant to grant an injunction in the absence of clear proof of irreparable injury.

(ii) *Anton Piller Orders*

The nature and purpose of the so-called Anton Piller orders was described as follows by the English House of Lords in the case of *Rank Film Distributors Ltd. v. Video Information Centre*:[182]

---

181 (1985), 6 C.P.R. (3d) 511 (Fed. T.D.).

182 [1981] 2 All E.R. 76 at 78 (H.L.).

> They [Anton Piller orders] are designed to deal with situations created by infringement of patents, trade marks and copyright, or more correctly with acts of piracy which have become a large and profitable business in recent years. They are intended to provide a quick and efficient means of recovering infringing articles and of discovering the sources from which these articles have been supplied and the persons to whom they are distributed before those concerned have had time to destroy or conceal them. Their essence is surprise. Because they operate drastically and because they are made, necessarily, *ex parte*, i.e., before the persons affected have been heard, they are closely controlled by the court. . . . They are only granted on clear and compelling evidence, and a number of safeguards in the interest of preserving essential rights are introduced. They are an illustration of the adaptability of equitable remedies to new situations.

The three conditions that were prescribed in the original *Anton Piller* case[183] for the making of such order were as follows:

> First, there must be an extremely strong prima facie case. Secondly, the damage, potential or actual, must be very serious for the plaintiff. Thirdly, there must be clear evidence that the defendants have in their possession incriminating documents or things, and that there is a real possibility that they may destroy such material before any application *inter partes* can be made.

Somewhat less stringent tests for the granting of an Anton Piller order were adopted by the English Court of Appeal in the case of *Yousif v. Salama*.[184] In that case, the court considered only two questions, first, whether the documents sought to be seized were essential to the plaintiff's case, and second, whether such documents were at serious risk because they might be dishonestly destroyed. The court noted that it was difficult to form any confident view as to the merits of the application since the evidence was inevitably one-sided and the defendants did not have an opportunity to answer it.

In the case of *Nintendo of America, Inc. v. Coinex Video Games Inc.*,[185] the court referred to the tests set out in the Anton Piller case as well as those in the *Yousif* case. Heald J. found that the plaintiff had made out an extremely strong *prima facie* case in view of the defendant's admissions that the plaintiff's game boards had been copied. In any event, the plaintiff had satisfied the less stringent tests adopted in the *Yousif* case. While granting an Anton Piller order to provide for the preservation of all materials and documents relating to the alleged infringement of copyright in the plaintiff's audio-visual games, the order was made subject to a number of conditions. These conditions included the following:

---

183 *Anton Piller K.G. v. Mfg. Processes Ltd.*, [1976] 1 All E.R. 779 at 784 (C.A.).

184 [1980] 3 All E.R. 405 (C.A.).

185 (1982), 69 C.P.R. (2d) 122 (Fed. C.A.).

a) the order and copies of supporting material were to be served forthwith;
b) the plaintiff was to undertake to be bound by any order of the court as to damages and provide a surety bond in the amount of $75,000.00 as security for the undertaking as to damages;
c) the documents or things seized were to be transmitted by the plaintiff's solicitor to the sheriffs in the defendants' jurisdictions to be utilized solely for the purpose of civil proceedings against the defendants or other persons in relation to the plaintiff's copyright and the subject audio-visual games;
d) entry into the defendants' premises on behalf of the plaintiff was restricted to not more than four persons;
e) the right of entry and removal was restricted to documents and materials directly related to the plaintiff's games; and
f) the defendants would be permitted to move the court to vary or discharge the order, or to increase the amount of security ordered on 24 hours notice to the plaintiff's solicitor.

It was noted that the foregoing safeguards would fully protect the rights and interests of the defendants while enabling the plaintiff at the same time to proceed with its action for copyright infringement.

By contrast to the Nintendo case, the Federal Court refused to grant an Anton Piller order in the case of *Chin-Can Communication Corporation v. Chinese Video Centre Ltd.*[186] The Court found that the plaintiff had failed to establish the second and third conditions set out in the original *Anton Piller* case, namely, that there was potential or actual damage that was very serious to the plaintiffs, and that there was clear evidence that the defendants had incriminating documents or things in their possession and there was a real possibility that such materials may be destroyed before an application *inter partes* could be made. As to the condition that the damage must be very serious for the applicant, the court noted that there was only hearsay evidence before the court and no indication as to the seriousness or amount of loss that the plaintiff might suffer. With respect to the possibility of the destruction of incriminating materials, the court noted that it was necessary to adduce some "convincing, concrete, factual evidence and not a mere statement of opinion or belief by a witness to justify a finding on the issue of real or grave danger that the evidence will be destroyed or removed."[187]

The conditions prescribed in the *Anton Piller* case for the granting of an order were applied somewhat less strictly in the case of *Aldrich v. Struk*,[188] Ruttan J. found that the plaintiff met the first condition prescribed in the *Anton Piller* case in that the plaintiffs had a strong claim to right

186 (1983), 70 C.P.R. (2d) 184 (Fed. T.D.).
187 *Ibid.*, at 188.
188 (1984), 8 C.P.R. (3d) 369 (B.C.S.C.).

of copyright in the videotapes in question and the defendants knew that they had no right of copyright themselves. With respect to the second condition, that is, proof of actual or potential damage of a serious nature, Ruttan J. noted that the "modern interpretation" of this requirement was that it was not necessary to prove actual financial damage, but it was sufficient for the plaintiff to show potential damage by an inability to prove its case if the subject-matter were to disappear. Finally, with respect to the possibility of destruction, Ruttan J. found a presumption that destruction would occur based on the defendant's apparent conduct of dishonesty. It was noted that such destruction would result in very heavy losses to the plaintiff.

The Anton Piller order raises questions as to whether such an order may violate a defendant's privilege against self-incrimination. In the case of *Rank Film Distributors Ltd. v. Video Information Centre*,[189] it was held that the requirements of the order compelling the defendants to supply information and disclose documents should be deleted since such requirements would infringe the defendant's privilege against self-incrimination, and could lead to prosecution under the Copyright Act, 1956 and the Theft Act, 1968. The decision in the *Rank Film* case was subsequently abrogated by section 72 of the Supreme Court Act, 1981 (U.K.) in respect of actions involving intellectual property infringement.

In *Chin-Can Communications Corp. v. Chinese Video Centre Ltd.*,[190] Addy J. noted that the Anton Piller order raised a very serious question of self-incrimination in view of the penal provisions of sections 25 and 26 of the Copyright Act and in view of the Canadian Charter of Rights and Freedoms. He stated that a valid argument could be made against any general disclosure order being made in the future, even where a general right to search is allowed under an Anton Piller order. However, in the case of *Dominion Citrus & Drugs Ltd. v. Laconte Meat Market*,[191] Master Donkin summarily dismissed the defendant's argument that an application to bring material to an examination for discovery and to answer questions following the issuance of an Anton Piller order could amount to an "unreasonable search and seizure" for purposes of section 8 of the Charter.[192]

The question of whether the seizure of documents under an Anton Piller order may result in an involuntary waiver of privilege to such documents was considered by the Ontario Supreme Court in the case of

189 [1981] 2 All E.R. 76 (H.L.).
190 *Supra*, footnote 186.
191 (1983), 33 C.P.C. 105 (Ont. Master).
192 See J. Berryman, "Anton Piller Injunctions: An Update" (1985), 2 I.P.J. 49; F.S. Freedman, "The Use and Development of Anton Piller Orders in Canada as a Remedy Against Piracy and Counterfeiting" (1983), 43 R. du B. 871.

Sunwell Engineering Co. v. Mogilevsky.[193] In that case, Gray J. held that the doctrine of involuntary waiver of privilege applied to the inadvertent disclosure of documents but did not apply where the documents were obtained pursuant to a court order. Accordingly, the defendant was entitled to claim privilege for letters written by its patent attorney which passed on a solicitor's opinion notwithstanding that such letters had been seized pursuant to an Anton Piller order. Gray J. found it unnecessary to consider whether the defendant's rights had been infringed under section 8 of the Charter.

(b) Permanent Relief

(i) *Damages and an Accounting for Profits*

Subsection 20(4) of the Copyright Act provides as follows:

> Where any person infringes the copyright in any work that is protected under this Act, such person is liable to pay such damages to the owner of the right infringed as he may have suffered due to the infringement, and in addition thereto such part of the profits that the infringer has made from such infringement as the court may decide to be just and proper; and in proving profits the plaintiff shall be required to prove only receipts or revenues derived from the publication, sale or other disposition of an infringing work, . . . and the defendant shall be required to prove every element of cost that he claims.

The distinction between an award of damages and an account for profits is stated as follows in Halsbury:[194]

> The principle underlying relief at law is that the plaintiff has suffered loss by defendant's breach of contract or wrongful conduct, and damages are awarded for the purpose of making good this loss. The principle underlying relief in equity is that the defendant has improperly received or withheld property, or profits from property (such profits or property belonging to the plaintiff) and he is required to restore the property or to account for the profits. Thus at law the extent of the remedy is measured by the loss to the plaintiff which is covered by the damages awarded; in equity the extent of the remedy is measured by the gain to the defendant which is ascertained by directing an account against him.

In *Apple Computer Inc. v. MacKintosh Computers Ltd.*,[195] Reed J. made an order for an accounting of profits in respect of the sale of computers incorporating copies of the plaintiff's programs based on the profit arising

193 (1986), 8 C.P.C. (2d) 14 (Ont. H.C.); leave to appeal to Ont. Div. Ct. refused (1986), 8 C.P.C. (2d) 14n (Ont. Div. Ct.).

194 4th ed., Vol. 16, p. 869, at para. 1297.

195 See *Nat. Film Bd. v. Bier* (1970), 63 C.P.R. 164 at 169 (Ex. Ct.); *T.J. Moore Co. v. Accessories de Bureau du Que. Inc.* (1973), 14 C.P.R. (2d) 113 at 125 (Fed. T.D.).

from the sale of the computers as a whole. She indicated that the amount of the profit could not be divided between that attributable to the programs and that attributable to the other components of the computers.

In an action for infringement, it is not necessary to prove specific loss. In *Smith, Kline & French Canada Ltd. v. Frank W. Horner Inc.*,[196] Walsh J. stated that

> the mere difficulty or even impossibility of calculating the quantum of damages by the use of any mathematical calculations does not justify a finding that no damages can be awarded when a finding of fault will result in entitlement to such damages, and the same would apply to an accounting of profits. The court must simply do the best it can under the circumstances and fix a global amount.

Similarly, in the case of *Webb and Knapp (Canada) Ltd. v. Edmonton*,[197] it was held that the fact that it was difficult to assess damages for breach of copyright was not a ground for awarding only nominal damages. Thus, where no specific loss can be proved, the amount of damages awarded is not necessarily small.[198]

In the case of *Schauenburg Industries Ltd. v. Borowski*,[199] it was held that damages in respect of infringement of copyright should be calculated based on the plaintiff's loss of profits resulting from sales made by the defendant during the period that the copyright infringement continued. It was noted that such sales would probably have been made by the plaintiff were it not for the defendant's wrongful conduct.

The court also found in the *Schauenburg* case that the plaintiff was entitled to punitive damages in view of the defendant's flagrant behavior and callous disregard for the plaintiff's rights. The courts will also take into account any kind of fraud or malicious conduct in awarding punitive damages.[200] On the other hand, where the infringement is an isolated act and the defendant has acted in good faith such as by acknowledging the original author and source in the infringing work, such conduct may be a mitigating factor which will affect the award of damages and preclude exemplary damages from being granted.[201] Punitive damages may also not be awarded where the defendant is acting in a mistaken belief that he is not acting in infringement of copyright.[202]

---

196 (1982), 68 C.P.R. (2d) 42 at 53 (Fed. T.D.).

197 (1970), 63 C.P.R. 21 (S.C.C.).

198 See *Underwriters' Survey Bureau Ltd. v. Massie & Renwick Ltd.*, [1942] Ex. C.R. 1 (Ex. Ct.).

199 (1979), 50 C.P.R. (2d) 69 (Ont. H.C.).

200 *Op. cit.*, footnote 198.

201 *Zamacois v. Douville*, [1943] 2 D.L.R. 257 (Ex. Ct.).

202 *Bayliner Marine Corp. v. Doral Boats Ltd.* (1985), 5 C.P.R. (3d) 289 (Fed. T.D.); reversed on other grounds (1986), 67 N.R. 139 (Fed. C.A.).

(ii) *Recovery of Infringing Copies*

Section 21 of the Copyright Act provides as follows:

> All infringing copies of any work in which copyright subsists, or of any substantial part thereof, and all plates used or intended to be used for the production of such infringing copies, shall be deemed to be the property of the owner of the copyright, who accordingly may take proceedings for the recovery of the possession thereof or in respect of the conversion thereof.

It would appear that a plaintiff would not be entitled to receive an accounting for profits earned by the infringer and damages for conversion at the same time.[203]

In determining whether a plaintiff is entitled to the recovery of infringing copies, the court will take into account the entire circumstances of the infringement and may refuse to grant either such recovery or damages in respect of the conversion if such a remedy is considered "inappropriate and . . . unduly severe in the circumstances."[204] Moreover, a court will not compel a defendant to recover and to deliver up infringing copies of a work that were sold to persons who are not named as parties to the action.[205]

(iii) *Permanent Injunction*

Where copyright infringement is established after a full trial of the action, the plaintiff will be entitled to a permanent injunction. In some cases, such an injunction may be the sole remedy granted to the plaintiff based on the provisions of section 22 of the Copyright Act which provides as follows:

> Where proceedings are taken in respect of the infringement of the copyright in any work and the defendant in his defence alleges that he was not aware of the existence of the copyright in the work, the plaintiff is not entitled to any remedy other than an injunction in respect of the infringement if the defendant proves that at the date of the infringement he was not aware, and had no reasonable ground for suspecting that copyright subsisted in the work; but if at the date of the infringement the copyright in the work was duly registered under this Act, the defendant shall be deemed to have had reasonable ground for suspecting that copyright subsisted in the work.

---

203 *Black v. Imperial Book Co.* (1904), 8 O.L.R. 9 at 16 (Ont. C.A.); affirmed (1905), 35 S.C.R. 488 (S.C.C.); leave to appeal to the Privy Council refused (1905), 21 T.L.R. 540 (P.C.).

204 *Bayliner Marine Corp. v. Doral Boats Ltd., op. cit.*; see also *Apple Computer Inc. v. MacKintosh Computers Ltd., op. cit.*, in which the court's order for delivery up was limited to the devices containing copies of the infringing programs, and not the other components of the computers in which the programs were contained.

205 *R. v. James Lorimer & Co.* (1984), 77 C.P.R. (2d) 262 (Fed. C.A.).

In the case of *Bulman Group Ltd. v. "One-Write" Accounting Systems Ltd.*,[206] Collier J. described this section as "a somewhat odd and ill-worded provision." The defendant contended in that case that he was not aware that copyright subsisted in business forms which were not registered under the statute and that he had no reasonable grounds for suspecting that copyright did subsist. He argued further that, since there appeared to be a practice among companies in that line of business to copy their competitors' forms, he assumed that copyright did not subsist in such forms. Collier J. noted that this was an "unwarranted assumption" and that "a wrong assessment of the legal and factual position cannot be an excuse to avert the remedy of damages and an accounting of profits, as well as the other relief claimed by the plaintiff."[207]

(c) Presumptions and Procedural Matters

Subsection 20(3) provides that where the defendant in a copyright infringement action puts in issue either the existence of copyright or the plaintiff's title, the work shall be presumed to be one in which copyright subsists and the author shall be presumed to be the owner of the copyright, unless in each case the contrary is proved. Moreover, when any such question is at issue, and no grant of copyright or assignment of copyright has been registered, any person whose name appears on a work as purporting to be that of the author is presumed to be the author unless the contrary is proved. This section should be read with subsection 36(2) which provides that a certificate of registration of copyright in a work is evidence that copyright subsists in the work and that the person registered is the owner of such copyright.

Subsection 20(6) of the Copyright Act provides that the Federal Court has concurrent jurisdiction with provincial courts to hear and determine all civil actions, suits or proceedings that may be instituted for violation of any of the provisions of the Act or to enforce the civil remedies provided thereby.

Reference should be made to section 24 of the Copyright Act which provides that an action in respect of infringement of copyright shall not be commenced after the expiration of three years following the infringement. In *Warner Bros. Seven Arts Inc. v. CESM — TV Ltd.*,[208] it was noted that the prescription in section 24 is against the right of the plaintiffs to be awarded damages but that the right to an injunction is not affected.

---

206 (1982), 16 B.L.R. 16 (Fed. T.D.).
207 *Ibid.*, at 28.
208 (1971), 65 C.P.R. 215 (Ex. Ct.).

(d) Summary Remedies

Section 25 of the Copyright Act provides that where a person knowingly makes for sale or hire, sells or lets for hire, offers for sale or hire, distributes for the purpose of trade (or to such an extent as to affect prejudicially the owner of the copyright), exhibits in public by way of trade or imports for sale or hire into Canada any infringing copy of a work in which copyright subsists, such person is guilty of an offence under the Act and is liable on summary conviction to a fine not exceeding $10.00 for every copy dealt with, and not exceeding $200.00 in respect of the same transaction, or in the case of a second or subsequent offence, either to such fine or to imprisonment for a term not exceeding two months.

In *R. v. Comox Valley Video Stop Ltd.*,[209] it was held that the word "transaction" as used in section 25 referred not to every infringing work, but rather to the "selling, letting for hire . . . any infringing copy."[210] Accordingly, the maximum penalty permitted by statute in respect of the offering for sale of 282 infringing copies of a copyright work was $200.

Because knowledge is an essential ingredient of an offence under section 25, it must be proved that the accused knew or ought to have known that he was engaged in an activity that would likely be an infringement of copyright.[211]

14. PROPOSALS FOR REVISION OF THE CANADIAN COPYRIGHT ACT

The wholesale revision of Canada's Copyright Act has been considered in numerous studies and reports published during the past three decades. In spite of this activity, no amendments had been introduced in Parliament as of late-1986. It is expected, however, that legislation will be introduced in the foreseeable future. To understand the background of such legislation, this section will review the history of proposals for a revision of the copyright law, particularly in relation to computer-related subject matter.

The first proposal for a major revision of the Copyright Act was a Report of a Royal Commission on Patents, Copyright, Trade Marks and Industrial Design which was issued in 1957.[212] Not surprisingly, the Report did not deal with programs or other computer-related subject-matter. However, the topic was addressed in a Report of the Economic Council

---

209 (1985), 4 C.I.P.R. 245 (B.C. Co. Ct.).

210 *Ibid.*, at 247.

211 See *Nat. Breweries Ltd. v. Paradis Ltd.*, [1925] S.C.R. 666 at 669-70 (S.C.C.).

212 *Report on Copyright* (Ottawa: Queen's Printer, 1957) (commonly referred to as the *Ilsley Commission Report*).

of Canada published in 1971.[213] The Economic Council recommended that Canada not take "any sort of world lead in extending patent or copyright protection to computer programs at this time."[214] The Council noted that even if other countries granted such protection, there might be "good practical reasons" for Canada not to do so. The Council noted in this regard the need for caution to ensure there were no hidden extensions of basic copyright principles such as to lead to the protection of ideas.

In a Report published under the auspices of the Department of Consumer and Corporate Affairs in 1977,[215] the authors noted that there was considerable uncertainty in the international community at that time as to whether copyright law was an appropriate form of protection for computer programs. It was recommended that computer programs *per se* not be protected by copyright, but that where they fell under existing categories of protected material, the protection attached to those categories should extend to the programs contained in such materials. It was further suggested that the Act include a provision that nothing in the Act would prevent the use of a computer program to operate a computer.

The copyright status of computer programs and other computer works was considered by a Task Force on Copyright Law Revision established by the Department of Communications which issued its final report in 1983.[216] The Task Force noted the uncertainty that existed under the current law as to the copyright status of programs, particularly programs expressed in object code form and embodied on computer-readable media. The Task Force considered but rejected the proposal to establish a separate statutory regime to deal expressly with computer programs. Instead, it recommended that programs be protected under the Copyright Act regardless of their form of expression. It was also recommended that the rights that should be granted to the creators of computer programs include the right to reproduce the programs in any form whatever, the right to publish programs, the right to "adapt" programs (i.e., to "translate" a program from one programming language to another, to convert a program from source code to object code form and vice versa and to add enhancements or new features), and the right to authorize any of the foregoing acts. It was further suggested that a person who lawfully obtains a copy of a copyright program be permitted to copy the program as part of a computer's temporary memory in conjunction with its use and to make a copy for archival purposes. Finally, it was recommended that copyright protection be granted

213 Economic Council of Canada, *Report on Intellectual and Industrial Property* (Ottawa: Information Canada, January, 1971).

214 *Ibid.*, at 103.

215 Keyes and Brunet, *Copyright in Canada — Proposals for a Revision of the Law* (Consumer and Corporate Affairs Canada, 1977).

216 Department of Communications, *Task Force on Copyright — Revision of the Copyright Law of Canada*, John D. Hylton, Q.C., Chairman, Final Report (January 14, 1983).

to an original compilation of material contained in a computer information storage and retrieval system as well as works created by the use or with the assistance of a computer.

In May, 1984, a White Paper prepared jointly by the Department of Consumer and Corporate Affairs and the Department of Communications was released by the Canadian Government.[217] The White Paper proposed to create a specialized form of statutory protection for computer programs. Under the proposed amendments, a computer program would be defined as "a set of instructions intended to operate a machine having information processing capabilities". It was proposed that the Act recognize two categories of computer programs, namely, programs in "human-readable" and "machine-readable" form. A "machine-readable" program would be defined as a "computer program that is not intended for human comprehension", but not including any instructions intended to make the program compatible with a particular machine or device.

The White Paper proposed that "human-readable" programs enjoy the same term of protection (that is, life of the author plus fifty years), and be subject to the same rights in favour of authors as other literary works, and that they be protected by the international copyright conventions. Programs in "machine-readable" form, on the other hand, would have a term of protection of only five years, and would not be subject to the international copyright conventions. Moreover, the only rights that would be accorded to a machine-readable program would be to publish the work if it were unpublished, and to make another machine-readable program or a human-readable program based upon, and identical with or substantially similar to, the protected program. Copyright in a machine-readable program would not include a moral right, a public performance right, a broadcast or cable transmission right, a market segregation right or a right to use the program.

The White Paper proposed that in order to qualify for copyright protection, a computer program would have to be marked with the name of the copyright owner and the year of publication. It was proposed that regulations be adopted specifying the nature of the markings required where an additional term of protection was claimed for programs which have been modified or updated.

The proposals in the White Paper, and particularly the proposal to create a different legal regime for "machine-readable" programs, gave rise to a number of questions and were widely criticized by legal writers and representatives of the computer industry.[218]

---

217 Government of Canada, *From Gutenberg to Telidon: A White Paper on Copyright* (Supply and Services Canada, 1984).

218 See, *e.g.*, Proceedings before the Subcommittee on the Revision of Copyright, Issue No. 17, June 17, 1985 (Supply and Services Canada, 1985).

Most of the objections made to the proposals in the White Paper indicated that the proposed distinction between human-readable and machine-readable programs was impracticable and unworkable. It was also not clear what was meant by the term "not intended for human comprehension", as defining a machine-readable program; for example, whether it referred to the ability of an ordinary layman as opposed to a skilled programmer to understand the program and whether the intent of the creator or an objective test should prevail.

It was also not clear whether the proposals in the White Paper were consistent with Canada's international copyright treaty obligations. In particular, it was questionable whether the implementation of the proposal that the right to convert a human-readable program into machine-readable form last only for five years would be consistent with Article IV of the Universal Copyright Convention which requires protection of a work for the life of the author plus 25 years after his death, or with Article 8 of the Berne Convention which requires that the "translation" right subsist during the whole term of copyright in the original work.

The public policy rationale underlying the proposal to limit the term of protection of "machine-readable" programs to five years was also questionable. The White Paper suggested that such term should encourage the dissemination of programs and avoid the abuse of monopoly power. However, if a work has economic value for a greater period of time than five years, it should be entitled to protection after the expiration of that period of time. On the other hand, if a work has a short life span in terms of economic value, it may not matter that copyright protection continues beyond that time. The creator will not likely assert any claim for damages or try to prevent copying of the work.

After the 1984 election and the coming into power of the new Conservative Government, the White Paper was referred to a special Parliamentary Subcommittee on the Revision of Copyright which was established in 1985. The Subcommittee invited submissions on the White Paper and held public hearings, a full day of which was devoted to computer programs and related works.[219]

In his appearance before the Parliamentary Subcommittee on May 30, 1985, the Honourable Michel Cot , the Minister of Consumer and Corporate Affairs, indicated that the proposals in the White Paper did not represent the policy of the then Canadian Government. He announced that the Government was considering instead the protection of computer programs in all forms as literary works with the full traditional term of copyright protection. The Minister also indicated that the enactment of the Semiconductor Chip Protection Act by the United States made it

219 *Ibid.*

incumbent upon Canada to pass similar legislation to protect the Canadian semiconductor chip industry.

On October 10, 1985, the Parliamentary Subcommittee issued its report to the House of Commons.[220] The report included some 137 recommendations for changes to the copyright law, including a number of recommendations for changes relating to computers and computer programs.

In its report, the Subcommittee rejected the recommendations made in the White Paper that human-readable and machine-readable programs be granted different rights under the Act and be subject to different terms of protection. The Subcommittee recommended, instead, that all computer programs be protected as a separate category of subject-matter with the full regime of protection and for the same term as other works (that is, life of the author plus fifty years). Such protection would be extended to foreign nationals on the basis of reciprocity.

The Report recommended further that the right of an author of a computer program to authorize reproduction be made subject to an exception to permit the making of a back-up copy. It was further suggested that the Government study the possibility of providing an exception to permit the reproduction of a substantial part of a pre-existing program as a non-substantial part of another program. It was contended that the computer industry accepts as "normal practice" that a substantial part of one program may be used in preparing another program different from the original.

Among the other recommendations made by the Subcommittee were the following:

(a) that a new right be created involving the introduction or input of a pre-existing work into a computer; the new right would not extend to the actual display of the protected work on a computer terminal;
(b) that a new definition of "fixation" be adopted to include all means capable of capturing a work, including capture in computer media, but not in a medium as "volatile" as a computer's main storage or display screen;
(c) that copyright in a work created with the assistance of a computer be owned by the person responsible for the making of the work and not the owner of the copyright in the program used;
(d) that ownership of copyright in compilations contained in computer data banks be vested in the individual or entity primarily responsible for the arrangements undertaken for making the compilations.

The Subcommittee recommended further that new legislation be adopted outside the Copyright Act to protect mass works fixed in

---

220 *Supra*, footnote 161.

semiconductor chips. Such legislation would protect the three-dimensional pattern of circuits and components resulting from the vertical layering of several masses. The new legislation would provide for the protection of the mass works of foreign nationals on a reciprocal basis.

In February, 1986, the Canadian Government announced its response to the various recommendations set out in the Subcommittee's Report.[221] The Government expressed its general agreement with the Subcommittee's recommendations relating to computer programs and related subject-matter, subject to certain qualifications. First, the Government indicated that the issue of whether copyright should include a right of display would be considered further. Second, the Government would further consider the definition of fixation. Third, the Government indicated that although it agreed in principle that programs should benefit from the full regime of protection, foreign computer programs would be protected on the basis of the national treatment principle, rather than on the basis of reciprocity. Fourth, the Government rejected the proposal to create an exception to permit the reproduction of a substantial part of a pre-existing program as a non-substantial part of another program. Finally, the Government indicated that mask works fixed in semiconductor chips would be protected under the Copyright Act but that they would be distinguished from other works.

In view of the policy stand taken by the Canadian Government in favour of copyright protection for computer programs and related works, it is likely that the principles set forth in the Subcommittee's Report, as qualified in the Government's response, will be reflected in any legislation introduced in Parliament. At the time of writing, however, the timetable for the introduction or eventual enactment of such legislation remains uncertain.

221 Government of Canada, Government Response to the Report of the Sub-Committee on the Revision of Copyright (Supply and Services Canada, 1986).

5

Computer Technology and the Common Law Protection of Trade Secrets and Confidential Information in Canada

1. INTRODUCTION

The common law doctrine of trade secrecy is one of the most important forms of protection available for computer software and know-how in Canada. Trade secrecy protection is based on the principle that information and know-how which have economic value and which are treated as confidential by their holder are entitled to protection from appropriation by others. In the field of computer technology, trade secrecy protection enables creators to prevent their competitors from acquiring the benefits of valuable technology without incurring the developers' research and development costs.

Trade secrecy law is of value to the computer industry because it represents a means of protecting underlying ideas, algorithms and processes and not just their expression such as in the form of computer programs. The protection of information under trade secrecy law is also not dependent upon any registration procedure or compliance with other formalities.

Although trade secrecy law is of considerable importance to the computer industry, there are several difficulties with its application. One such difficulty is the need to distinguish between information which is truly unique and thus deserving of trade secret protection, and information which is common knowledge to those in the computer field. A related difficulty is the need to distinguish between information which has been misappropriated in breach of a confidential relationship and that which has been developed independently. A further difficulty arises in respect of mass-distributed software: because such software is subject to widespread dissemination and because licence agreements are often not executed by end users, it is questionable whether confidentiality obligations imposed on such users are truly binding.

In general, trade secrets are protected in Canada based on an action for breach of a confidential relationship, however the relationship arises.

In order to succeed in such an action for breach of confidence, three conditions must be met:

> First, the information itself, in the words of Lord Greene, M.R. in the Saltman case[1] . . . must 'have the necessary quality of confidence about it'. Secondly, that information must have been imparted in circumstances importing an obligation of confidence. Thirdly, there must be an unauthorized use of that information to the detriment of the party communicating it.[2]

In order to render the cause of action complete, all three conditions must be met, as indicated by the following statement from the decision of the Alberta Court of Appeal in *Ridgewood Resources Ltd. v. Henuset*:[3]

> During the course of argument it was contended that, though the information is not secret, the circumstances of disclosure may nevertheless produce a confidential relationship which makes the recipient liable for subsequent use to the detriment of the giver of the information. In my view that is not a correct statement of the law. Both elements, secrecy of the information and the circumstances of confidential disclosure, must co-exist. In addition, to render the cause of action complete, there must, of course, be unauthorized use of the information to the detriment of the party giving the information.

The Court in the *Ridgewood* case went on to find that "the communication of information which is in the public domain, though not known to the recipient, does not (apart from contract) make the recipient liable for its use."[4] In addition to the exception with respect to contractual obligations, the misuse of information which is otherwise in the public domain may also be restrained in other circumstances, such as in the relationship between a fiduciary and his principal.[5]

The following sections of this chapter will consider the meaning of a trade secret and how it may be distinguished from "confidential information", the conditions for the protection of such information, the extent to which computer software and related information and technology may qualify for such protection, the legal basis for the protection of trade secrets, and the remedies that may be available in respect of actions for breach of the rights of the holders of such information.

2. THE MEANING Of A "TRADE SECRET"

An Ontario court has adopted the following definitions based on

1 *Saltman Enrg. Co. v. Campbell Enrg. Co.* (1963), 65 R.P.C. 203 at 215.

2 *Coco v. A.N. Clark (Engineers) Ltd.*, [1969] R.P.C. 41 at 47.

3 (1982), 18 Alta. L.R. (2d) 68 at 75-76 (Alta. C.A.); leave to appeal to the Supreme Court of Canada refused (1982), 43 N.R. 90 (S.C.C.).

4 *Ibid.*, at 69.

5 See 129-131, *infra*.

various American authorities as to what constitutes a trade secret:

1st 'A trade secret . . . is a property right, and differs from a patent in that as soon as the secret is discovered, either by an examination of the product or any other honest way, the discoverer has the full right of using it . . .'

2nd 'A trade secret is a plan or process, tool, mechanism or compound known only to its owner and those of his employees to whom it is necessary to confide it . . .'

3rd 'The term "trade secret", as usually understood, means a secret, formula or process not patented, but known only to certain individuals using it in compounding some article of trade having a commercial value, and does not denote the mere privacy with which an ordinary commercial business is carried on.'

4th 'A trade secret may consist of any formula, pattern, device, or compilation of information which is used in one's business, and which gives him an opportunity to obtain an advantage over competitors who do not know or use it. A trade secret is a process or device for continuous use in the operation of the business. The subject matter of a trade secret must be secret.'[6]

The proper subject-matter for trade secrecy protection was described as follows in a leading text:

> The subject-matter capable of protection may be an industrial secret like a secret machine, process, or formula, or it may be industrial know-how (an increasingly important ancillary of patented inventions); it may be information of any sort; it may be an idea of a scientific nature, or of a literary nature (such as the plot of a story or the theme of a television series), or it may be a slogan or suggestion for a method of advertising; lastly, the subject-matter may be the product of work, or expenditure of money, or of trial and error, or the expenditure of time.[7]

The foregoing definitions indicate that virtually any type of information may be entitled to trade secrecy protection provided that the necessary conditions of protection are met. Some authorities have indicated, however, that a distinction should be made between a "trade secret" and a more general concept of "confidential information". For example, in *Schauenberg Industries Ltd. v. Borowski*,[8] it was held that drawings of a machine used in a manufacturing process represented an employer's confidential information even though they were not trade secrets. The distinction would appear to be that a trade secret consists of a specific plan, process or formula whereas "confidential information" may include general methods of doing business as well as "ephemeral" information

6 *R.L. Crain Ltd. v. Ashton*, [1949] O.R. 303 at 308-309 (Ont. H.C.); affirmed [1950] O.R. 62 (Ont. C.A.).

7 A.E. Turner, *The Law of Trade Secrets* (London: Sweet & Maxwell, 1962) 4.

8 (1979), 50 C.P.R. (2d) 69 (Ont. H.C.).

concerning a particular contract or transaction.[9]

The distinction between "true" trade secrets and more general confidential information may be of some significance in the context of employer-employee relationships. In *Faccenda Chicken Ltd. v. Fowler*,[10] it was held that although an employee could be restrained from disclosing or using any information received in confidence during the course of his employment, he could be prevented from disclosing or using only "true" trade secrets after the termination of his employment, either as an implied term of his employment or by means of a restrictive covenant. Among the factors which the Court indicated should be considered in determining whether information was to be classified as a true trade secret were the nature of the employment, the nature of the information itself, whether the employer impressed on the employee the confidentiality of the information and whether the relevant information could be readily isolated from other information. In applying these factors, the Court found in the circumstances in the *Faccenda* case that neither information about prices nor general sales information had the degree of confidentiality to justify an order preventing its disclosure or use by the employee following termination of his employment.

With the possible exception of employer-employee relationships, in general the courts have applied the same legal principles in protecting any information of a confidential character, whether or not it is strictly speaking a trade secret. Accordingly, for purposes of this text, the terms "trade secret" and "confidential information" are, unless otherwise noted, used interchangeably to refer to any information which meets the conditions for protection at common law as set forth in greater detail below.[11]

3. CONDITIONS FOR THE PROTECTION OF TRADE SECRETS: WHEN IS INFORMATION TRULY CONFIDENTIAL?

In order to be protected as a trade secret, it must be shown that any information is not publicly available or is not otherwise a matter of general knowledge within a particular industry or trade. In addition, it must be shown that the information was treated by its holder at all times as confidential, although the degree of secrecy required may vary according to circumstances.

The following is a summary of the implications of these principles for the protection of computer programs and data under Canadian law.

---

9 See Restatement of the Law of Torts, Article 759.

10 [1986] 1 All E.R. 617 (C.A.).

11 See D. Vaver, "Civil Liability for Taking or Using Trade Secrets in Canada" (1981-82), 5 Can. Bus. Law Journal 253 at 257-258.

(a) Distinction Between Confidential Information and Matters of Public Knowledge

Although information which is publicly available or which is generally known in an industry will not be entitled to trade secrecy protection, such protection may be extended to an original combination of generally known information. The principle was stated as follows by Lord Greene M.R. in *Saltman Engineering Co. v. Campbell Engineering Co.*:[12]

> The information, to be confidential, must, I apprehend, apart from contract, have the necessary quality of confidence about it, namely, it must not be something which is public property and public knowledge. On the other hand, it is perfectly possible to have a confidential document, be it a formula, a plan, a sketch, or something of that kind, which is the result of work done by the maker on materials which may be available for the use of anybody; but what makes it confidential is the fact that the maker of the document has used his brain and thus produced a result which can only be produced by somebody who goes through the same process.

Where a person receives information some of which is a matter of public knowledge and other parts of which are confidential, the court will restrain the use of that part which is confidential. As stated by Lord Denning M.R. in *Seager v. Copydex Ltd.*:[13]

> When the information is mixed, being partly public and partly private, then the recipient must take special care to use only the material which is in the public domain. He should go to the public source and get it: or at any rate, not be in a better position than if he had gone to the public source. He should not get a start over others by using the information which he received in confidence.

The distinction between confidential information and that which is generally available will depend on the nature of a particular business or trade and the manner in which information is disseminated within that business. In *Chevron Standard Ltd. v. Home Oil Co.*,[14] the Court refused to grant an injunction to restrain a former employee from using information acquired in the course of employment. It was noted that the defendant's new employer had collected a considerable amount of information from legitimate sources and it was at liberty to exploit information available in the industry market.

In a recent Ontario decision,[15] the Court refused to grant an injunction to restrain a computer programmer from using customer lists which the plaintiff claimed to represent its confidential information. The Court held

12 [1963] 3 All E.R. 413 at 415.

13 [1967] 1 W.L.R. 923 at 931-932 (C.A.).

14 [1980] 5 W.W.R. 624 (Alta. Q.B.); affirmed [1982] 3 W.W.R. 427 (Alta. C.A.).

15 *Canavest House Ltd. v. Lett* (1984), 2 C.P.R. (3d) 386 (Ont. H.C.).

that information concerning customers known to the employee only in virtue of his employment would be confidential. However, this was not the case where knowledge of the customers or potential customers was derived from general trade sources available to all competitors in the business. In the case of institutions known to the public such as banks and insurance companies, any person involved in the provision of programming services would know of their existence and of their interest in computer programs dealing with investment counselling.

In order to be protected as a trade secret, it is not necessary that the information be reduced to writing. However, as was noted in the case of *Fraser v. Thames Television Ltd.*,[16] the absence of any writing may make it more difficult to prove the confidential character of the information as well as the circumstances in which it was communicated:

> In my judgment there is no reason in principle why an oral idea should not qualify for protection under the law of confidence, provided it meets the other criteria I discuss below. Neither the originality nor the quality of an idea is in any way affected by the form in which it is expressed. No doubt both the communication and the content of an oral idea may be more difficult to prove than in the case of a written idea, but difficulties of proof should not affect the principle any more than in any other branches of the law where similar problems arise (e.g. contract and defamation).

It was held in the *Thames Television* case that an idea was entitled to protection when the content of the idea was clearly identifiable, original, of potential commercial attractiveness and capable of reaching fruition. By contrast, in *Promotivate International Inc. v. Toronto Star Newspapers Ltd.*,[17] it was held that information given by the plaintiff to the defendant was not confidential because it lacked any element of originality.

(b) Information to be Treated as Confidential by the Holder

In order for information to be protected as confidential, the holder must not disclose such information to other persons except under conditions of confidentiality. Thus, if information previously treated as confidential is published in specifications accompanying a patent application, it becomes a matter of "common knowledge" and can no longer be treated as a secret.[18] Although ancillary secrets connected to the patented invention may continue to be protected, the onus will be on the claimant to show the confidential character of such matters.[19] Information may also cease to be protected as confidential if it is required to be filed with governmental

16 [1984] 1 Q.B. 44 at 65.

17 (1985), 53 O.R. (2d) 9 (Ont. H.C.).

18 *Mustad & Son v. Allcock*, [1963] 3 All E.R. 416 (H.L.).

19 *Ibid.*

authorities for publication or if the information is used for trading purposes with other companies.[20]

Notwithstanding that confidential information becomes public, a person who has previously received it in confidence may be restrained from using it for a limited period of time after its public disclosure. This is a reflection of the so-called "springboard" principle which was explained as follows in *Terrapin Ltd. v. Builders Supply Co. (Hayes)*:[21]

> . . . a person who has obtained information in confidence is not allowed to use it as a springboard for activities detrimental to the person who made the confidential communication, and springboard it remains even when all the features have been published or can be ascertained by actual inspection by any member of the public . . . . The possessor of the confidential information still has a long start over any member of the public . . . the possessor of such information must be placed under a special disability in the field of competition to ensure that he does not get an unfair start.

The measures required to be taken to protect confidential information will depend on the circumstances, and in some cases, relative or partial secrecy will be sufficient. For example, in the case of *Com-Share v. Computer Complex*,[22] the court held that the disclosure of software in isolated instances to persons who were not in the same business as the plaintiff did not result in a loss of protection for the software as against the plaintiff's competitors to whom no such disclosure had been made.[23]

The persons to whom confidential information is disclosed need not be subject to express covenants of confidentiality, and information provided for a specific purpose pursuant to an oral agreement may be protected.[24] However, as was noted in *Consolidated Textiles Ltd. v. Central Dynamics Ltd.*,[25] "Where there is some doubt as to whether information is in fact confidential the plaintiff is obviously in a better position when the defendant had been advised that it was confidential." In *Molnar Lithographic Supplies Ltd. v. Sibatory*,[26] it was held that a manufacturing process was not entitled to protection as a trade secret as against the plaintiff's former employee on the grounds, *inter alia*, that there was not "sufficient evidence to establish to the learned trial Judge's satisfaction that the plaintiff had intimated to the defendant that the process involved was a secret one."[27]

---

20 *Mobil Oil Can. Ltd. v. Cdn. Superior Oil*, [1979] 4 W.W.R. 481 (Alta. Q.B.); addendum [1980] 1 W.W.R. 453 (Alta. Q.B.).

21 [1960] R.P.C. 128 at 130 (C.A.); see also *Schauenberg Industries v. Borowski* (1979), 50 C.P.R. (2d) 69 at 79-80 (Ont. H.C.).

22 338 F. Supp. 1229 (1971); affirmed 458 F. 2d 1341 (1972).

23 See *Franchi v. Franchi*, [1967] R.P.C. 149 at 153.

24 *Breeze Corps. v. Hamilton Clamp & Stampings Ltd.*, [1962] O.R. 29 (Ont. H.C.).

25 [1974] 2 F.C. 814 at 826-827 (Fed. T.D.).

26 (1974), 14 C.P.R. (2d) 197 (Ont. C.A.).

27 *Ibid.*, at 203-204.

A court is more likely to treat information as a trade secret if the holder has taken various measures to prevent its disclosure to third parties. In *Consolidated Textiles Ltd. v. Central Dynamics Ltd.*,[28] the court found that a system for converting a fly-shuttle loom to a water-jet loom was entitled to protection as a trade secret. The court noted that the project for the development of the process was initially carried out in strict secrecy in a special part of the plaintiff's plant where only key personnel involved in the actual project were allowed. Moreover, except for key personnel, persons working on the project were not informed of the relationship between what they were doing and the proposed finished project. Visitors were required to sign in and staff working on the project were warned not to discuss it with other persons. Persons to whom experimental parts were distributed were advised orally of the need for secrecy and purchase orders were marked confidential.

In *International Tools Ltd. v. Kollar*,[29] the court found that a manufacturing process was a trade secret where the process was regarded by the parties involved as secret, where the manufacture was carried out in a separate room to which access by unauthorized persons was prohibited, where the manufacturing instructions were kept in that room and the items of manufacture were accounted for, and where persons who were working on the process were warned of its secret character. On the other hand, the court held that a separate manufacturing process could not be treated as a secret where the manufacture was carried out in an outer shop, where no effective measures were taken to prevent some 250 employees from observing the process or to make such employees aware that it was a secret process, and where the process was one which could be easily imitated.

Other authorities have adopted somewhat less stringent tests as to when an owner may claim protection for information as secret or confidential. In *R. v. Stewart*,[30] Cory J.A. adopted the following statement by Megarry V.C. in *Thomas Marshall (Exports) Ltd. v. Guinle*,[31] setting out the tests to be met:

> First, I think that the information must be information the release of which the owner believes would be injurious to him or of advantage to his rivals or others. Second, I think the owner must believe that the information is confidential or secret, i.e., that it is not already in the public domain. It may be that some or all of his rivals already have the information: but as long as the owner believes it to be confidential I think he is entitled to try and protect it. Third, I think that the owner's belief under the two previous heads must be reasonable. Fourth, I think that the information must be judged in

---

28 *Supra*, footnote 25 at 828-829.
29 [1966] 2 O.R. 201 (Ont. H.C.); affirmed [1968] 1 O.R. 669 (Ont. C.A.).
30 (1983), 42 O.R. (2d) 225 at 245 (Ont. C.A.).
31 [1978] 3 All E.R. 193 at 209-210.

the light of the usage and practices of the particular industry or trade concerned. It may be that information which does not satisfy all these requirements may be entitled to protection as confidential information or trade secrets: but I think that any information which does satisfy them must be of a type which is entitled to protection.

4. COMPUTER SOFTWARE AND TECHNOLOGY AS A SUBJECT-MATTER OF TRADE SECRECY PROTECTION

(a) The Protection of Software as Trade Secrets

In general, any process or technology resulting from research and development efforts may qualify for protection as a trade secret. In *Polyresins Ltd. v. Stein-Hall Ltd.*,[32] Galligan J. made the following observations:

> . . . it appears obvious to me that when private corporations involved in highly competitive commercial enterprises carry out product research and development projects, they are doing so for their own private benefit to improve their products so that their competitive position in the industry will be improved or maintained. Processes resulting from such research which give them new products or which improve existing ones for the purpose of maintaining or improving their competitive position in the market generally must be trade secrets.

In *Northern Office Microcomputers (Pty.) Ltd. v. Rosenstein*,[33] Marais A.J. held that it seemed to be plain "that computer programmes may qualify for protection which the common law accords to confidential matter." However, the broad concept of a computer program attuned to the needs of a particular professional group, that is, dentists and doctors, could not be protected as confidential. The court noted in this regard that the plaintiff could not be prevented from applying his mind to the development of his own programs designed for doctors and dentists, nor from using his recollection based on his previous employment of things to be included in such programs. At the same time, the defendant could be restrained from making a copy of the plaintiff's programs, from making any translation of such programs into any other computer language, or from using or permitting any other person to use any copy of the plaintiff's programs.

In *Canavest House Ltd. v. Lett*,[34] the plaintiff sought an injunction to restrain its former employee from copying or otherwise using certain computer programs developed in the course of his employment. The Court

32 (1971), 5 C.P.R. (2d) 183 at 185 (Ont. H.C.).

33 [1981] F.S.R. 124 at 138 (Sup. Ct. of S.A.).

34 (1984), 2 C.P.R. (3d) 386 (Ont. H.C.).

found that because the plaintiff had "ownership" in the programs, the defendant should be enjoined from providing copies of the programs or of the notes and material leading to their creation to any other person. In addition, the defendant and any person having knowledge of the order were required to deliver to the plaintiff all copies of the software in their possession.

In his judgment, Callon J. did not deal expressly with whether the software was protected as a trade secret. However, the terms of the order preventing the defendant from providing copies of the software to third persons and requiring that all information concerning the software be kept confidential in all proceedings prior to trial indicate that the software was considered to be the plaintiff's confidential information.

In *Gemologists International Inc. v. Gem Scan International Inc.*,[35] the court noted that the plaintiff had implemented specific confidentiality and security procedures to protect its copyright interests in software programs and program manuals which it spent over four years developing. The court granted an injunction to restrain the plaintiff's former employees from using, copying or modifying the computer appraisal system including program disks, source code and documentation or from disclosing information concerning such system, and to require the delivery up of all documentation relating to the system and any modifications in the defendants' possession.

Different tests have been applied in various cases as to the extent of the protective measures that should be taken to protect computer software as a trade secret. In *Com-Share v. Computer Complex*,[36] the court found the existence of a trade secret based on such security measures as the stamping of each page of the source code listings with Com-Share's proprietary legend, the creation of passwords built into the system to prevent unauthorized access, and the locking up of the magnetic tapes when not in use. In *Jostens Inc. v. National Computer Systems*,[37] the court found that there was no trade secret in the absence of a trade secret policy by the plaintiff. The court held that employees must be informed about keeping developments secret so they are aware that information is not to be made readily available to those in the industry. By contrast, in *Lake Mechanical Systems Corp. v. Crandell Mechanical Systems Inc.*,[38] the court found that computer programs were protected as confidential information notwithstanding evidence indicating they were treated in a "haphazard" manner with the only security measure taken being to keep a spare copy in a separate location.

---

35 (1986), 7 C.I.P.R. 225 (Ont. H.C.).

36 *Supra*, footnote 22, at 1234.

37 318 N.W. 2d 691 (Minn., 1982).

38 (1985), 31 B.L.R. 113 (B.C. Chambers).

Some of the practical measures that may be taken to ensure that computer software is entitled to trade secrecy protection include the following:

1. limiting access to programs and computer areas on a need-to-know basis;
2. monitoring all software and document copying;
3. entering into restrictive covenants with key employees to prevent disclosure and to limit post-employment competition;
4. requiring all third parties having access to computer programs and their employees, subcontractors, and other related parties having access, to sign a restrictive non-disclosure agreement;
5. using software with a built-in "lock-out", "time bomb" or self-destruct feature activated upon copying;
6. using firmware as opposed to software where feasible to make copying more difficult;
7. licensing a machine-readable object code only and prohibiting disassembly, reverse engineering and decompilation;
8. destroying all obsolete, excess, and preliminary copies of proprietary materials to prevent their use or reproduction;
9. adopting appropriate physical security measures, including the use of employee badges, restricted areas, audit trails of all users accessing material and secure storage; and
10. encrypting high value material.[39]

(b) The Effect of Reverse Engineering for the Protection of Computer Programs under Trade Secrecy Law

As a general rule, to the extent that a product sold in the open market may be examined to reveal information about its method of manufacture or the process by which it is produced, the information will not be protected as a trade secret. The principle was stated as follows in *R.L. Crain Ltd. v. Ashton*:[40]

> A trade secret differs from a patent right in that as soon as the secret is discovered, either by an examination of the product, or in any other honest way, the discoverer has the full right of using it.

In *Breeze Corps. v. Hamilton Clamp & Stamping Ltd.*,[41] the Court indicated that a competing engineering company could disassemble a clamp manufactured by the plaintiff and design tools to produce the

39 R. Mantles, "Trade Secret and Copyright Protection of Computer Software" (1983-84), 4 Computer Law Journal 669 at 676.

40 [1949] 2 O.R. 303 at 309 (Ont. H.C.); affirmed [1950] O.R. 62 (Ont. C.A.).

41 [1962] O.R. 29 (Ont. H.C.).

component parts of the clamp. The provision of samples of a finished product and its component parts did not by itself represent a disclosure of the plaintiff's trade secret.

The susceptibility of computer software or semiconductor chip products embodying programs to reverse engineering may determine whether the underlying source programs are entitled to trade secrecy protection. As indicated in the Restatement of Torts,[42] one of the factors which determines whether information is to be treated as a trade secret is "the ease of difficulty with which the information could be properly acquired or duplicated by others." Thus, if the underlying source code or the algorithms of a computer program can be readily detected through a process of reverse engineering or decompilation, they may lack the unique character necessary to be treated as trade secrets.

The general rule that a product may be "reverse engineered" to detect an underlying secret process or method of manufacture may not apply in the case of computer programs which are marketed under licence arrangements which expressly prohibit such activities. Such licence arrangements are intended to provide for the preservation of all proprietary rights in favour of the program creators, and limit the uses that may be made of the programs by licensees. Such licences commonly prohibit licensees from transferring possession of the programs to third parties, and prohibit various unauthorized uses such as decompilation, disassembly or reverse engineering of the programs. Such licence arrangements would appear to give program creators an effective means of prohibiting reverse engineering provided that such arrangements are contractually binding.

(c) Trade Secrecy Protection for Mass-Distributed Software

The distribution of software in the mass market raises the question of whether the mere fact that such software is widely distributed precludes trade secrecy protection. This result would occur if the courts were to find that mass distribution itself is inconsistent with the required element of secrecy.

The case of *Data General Corp. v. Digital Computer Controls Inc.*[43] suggests that the mere number of disclosures should not cause trade secrecy protection to be invalidated where the programs are otherwise entitled to protection. In that case, the plaintiff sought trade secrecy protection for maintenance diagrams which were accessible by almost 6,000 people by the time of trial. The Court found that Data General had taken adequate measures to protect the confidentiality of the diagrams, stating that "dissemination is not significant if in confidence."

---

42 Article 757, Comment B.

43 357 A. 2d 105 (1975).

The applicability of trade secrecy protection to mass-distributed software will turn on the application of the principle that relative, as opposed to absolute, secrecy is all that is required for trade secrecy protection. In *K-2 Ski Co. v. Head Ski Co.*,[44] the Court noted that:

> There are two common law doctrines on secrecy: (1) absolute secrecy and (2) relative secrecy. The better view . . . is the majority view of relative secrecy which has been adopted by the Restatement of Torts S.757 . . . [which] requires a substantial element of secrecy . . . and that reasonable measures under the circumstances be taken to protect the secret. . . . The necessary determination of "[w]hether such a degree of secrecy exists in a particular case is a question of fact" . . . and the trier of the fact must consider "the entirety of circumstances surrounding use" of the secret . . .[45]

Among the measures that the courts may consider in determining whether reasonable measures are taken to protect the secrecy of mass-distributed software are whether the object code only is distributed, how difficult it would be to decompile or disassemble the program in order to reconstruct the underlying source code, the degree to which the source code is truly unique rather than representing a compilation of commonly known routines and algorithms, and what protective devices and other practical measures are taken to prevent the unauthorized duplication or decompilation of the program.[46]

In addition to showing that reasonable measures are taken to protect the secrecy of mass-distributed software, it must also be shown that the restrictions on disclosure or use of the software, as set forth in end user licence agreements, are contractually binding on the licensees. Some questions as to the enforceability of such agreements may arise in the case of "shrink-wrap" licences which are not signed by the licensees.[47]

5. LEGAL BASIS FOR THE PROTECTION OF TRADE SECRETS: RELATIONSHIPS IN WHICH OBLIGATIONS OF CONFIDENCE WILL BE IMPOSED

Apart from the terms of an express or implied contract, confidential information is protected in Canada based on the nature of the relationship between the party disclosing such information and the recipient. Specifically, such a duty may arise as between an employer and employee or as between a fiduciary (such as an agent, officer or director) and his principal. Such an obligation may also arise based on principles of equity

44 506 F. 2d 471 (9th Cir., 1974).

45 *Ibid.*, at 473-474.

46 M. Gilburne and R. Johnston, "Trade Secret Protection for Software Generally and in the Mass Market" (1981-82), 3 Computer Law Journal 211 at 236.

47 See Part V, Chapter 17, *infra*.

to prevent the unauthorized disclosure or use of information provided to a recipient in circumstances where it was intended to be treated as confidential.

The following is an outline of the duties that may arise in the context of the foregoing relationships.

(a) Employer-Employee Trade Secret Relationships

The general rule that a creator will lose trade secrecy protection if his secrets are disclosed is not applicable to disclosures made by an employer to his employees. An employee has an implied obligation not to disclose or use for his own advantage trade secrets acquired by the employee in the course of employment.[48] Moreover, it is a well-established principle that "when a person is hired for the express purpose of developing an invention, whatever new material, method or process he discovers belongs to his employer."[49] Such obligations will be imposed as a matter of common law even in the absence of an express confidentiality agreement executed by the employee.[50] Moreover, even if the provisions of an employment agreement dealing with an employee's obligations of confidentiality and non-competition are found to be unenforceable, the courts may nevertheless enforce common law remedies.[51]

Although an employee will be bound not to disclose confidential information at common law, at least during the term of his employment, it is advisable not to rely solely on such implied obligations but to set out in a written employment agreement the scope of the employee's obligations. Such an agreement will be of assistance in defining the nature of the confidential information, will serve as a reminder to the employee of his obligations and may provide for the protection of confidential information after the termination of employment.

While an employer is entitled to protect his trade secrets, he cannot prevent an employee from using the general skill and knowledge acquired in the course of employment.[52] It has been said that a trade secret has an objective character, distinct from subjective matters such as a person's aptitudes, skill, dexterity or manual or mental ability.[53]

Although the strict standards applicable to employees' non-competition covenants have generally not been applied in determining the validity

---

48 *Cdn. Aero Service Ltd. v. O'Malley*, [1974] S.C.R. 592 at 606 (S.C.C.).

49 *Devoe-Holbein Inc. v. Yam* (1984), 3 C.P.R. (3d) 52 at 66 (Que. S.C.).

50 *Bee Chemical Co.v. Plastic Paint & Finish Specialties Ltd.* (1979), 47 C.P.R. (2d) 133 (Ont. C.A.); leave to appeal to Supreme Court of Canada refused, November 6, 1979.

51 *Investors Syndicate Ltd. v. Versatile Invts. Ltd.* (1983), 73 C.P.R. (2d) 107 at 111 (Ont. C.A.); *Creditel of Can. Ltd. v. Boothe* (1985), 6 C.P.R. (3d) 507 (Ont. H.C.).

52 *Genesta Mfg. Ltd. v. Babey* (1984), 48 O.R. (2d) 94 (Ont. H.C.).

53 See *Herbert Morris Ltd. v. Saxelby*, [1916] 1 A.C. 688 (H.L.).

of non-disclosure agreements, such agreements are nevertheless subject to a test of reasonableness.[54] Accordingly, a contract of employment should be drafted with only those restrictions required to protect the employer, taking into account an employee's particular responsibilities. A provision intended to protect an employer's trade secrets may be invalid as restraining competition if the specialized skill of a programmer means that his only opportunity for successful employment would be to work for his former employer's competitors.[55]

Employment agreements should also contain provisions for the release of an employee's non-disclosure obligations if the trade secret goes into the public domain or if the recipient obtains the information from another party who was entitled to disclose the trade secret. A confidentiality agreement may also contain a clause providing for the termination of the obligations of non-disclosure after a fixed period of time, which would be determined based on the estimated useful life of the trade secret. In a United States decision, it has been held that non-disclosure covenants are subject to the same restraints as are applicable to non-competition agreements, that is, that they must be reasonable as to both the geographical area and the period of time over which they extend.[56] In Canada, a non-disclosure covenant which is part of a more general covenant restricting the use of any information acquired in the course of employment or any solicitation of an employer's customers or other employees may be invalid if it contains no limitation as to time and no territorial restriction.[57]

(b) Fiduciary Relationships

The obligation not to disclose confidential information may also arise based on a fiduciary relationship. Such a relationship exists if a person occupies a position of trust in relation to his beneficiary. The basic obligation of a fiduciary is not to put himself in a position where his duty

---

54 *Elsley v. J.G. Collins Ins. Agencies Ltd.* (1978), 36 C.P.R. (2d) 65 at 71 (S.C.C.); *Triplex Safety Glass Co. v. Scorah*, [1938] Ch. 211 at 217.

55 *Genesta Mfg. Ltd. v. Babey, supra*, footnote 52.

56 *Disher v. Fulgoni*, 464 N.E. 2d 639 (Ill. App., 1984).

57 *Nat. Financial Brokerage Center Inc. v. Investors Syndicate Ltd.* (1986), 9 C.P.R. (3d) 497 (Que. C.A.). In *Nelson Burns & Co. v. Grathorn Indust. Ltd.* (1983), 42 O.R. (2d) 705 (C.A.); affirmed (1986), 56 O.R. (2d) 604 (S.C.C.), the court found that a covenant by an employee not to use information acquired in the course of employment either during the continuance of his employment "or at any time thereafter" was void as unreasonable since it meant that the employee might have a life-long obligation not to use knowledge and experience gained from the only business he knew. See also *Faccenda Chicken Ltd. v. Fowler*, [1986] 1 All E.R. 617 (C.A.), in which it was held that an employee may not be restrained from disclosing or using, after the termination of his employment, information which is of a confidential character but is not truly a trade secret.

and his personal interest conflict. Accordingly, the fiduciary must not disclose or use trade secrets for his personal gain.

Senior management personnel may owe duties to their companies both as fiduciaries and as employees. It is clear, however, that the courts will impose higher standards on persons in fiduciary positions than ordinary employees, as indicated by the following extract from the case of *Helbig v. Oxford Warehousing Ltd.*:[58]

> . . . once an employee reaches a certain level, he can stand in a fiduciary position to his employer and must be considered apart from the usual master and servant relationship. As a fiduciary, he then owes a clear duty to his company to use all his energy, ability and imagination in the best interest of the company, and if he is successful in developing a business opportunity, he does so for the benefit of the company exclusively. Where, as here, he insists that he can promote his interests and those of the company in tandem, he and his company are in a position of fundamental conflict and management is entitled to terminate his services without further notice. The question of who is right and who is wrong has no application. One party must give way and management has the right to insist that it be the employee executive.

A fiduciary may be bound not to disclose or use information belonging to his principal even if the information is not confidential. In *Investor's Syndicate Ltd. v. Versatile Investments Ltd.*,[59] the Court held that "the liability for breach of fiduciary duty is not dependent on a finding that the information was confidential."[60] The Court in that case cited with approval the finding by Lord Denning M.R. in *Boardman v. Phipps*[61] that an agent may be in breach of his fiduciary duty if he turns to his own use so as to make a personal profit any "information or knowledge which he has been employed by his principal to collect or discover, or which he has otherwise acquired, for the use of his principal." The obligation of a fiduciary not to make any "unfair use" of any information acquired in the course of performing his duties means that for some period of time at least, he may be restrained from soliciting business from customers of his former employer.[62]

In *International Corona Resources Ltd. v. Lac Minerals Ltd.*,[63] R.E. Holland J. stated that "the law regarding the obligations imposed by the delivery of confidential information is distinct and independent from the

---

58 (1985), 6 C.P.R. (3d) 460 at 469 (Ont. C.A.); leave to appeal to the Supreme Court of Canada refused (1985), 52 O.R. (2d) 754 (S.C.C.).

59 (1983), 73 C.P.R. (2d) 107 (Ont. C.A.).

60 *Ibid.*, at 113.

61 [1965] 1 All E.R. 849 at 856 (C.A.); affirmed [1966] 3 All E.R. 721 (H.L.).

62 *Alberts Ltd. v. Mountjoy* (1977), 36 C.P.R. (2d) 97 (Ont. H.C.); *G.T. Precision Welding (Ont.) Ltd. v. Nelligan* (1984), 3 C.P.R. (3d) 511 (Ont. H.C.).

63 (1986), 53 O.R. 737 (Ont. H.C.).

law dealing with the imposition of fiduciary duties."[64] It was noted in the circumstances of that case, however, that there was an intertwining of the two sets of legal principles.

There are no clearly defined rules as to the proper categories of persons to whom fiduciary duties attach. Although directors, officers and senior employees of a company are subject to fiduciary duties, less senior employees and technical personnel may not be subject to such strict duties. For example, in *Genesta Manufacturing Ltd. v. Babey*,[65] Barr J. expressed "grave doubts" as to whether a plant manager whose responsibilities had been reduced so that no employees were reporting to him at the time of the alleged breach of confidence owed a fiduciary duty to his employer. On the other hand, fiduciary duties may be owed between prospective partners or parties who have entered into negotiations toward a joint venture or other business relationship before a partnership, joint venture or other arrangement is finalized.[66]

(c) Other Confidential Relationships

The obligation not to disclose or use confidential information is not dependent upon the existence of a contract or a specific legal relationship between the party disclosing the information and the recipient. Instead, such obligation will be imposed whenever confidential information is disclosed in circumstances where it is apparent that an obligation of confidence was intended. The principle was set forth as follows by Lord Denning M.R. in the case of *Seager v. Copydex*:[67]

> The law on this subject does not depend on any implied contract. It depends on the broad principle of equity that he who has received information in confidence shall not take unfair advantage of it. He must not make use of it to the prejudice of him who gave it without obtaining his consent.

The equitable doctrine of confidence means that if information is disclosed in the course of conducting negotiations for the establishment of a joint venture or partnership, the recipient will be bound not to use such information even if a contract is not made.[68] Other circumstances in which the disclosure of confidential information may be restrained are where confidential information is provided by a consultant to a potential

---

64 *Ibid.*, at p. 771. The common law duties of a fiduciary may also co-exist with a non-competition covenant: *Wallace Welding Supplies Ltd. v. Wallace* (1986), 9 C.P.R. (3d) 514 (Ont. H.C.).

65 (1984), 48 O.R. (2d) 94 (Ont. H.C.).

66 *United Dom. Corp. v. Brian Pty. Ltd.* (1985), 60 A.L.R. 741 (H.C. Aust.); *Int. Corona Resources Ltd. v. Lac Minerals Ltd.*, *op. cit.*, at 776-777.

67 [1967] 1 W.L.R. 923 at 931 (C.A.).

68 See *Coco v. A.N. Clark (Engr.) Ltd.*, [1969] R.P.C. 41.

client, an invention is disclosed by an inventor to a manufacturer, an idea for a television program is disclosed to a television company, or where a franchise scheme is disclosed to prospective franchisees.[69]

Although the existence of a contract is not necessary for the protection of confidential information, the absence of an express agreement may make it more difficult for a plaintiff to demonstrate that the information was disclosed in circumstances in which it was intended to be treated as confidential.[70] The courts may also be reluctant to impose an obligation of confidentiality if the relationship is that of independent contractors rather than that of employer and employee,[71] although in at least one case, it was held that a person who represents another person or acts on his behalf may be an agent even if he is an independent contractor.[72] Similarly, in the absence of an agreement to the contrary, confidential information acquired by members of a partnership in the course of carrying on business may be used by the partners in competition with one another if the partnership is dissolved.[73]

Although most cases involving claims for breach of confidence in the absence of a contract or a specific legal relationship would appear to be founded in equity, other cases suggest that such a claim may be founded in tort law or be based on concepts of property. For example, in *MacDonald v. Vapour Canada Ltd.*,[74] Laskin C.J.C. suggested that the prescriptions set forth in paragraph 7(*e*) of the Trade Marks Act were "simply a formulation of the tort of conversion, perhaps writ large and in a business context." In other cases, confidential information has been characterized as the "property" of the person who expends time and effort in order to compile or collect the information. For example, in *R. v. Stewart*,[75] it was held that confidential information was "property" for purposes of the theft and fraud provisions of the Criminal Code.

In the context of other relationships, the Supreme Court of Canada has applied the doctrine of unjust enrichment so as not to "allow any man unjustly to appropriate to himself the value earned by the labours of

---

69 See Bereskin, "Trade Secrets and Intellectual Property: Identifying the Dangers — Closing the Loopholes", Paper presented at Conference on Protection of Trade Secrets and Confidential Information, Institute for International Research, Toronto, June, 1985, at 21-23.

70 *Wildwood Farm Services Int. (1975) Inc. v. T. Clay Mfg. Ltd.* (1978), 44 C.P.R. (2d) 157 (Ont. H.C.).

71 *Monarch Messenger Services Ltd. v. Houlding* (1984), 2 C.P.R. (3d) 235 (Alta. Q.B.).

72 *Investors Syndicate Ltd. v. Versatile Invts. Ltd.* (1983), 73 C.P.R. (2d) 107 (Ont. C.A.).

73 *Sinclair v. Ridout*, [1955] O.R. 167 (Ont. H.C.); *Nufort Resources Inc. v. Eustace* (1985), 29 B.L.R. 282 at 309 (Ont. H.C.).

74 (1976), 66 D.L.R. (3d) 1 at 13 (S.C.C.).

75 (1982), 38 O.R. (2d) 84 (Ont. H.C.); affirmed (1983), 43 O.R. (2d) 225 (Ont. C.A.).

another."[76] This doctrine would also apply to third parties who receive confidential information with notice of its confidential character. However, as noted in greater detail below, the position of a person who acquires confidential information for value without notice has not been settled in Canada.

6. THE PROTECTION OF TRADE SECRETS IN THE ABSENCE OF A CONFIDENTIAL RELATIONSHIP: INDUSTRIAL ESPIONAGE AND THIRD PARTY RECIPIENTS

The principles of equity reviewed in the foregoing section would appear wide enough to impose an obligation of confidence on any person who acquires information that he either knows or should know was obtained in breach of a confidential relationship. In an early English case,[77] the Court issued an injunction restraining the defendant from exhibiting or publishing drawings made by the plaintiff. The Court indicated that in the absence of any explanation as to the manner in which the defendant obtained possession of the etchings, it was bound to assume that such possession had "its foundation in a breach of trust, confidence or contract." The Court noted further that if the defendant availed himself surreptitiously of the information, which he could not have had except from a person guilty of a breach of contract in communicating it, he could not be permitted to avail himself of that breach of contract.[78]

Although any party who surreptitiously receives information disclosed in confidence will be fixed with liability, it is not clear whether such liability will be imposed on an "innocent" third party. It is also unclear whether an action for breach of confidence will succeed against a person who obtains trade secrets by means of industrial espionage but who has no confidential relationship, directly or indirectly, with the holder of the secrets.

In *Fraser v. Thames Television*,[79] Hirst J. noted that "in order to be fixed with an obligation of confidence, a third party must know that the information was confidential; knowledge of a mere assertion that a breach of confidence has been committed is not sufficient." In *International Tools Ltd. v. Kollar*,[80] the Court granted an injunction restraining use or disclosure of the plaintiff's trade secrets against not only the plaintiff's former employees but also the employees' partner to whom the information had

76 *Rathwell v. Rathwell*, [1978] 2 S.C.R. 436 at 455 (S.C.C.); see also *Pettkus v. Becker*, [1980] 2 S.C.R. 834 (S.C.C.); and *Lake Mechanical Systems Corp. v. Crandell Mechanical Systems Inc.* (1985), 31 B.L.R. 113 (B.C. Chambers).

77 *Prince Albert v. Strange* (1849), 47 E.R. 1302.

78 *Ibid.*, at 1311.

79 [1984] 1 Q.B. 44 at 65.

80 [1968] 1 O.R. 669 (Ont. C.A.).

been communicated. The Court noted that the partner must have acquired the information in one of two ways, that is, either as a purchaser for value without notice of any obligation affecting it, or from the plaintiff's former employees in circumstances which resulted in there being an equity against him. In the absence of evidence indicating that he was a purchaser for value, the Court concluded that the partner could be restrained from using the information brought to him by the plaintiff's former employees.

In *Tenatronics Ltd. v. Hauf*,[81] the individual defendant left the plaintiff's employment and accepted a position with the defendant company. The latter then commenced using circuit diagrams which the Court found to be the plaintiff's confidential information. The Court enjoined the defendant company from copying, manufacturing or marketing copies of the plaintiff's diagrams. Referring to the test applied in the case of *International Tools v. Kollar*, the Court noted that there was no doubt that the equity was against the defendant company. The Court also quoted with approval the following statement by Lord Denning M.R. in the case of *Saltman Engineering Co. v. Campbell Engineering Co.*:[82]

> If a defendant is proved to have used confidential information, obtained directly or indirectly from a plaintiff, without the consent, express or implied, of the plaintiff, he will be guilty of an infringement of the plaintiff's rights.

In at least one Canadian case, it has been suggested that even an "innocent" third party may be enjoined from using trade secrets communicated to him by another person in breach of a confidential relationship. In *Polyresins Ltd. v. Stein-Hall Ltd.*,[83] Galligan J. found that the plaintiff's former employees made improper use of the plaintiff's processes and formulae which came to their knowledge as a result of their employment and which they knew to be secret. Moreover, as a result of such use, the corporate defendant gained the benefits of the trade secrets to which it was not entitled and which came to it improperly. Based on these facts, Galligan J. concluded that in his opinion an injunction would lie "against a third person who has acquired information to which he is not entitled even assuming he does not have notice of a breach of duty on the part of the person who imparted it to him."[84]

An action may also be asserted against a party who is not in a confidential relationship with the disclosing party, or against an "innocent"

81 [1972] 1 O.R. 329 (Ont. H.C.).

82 (1948), 65 R.P.C. 203 at 206.

83 [1972] 2 O.R. 188 (Ont. H.C.).

84 *Ibid.*, at 195. See also *Butler v. Bd. of Trade*, [1971] 1 Ch. 680; and *I.C.L. Enrg. Ltd. v. Associated Corrosion Contractors Ltd.* (1985), 10 C.P.R. (3d) 313 (B.C.S.C.). In the latter case, it was held that once it was determined that the defendants used the plaintiff's design plans without its consent or authorization, it was irrelevant whether such use was made deliberately or unknowingly.

third party who receives confidential information based on a claim for unjust enrichment. This doctrine has been applied in cases where a person receives a benefit from another person if the court finds it would be unjust for the recipient to retain such benefit. In *Pettkus v. Becker*,[85] the Supreme Court applied such doctrine in circumstances where there was "an enrichment, a corresponding deprivation and absence of any juristic reason for the enrichment." This decision suggests that at least in some circumstances the courts will apply an objective test to prevent the use of confidential information belonging to another person whether or not the recipient knew such information was confidential and whether or not there existed any confidential relationship.

7. REMEDIES FOR BREACH OF TRADE SECRET RIGHTS

(a) Injunction

The most commonly sought remedy in respect of the infringement of a plaintiff's trade secrets is an injunction to restrain the defendant from disclosing the trade secrets to others or from using them for his own benefit or purposes. Unlike covenants in restraint of competition, an injunction to restrain the improper use of trade secrets in violation of common law rights need not be subject to any limitation as to time.[86] However, in one case the injunction was limited to such period of time as the trade secret "remains a secret exclusively known to the plaintiff or to those to whom it has been voluntarily disclosed under conditions imposing an obligation of confidentiality upon them."[87]

Where confidential information together with information which is publicly available are used as a "springboard" by a defendant to get a start over others in developing a new product, a court may refuse to grant an injunction and find damages to be a sufficient remedy depending on the worth of the confidential information in saving the defendant time and trouble.[88] On the other hand, a court may grant an injunction even in respect of confidential information which is subsequently made public, for a limited period of time corresponding to the time that a defendant had a "long start" over other members of the public as a result of making improper use of the information at the time it was confidential.[89] The "long

85 [1980] 2 S.C.R. 834 at 848 (S.C.C.).

86 *Bee Chemical Co. v. Plastic Paint & Finish Specialties Ltd.* (1979), 47 C.P.R. (2d) 133 (Ont. C.A.); leave to appeal to Supreme Court of Canada dismissed on November 6, 1979. But see footnotes 56 and 57 and accompanying text.

87 *Int. Tools Ltd. v. Kollar, supra*, footnote 80, at 676.

88 *Seager v. Copydex*, [1967] 1 W.L.R. 923 at 931-932 (C.A.).

89 *Terrapin Ltd. v. Builders' Supply Co. (Hayes)*, [1960] R.P.C. 128 (C.A.); *Schauenberg Indust. Ltd. v. Borowski* (1979), 50 C.P.R. (2d) 69 at 79 (Ont. H.C.).

start" period will depend on the length of time that would have been required to develop any competing product in the absence of misuse of the trade secret.[90]

An interlocutory injunction may be granted to restrain the disclosure or use of trade secrets pending trial of the action. The tests for the granting of such an injunction would appear to be those set forth in the case of *Yule Inc. v. Atlantic Pizza Delight Franchise (1968) Ltd.*:[91]

> It is sufficient if the applicant establishes that there is a substantial issue to be tried; that it has demonstrated a prima facie case, in the sense that the applicant has a legal right which it is attempting to protect pending trial; that the failure to grant the relief will result in a threatened harm to the applicant which may not be adequately compensable by way of damages and finally, that the preponderance of convenience must be in favour of the applicant . . .

A more onerous test would appear to be applicable if an injunction is sought in the context of an employer-employee relationship. In *Canavest House Ltd. v. Lett*,[92] it was held that an employer seeking an injunction against his former employee that goes straight to the defendant's means of livelihood must establish a strong *prima facie* case. The principles applicable in such a case are the same as those applicable to the enforcement of restrictive covenants in an employment contract.[93]

On the question of proof of "irreparable harm", in the case of *Polyresins Ltd. v. Stein-Hall Ltd.*,[94] Galligan J. noted that loss of competitive advantage would lead to irreparable harm to the plaintiff:

> It is hard to imagine how the plaintiff can fail to suffer irreparable harm if its secret is not now protected as much as is possible. Once the secret is lost the competitive advantage is lost with respect to the products involved and may never be regained. Such a loss would, in my opinion, be impossible to measure in damages.[95]

In *Eli Lilly Canada Inc. v. Shamrock Chemicals Ltd.*,[96] McCart L.J.S.C. noted that disclosure of the plaintiff's trade secrets would affect its affiliated companies throughout the world, resulting in damages that would be virtually impossible to calculate and that the defendant would be highly unlikely to be able to pay.

With respect to "balance of convenience", a court will take into

---

90 *Genesta Mfg. Ltd. v. Babey* (1984), 48 O.R. (2d) 94 at 105-106 (Ont. H.C.).

91 (1977), 34 C.P.R. (2d) 284 at 286 (Ont. H.C.); affirmed (1977), 35 C.P.R. (2d) 273 (Ont. Div. Ct.).

92 (1984), 2 C.P.R. (2d) 386 (Ont. H.C.).

93 *Cantol Ltd. v. Brodi Chemicals Ltd.* (1978), 42 C.P.R. (2d) 111 (Ont. H.C.).

94 (1971), 5 C.P.R. (2d) 183 (Ont. H.C.).

95 *Ibid.*, at 191.

96 (1985), 6 C.I.P.R. 5 (Ont. H.C.).

account the stage of the development of the plaintiff's and defendant's products and their relative positions in the marketplace. In *Eli Lilly Canada Inc. v. Shamrock Chemicals Ltd.*,[97] the Court noted that the plaintiff had a substantial and established position in the Canadian and world markets, whereas the defendants' competing product was in its infancy and was not established in the Canadian market.

If an action for an injunction is brought in respect of a breach of contract for which liquidated damages are stipulated, the plaintiff must choose between the liquidated damages and an injunction as the remedy for that breach. If the plaintiff elects to take an injunction and not the liquidated sum, he may recover damages in respect of any breach up to the date of the injunction or the date that the plaintiff should have brought the injunction if he had not delayed. In addition, the plaintiff may obtain an injunction in respect of future breaches and liquidated sums set forth in the contract in respect of any breach of contract committed up to the date of the injunction.[98]

(b) Damages

Damages in an action for breach of confidence will generally be assessed based on the losses actually suffered, subject to equitable factors which the Court may consider. The principles which are applicable would appear to be the same as those that apply in an action for breach of a restrictive covenant. These were expressed as follows in *Elsley v. J.G. Collins Ins. Agencies Ltd.*:[99]

> . . . the amount of compensation which would satisfy the loss suffered, and which the Court considers it just and equitable be paid, usually happens to be equivalent to the amount of legal damages which would be appropriate. The award is still governed, however, by general equitable considerations which would not apply if the plaintiff were seeking damages at law rather than in equity. These considerations might serve, for example, to reduce the amount, due to such factors as delay or acquiescence. In addition, if the parties have agreed on a set amount of damages at law, or a maximum amount, it would be unconscionable, in my opinion, to allow recovery of a greater amount of damages in equity.

Where breach of a confidential relationship gives rise to a "spring-board" period during which a defendant gains a head start over other competitors, damages may be granted based on the loss of profits that the plaintiff would have made during such period in the absence of the

---

97 *Ibid.*

98 *Elsley v. J.G. Collins Ins. Agencies Ltd.* (1978), 36 C.P.R. (2d) 65 (S.C.C.).

99 *Ibid.*, at 79.

defendant's wrongful conduct.[100]

Where the Court finds that the defendant has displayed "flagrant behavior and a callous disregard" of the plaintiff's rights in breach of fiduciary duties, an award for punitive damages may be made.[101] However, such damages will be granted only if the plaintiff's conduct can be characterized as "high handed, malicious . . . showing a contempt of the plaintiff's rights" or showing "malice, wantonness, insult and persistent repetition."[102] In the *Genesta* case,[103] the Court noted that the defendant's state of mind at the time of the conduct complained of, rather than the violence of the reaction of the offended party, must determine whether an award for punitive damages should be made. In *Fern Brand Waxes Ltd. v. Pearl*,[104] the Ontario Court of Appeal refused an award of punitive damages in an action for breach of fiduciary duty on the grounds that: "No claim is made in tort. No precedent exists to justify an award of punitive damages in this type of action." It has been suggested that this decision failed to take into account the broad discretion of the court to award damages to an action for breach of an equitable obligation, and that there is no reason in principle to prevent a court from awarding damages in the case of any wrong.[105]

(c) Accounting for Profits

A plaintiff in an action for breach of confidence may also be entitled to an accounting for profits earned by the defendant. In an action for breach of fiduciary duty, it was held that an agent will be held accountable where he uses information or knowledge which he has been employed by his principal to collect, or which he has otherwise acquired for the use of his principal, so as to make a profit.[106] Such profits will be determined based on the difference between the sale price received by the defendant for any products sold in breach of the plaintiff's rights and the costs incurred by the defendant, including the costs of manufacture, research and development costs, and promotional expenses.[107]

A plaintiff, in general, will be required to elect between his claim for damages and an accounting for profits earned by the defendant. The need to make an election has been described as being "in the end a doctrine

---

100 *Schauenberg Indust. Ltd. v. Borowski* (1979), 50 C.P.R. (2d) 69 at 79 (Ont. H.C.).

101 *Ibid.*, at 80.

102 *Denison v. Fawcett*, [1958] O.R. 312 at 319-320 (Ont. C.A.); *Genesta Mfg. Ltd. v. Babey*, *supra*, footnote 90, at 104.

103 *Ibid.*

104 [1972] 3 O.R. 829 at 837 (Ont. C.A.).

105 See Vaver, "Civil Liability for Taking or Using Trade Secrets in Canada" (1981-82), 5 Can. Bus. Law Journal 253 at 292-296.

106 *Boardman v. Phipps*, [1966] 3 All E.R. 721 (H.L.).

107 *Peter Pan Mfg. Corp. v. Corsets Silhouette Ltd.*, [1964] 1 W.L.R. 96.

based on simple considerations of common sense and equity,"[108] that is, so as to avoid the possibility of double recovery. However, American courts permit both an award for damages and an account, and it has been suggested that there is little reason why a plaintiff should be precluded from obtaining both remedies, subject to the discretion of the court so as to prevent an overlap.[109]

(d) Other Remedies: Delivery Up and Inspection

A court may order delivery up of any documents or materials containing any confidential information belonging to the plaintiff.[110] A constructive trust may also be imposed to provide a tracing remedy for any property that comes into the hands of a person in breach of a confidential relationship.[111] A trust relationship may also be imposed to require delivery up of any machine manufactured by use of the plaintiff's secret design plans.[112]

A court may also make an order for the appointment of independent experts to inspect premises belonging to the defendant to enable the plaintiff to ascertain whether there has been any violation of its rights.[113]

In an action for appropriation of trade secrets, the Court may order the exclusion of the public from the hearing so as to prevent any public disclosure of the secrets.[114] The Court may also order all documents and affidavit evidence, oral evidence and other material containing the plaintiff's confidential information to be sealed and kept confidential in all proceedings prior to trial.[115] Conditions may also be imposed on the conduct of an examination for discovery and other proceedings in order to protect a litigant's trade secrets. In *Big Country Gas Co-operative Ltd. v. Plastex Pipeline Systems Ltd.*,[116] the Court ordered that the defendant Dow Chemical was not required to disclose any of its secret formulae and processes unless and until the plaintiffs and their solicitors gave written

---

108 *Johnson v. Agnew*, [1980] A.C. 367 at 398 (H.L.).

109 See Vaver, *supra*, footnote 105 at 296-300.

110 *Canavest House Ltd. v. Lett* (1984), 2 C.P.R. (3d) 386 (Ont. H.C.); and *Gemologists Int. Inc. v. Gem Scan Int. Inc.* (1986), 7 C.I.P.R. 225 (Ont. H.C.).

111 *Pre-Cam Exploration & Dev. Ltd. v. McTavish*, [1966] S.C.R. 551 (S.C.C.); see also *Indust. Furnaces Ltd. v. Reaves*, [1970] R.P.C. 605 at 627; and *Int. Corona Resources v. Lac Minerals*, *supra*, footnote 63. In the latter case, the court ordered the delivery up of a valuable mining property that was acquired by the defendant based on the improper use of confidential information.

112 *I.C.L. Enrg. Ltd. v. Associated Corrosion Contractors Inc.* (1985), 10 C.P.R. (3d) 313 (B.C.S.C.).

113 *Terrapin Ltd. v. Tecton Structures Ltd.* (1967), 59 W.W.R. 374 (B.C. Chambers); affirmed (1968), 64 W.W.R. 129 (B.C.C.A.).

114 *Eli Lilly Can. Inc. v. Shamrock Chemicals Ltd.* (1985), 6 C.I.P.R. 5 (Ont. H.C.).

115 *Canavest House Ltd. v. Lett*, *supra*, footnote 110 at 395.

116 (1986), 10 C.P.R. (3d) 421 (Alta. Q.B.).

undertakings not to disclose any such information except to experts retained by them. In addition, they were required to ensure that any such experts were not associated with any competitor of Dow Chemical and that the experts gave similar written undertakings. The undertakings were also to provide that all written materials concerning any secret information would be destroyed immediately upon the conclusion of the litigation.

8. CONCLUSIONS

The foregoing review demonstrates that the law relating to trade secrecy protection can be of considerable importance in protecting computer programs and other technology and know-how in the computer industry. At the same time, some questions arise as to the doctrinal basis underlying the protection of trade secrets, the determination of when information is truly secret, whether specific trade secrets should be distinguished from general confidential information, the rights and obligations of the parties to a confidential relationship and how trade secrets should be protected against persons who are not in a confidential relationship with the person who develops or holds the secrets.

In other jurisdictions, the common law principles relating to the protection of trade secrets have been codified in statutory form. For example, in the United States, a Uniform Trade Secrets Act was approved in 1979 by the National Conference of Commissioners on Uniform State Laws and was subsequently enacted with variations in a number of states. The adoption of a new statutory tort for misappropriation of trade secrets has also been recommended by the Law Reform Commission of the United Kingdom. A draft report recommending the adoption of similar legislation by the Canadian provinces was published by the Alberta Institute of Law Research and Reform in 1984.[117] A further joint report by the Alberta Institute of Law Research and Reform and a federal/provincial task force on trade secrets was also released in late 1986.[118] However, to date, no specific legislation has been introduced in Canada to codify the principles relating to trade secrecy protection.

---

117 *Protection of Trade Secrets*, Report for Discussion No. 1, Institute of Law Research and Reform, Edmonton, Alberta, February, 1984.

118 *Trade Secrets*, Report No. 46, Institute of Law Research and Reform, Edmonton, Alberta and a Federal Provincial Working Party, July 1986. Among the recommendations of the Report are that a Trade Secrets Protection Act be enacted by the common law provinces and that new provisions be included in the Criminal Code dealing with the improper acquisition, disclosure or use of trade secrets.

6

Canadian Patent Law and its Application to Computer-Related Subject-Matter

1. INTRODUCTION: THE BASIC PRINCIPLES OF CANADIAN PATENT LAW

The Canadian Patent Act[1] provides protection for any new "invention". An invention is defined as "any new and useful art, process, machine, manufacture or composition of matter, or any new and useful improvement in any art, process, machine, manufacture or composition of matter."[2] This definition should be read with subsection 28(3) of the Act which provides that "no patent shall issue ... for any mere scientific principle or abstract theorem."

Patent protection is conditional upon the successful prosecution of an application and the issuance of letters patent by the Patent Office. Patent protection is also limited as to time, being a term of 17 years.[3] During this time, the owner of the patent enjoys a statutory monopoly entitling him to "the exclusive right, privilege and liberty of making, constructing, using and vending to others to be used the said invention."[4]

As a condition of the grant of such a monopoly, the inventor is required to provide full disclosure of the subject-matter of the invention. Such disclosure is made in the specification set out in the application to the Patent Office which include detailed claims as to the scope of the invention.[5]

In order for a patent to be issued, certain requirements prescribed both by statute and by case law must be met. First, the subject-matter of the claim must be original in that it must be derived from the inventor rather than from another person. Second, the claim must be novel, that is, it must have some unique features and be different from the prior art in some material respect. Third, the invention must have "utility", that

1 R.S.C. 1970, c. P-4.

2 *Ibid.*, s. 2.

3 *Ibid.*, s. 48

4 *Ibid.*, s. 46.

5 *Ibid.*, ss. 35-36.

is, a person skilled in the particular art must be able to put the invention into practice based on the information disclosed to the Patent Office. Fourth, the invention must not be obvious, that is, the unique features of the claim must not be apparent to other persons skilled in the particular field.

2. ADVANTAGES AND DISADVANTAGES OF PATENT PROTECTION FOR COMPUTER-RELATED SUBJECT-MATTER

The principal advantage of patent protection is that it provides the owner of the patent with the exclusive right to manufacture, distribute or sell the invention during a period of 17 years. The patent protects not only the "preferred embodiment" or particular applications of an invention but the entire inventive concept. By contrast, copyright law protects only the form of expression of literary and certain other works. A patent can be enforced against any person, including a person who develops the same invention through his independent efforts. In this regard, patent protection is advantageous to trade secrecy protection which, in Canada, generally depends upon the relationship between the owner of the trade secret and any person to whom the information is disclosed.

The principal disadvantage of relying on patent law in order to protect computer-related subject-matter is that the prosecution of a patent application is very expensive and time-consuming. The time required to obtain a patent may be in excess of two years from the initial application. In many cases, computer programs will be obsolete or cease to be of significant commercial value by the time that a patent is issued. The need to provide a precise description of the subject-matter of the invention means that considerable technical skill and expert legal advice are required to prepare the application and accompanying documentation. Even after a patent is issued, enforcement may be difficult since any alleged infringer is likely to argue that the patent is invalid. A claim for patent protection may also be inconsistent with protection of the subject-matter as a trade secret in that it is a condition of the issuance of the patent that full disclosure be made of the scope of the invention. Patent protection may also be inconsistent with copyright protection insofar as recent case law suggests that the Patent Act, the Copyright Act and the Industrial Design Act should not be interpreted so as to provide overlapping protection.[6]

3. APPLICATION OF PATENT ACT TO COMPUTER PROGRAMS AND OTHER COMPUTER-RELATED SUBJECT-MATTER

In order to support a claim for patent protection, computer software

6 See *Rucker Co. v. Gavel's Vulcanizing Ltd.* (1985), 6 C.I.P.R. 137 (Fed. T.D.).

must come within the statutory definition of an invention, and must meet the tests of originality, novelty, utility and unobviousness set forth above. In addition, it must be shown that the subject-matter of the invention is not a "mere scientific principle or abstract theorem".[7] Although the latter limitation means that the Patent Act is of limited assistance in protecting computer programs as such, the case law has indicated that protection may extend either to a novel apparatus that is programmed in a certain manner where the novelty lies in the apparatus itself, or to a method or process carried out with a novel apparatus. Recent cases further suggest that a patent may be obtained for a novel method or process that is a part of a control or other computer system, or that otherwise produces a useful result, even if a computer program is essential to the implementation of that method or process.

One of the earliest Canadian cases which considered the patentability of computer-related subject-matter was *Re Application Number 961,392 (Waldbaum's Application)*.[8] The Patent Appeal Board held in that case that computer programs or lists of instructions for operating a computer were not patentable. The Board defined the term "program" to mean a set of ordered steps or list of instructions specifying the internal changes of state of physical devices within a data processor. While finding that a program *per se* was not patentable, the Board found that a patent could be issued for a computer programmed to operate in a certain way, or for a method setting out the steps for controlling the operation of a computer or conditioning the computer's operation. Thus, the Board accepted that a patent could be issued for a known data processor or computer that was programmed in such a way that its memory could count the traffic density on telephone lines. The Board also found that methods for controlling the operation of a computer could be equated to methods of operating a machine and were also patentable. In addition, such protection could be granted to a process for conditioning the operation of a computer and a process defining a new use of a known computer.

In *Re Application No. 096,284*[9] the invention related to a process for analyzing data obtained from bore holes concerning the availability of hydrocarbons in soil formations. The invention was described as follows:

> [A] method of seismic exploration in which acoustic signals are generated, reflected from subsurface interfaces, and then detected. The detected acoustical signals are translated into electric signals which are then processed to a convenient form using automatic computing apparatus.[10]

---

7 *Ibid.*, subs. 28(3).

8 (1971), 5 C.P.R. (2d) 162 (Cdn. Patent Office).

9 (1978), 52 C.P.R. (2d) 96 (Pat. Appeal Bd.).

10 *Ibid.*, at 98.

The applicant claimed that the invention consisted of a process for the mechanical performance of a series of calculations so as to produce useful information.

The examiner's rejection of the application for a patent was upheld by the Patent Appeal Board. After reviewing developments since *Waldbaum's Application*, including decisions in other jurisdictions, the Board set out five propositions reflecting a stricter interpretation as to the application of the Patent Act to computer-related subject-matter:

1. Claims to a computer program *per se* are not patentable;
2. Claims to a new method of programming a computer are *not* patentable;
3. Claims to a computer programmed in a novel manner, *expressed in any and all modes*, where the novelty lies solely in the program or algorithm, are not directed to patentable subject-matter under s. 2 of the *Patent Act*;
4. Claims to a computing apparatus programmed in a novel manner, where the patentable advance is in the apparatus itself, are patentable; and
5. Claims to a method or process carried out with a specific novel apparatus devised to implement a newly discovered idea are patentable.[11]

In applying these criteria in the case at bar, the Patent Appeal Board found that the applicant's claims were directed to a method comprising a series of operational steps characterizing the performance of a method. The claims were not tied to a novel apparatus devised to implement the new idea, and the novelty laid solely in the program or algorithm. The Board concluded that a method for programming a computer is not directed to patentable subject matter.

The decision of the Patent Appeal Board was appealed to the Federal Court of Appeal.[12] Pratte J. on behalf of the Court noted that a mathematical formula could be assimilated to "a mere scientific principle or abstract theorem," the patentability of which was negated by subsection 28(3) of the Patent Act. Moreover, the "processes" referred to in section 2 of the Patent Act do not include mental operations and processes. Pratte J. concluded that, although the fact that a computer is used to implement a discovery does not change the nature of that discovery, the subject-matter of the within claim was not patentable on the following grounds:

> What the appellant claims as an invention here is merely the discovery that by making certain calculations according to certain formula, useful information could be extracted from certain measurements. This is not, in my view, an invention within the meaning of s. 2.[13]

Leave to appeal from the decision of the Federal Court of Appeal to the

---

11 *Ibid.*, at 111-112.

12 (1981), 56 C.P.R. (2d) 204 (*sub nom. Schlumberger Can. Ltd. v. Commr. of Patents*) (Fed. C.A.).

13 *Ibid.*, at 206.

Supreme Court of Canada was dismissed on October 20, 1981.[14]

Subsequent to the decision in the *Schlumberger Case*, the Patent Appeal Board considered a number of other applications involving computer-related subject-matter. In several cases, the Board accepted the patentability of the claimed inventions based on the finding that the invention involved a novel apparatus or a method or process carried out with a novel apparatus. In other cases, however, the Board has permitted the patentability of a computer-related invention such as a control system which involves a novel method or process even if a computer program is essential for its implementation. However, it would be premature to conclude based on these decisions that a patent will now be issued for a computer program itself.

In *Re Application for Patent of Lelke (Now Patent No. 1,113,194)*,[15] the applicant sought to obtain a patent for a system of laying out advertising copy by computer. Data assembled by a microcomputer were displayed at work stations on CRT display units. Following an initial rejection, the application was amended to include a restriction that an essential feature of the invention was a bit image memory used to refresh the visual display. The Patent Appeal Board held that the amendment was sufficient to overcome the examiner's objections. The decision appears to reflect the principles adopted by the Patent Appeal Board in 1978 insofar as an apparatus was an essential element of the patentable invention.

The principles set out in the *Schlumberger Case* were applied in the case of *Re Application for Patent No. 178,570*,[16] as a basis for denying patent protection for a "Securities Valuation System". The purpose of the invention was to compute and display the current value of an investment portfolio based on the changing values of the holdings. The information was generated from a central data base connected by telephone lines to various remote user terminals on which updated information was displayed. The application described a method of operating a general purpose computer when restricted to use for gaining access to data bases by means of a particular sub-program and the computer system itself when organized for such purpose.

In its decision, the Patent Appeal Board noted that the Patent Act contained no special provision so as to exclude inventions involving computers. The Board also held that the exclusion in subsection 28(3)

---

14 (1981), 63 C.P.R. (2d) 261 (S.C.C.).

15 (1981), 72 C.P.R. (2d) 139 (Pat. Appeal Bd.). In several other cases, claims have been allowed for a novel combination or arrangement of known computer apparatus: see *Re Application for Patent of Gen. Electric Co.* (Now Patent No. 1,188,775), *infra*, footnote 22; *Re Application of Fujitsu Ltd.* (now Patent No. 1,200,911) (1985), 9 C.P.R. (3d) 475 (Pat. App., Bd.), and *Re Application of Vapor Can. Ltd.* (Now Patent No. 1,203,625) (1985), 9 C.P.R. (3d) 524 (Pat. App. Bd.).

16 (1983), 2 C.P.R. (3d) 483 (Pat. Appeal Bd.).

for a "mere scientific principle or abstract theorem" was not restricted to mathematical formulas. The fact that the process in question produced a new end result did not make the apparatus or method used to achieve such result patentable. Thus, the use of a computer system in order to re-calculate investment holdings to reflect changing values could not make such re-calculation new.

The Board set out the following test as being applicable to determine the patentability of computer-related subject-matter:

> In computer-related subject-matter unless the actual physical aspects or embodiments used are patentable or unless the inherent capabilities of a computer have been combined with another system, which is already on its own merits within a statutory field of invention and thereby produce either a new tangible result or an improvement to a tangible result, then, the board considers, it is very difficult to find a patentable invention. We take the view that a process or procedure for using a known computer to process information, without further integration of that information into some practical system, is not patentable subject-matter within s. 2 of the *Patent Act*. To state what we understand from *Schlumberger*: that is why computers were invented.

The Board continued that the *Schlumberger Case* did not stand for the proposition that only mathematical procedures were unpatentable and that non-mathematical procedures for organizing and operating computer systems should be patentable. Instead, what was in issue was the patentability of procedures merely to regulate the functioning of a computer. The Board went on to distinguish the decision of the United States Supreme Court in *Diamond v. Diehr*[17] in that the subject-matter which was found patentable in that case was a chemical manufacturing process made through a combination of the computer's inherent capabilities with other elements. In the case at bar, the process being claimed occurred entirely within the computer system and the end result was not a real change in the tangible thing but rather merely information. There was no further integration into a practical process as in the *Diehr* case. Since the true subject-matter of the applicant's claims involved computer programs, they were unpatentable even if they were new, useful or unobvious.

In at least two cases, the Patent Appeal Board found that inventions in the field of seismic exploration which involved the use of computers were patentable. In the case of *Re Application of Western Geophysical Co. of America (Now Patent No. 1,160,334)*,[18] the subject-matter related to seismic systems for use in marine or land surveys. The system involved the initiation of an impulse such as an explosive shot, with a record being made of the impulses received at sensors or detectors at spaced locations along the seismic cable extending from the shot point. Output from the

---

17 450 U.S. 175 (1981).

18 (1983), 3 C.P.R. (3d) 386 (Pat. Appeal Bd.).

sensors was routed to the data receiver control from where it could be processed.

The Board accepted an amended description provided by the applicant which included an apparatus. The Board found that the additional apparatus description was acceptable and that this was not like the *Schlumberger Case*, where the novelty laid only in a program or algorithm.

In the case of *Re Application for Patent of Societe National Elf Aquitaine (Now Patent No. 1,190,311)*,[19] the invention involved the use of reflected signals from acoustic vibrational waves which were processed to generate final seismograms. The Board found that the applicant's discovery was that seismograms of an improved signal-to-noise ratio could be obtained by the emission of signals divided into different band widths. This discovery represented a definite advantage over the existing method of seismic exploration. The application was not merely a description of a machine method of manipulating analytical data nor merely a mathematical theorem. Instead, its subject-matter was directed to an improvement in methods of seismic exploration which methods were patentable of themselves.

The *Schlumberger* decision was also distinguished in several cases involving process control systems used to monitor or control operations in an industrial plant.

In *Re Application No. 287,623 for Patent by Bartley and Giles (Patent No. 1,167,549)*,[20] the application related to a method and apparatus used in a process control system such as in electrical analog, digital, electronic, pneumatic, mechanical, hydraulic and other control system formats. The examiner rejected the claims on the grounds that they were directed to an algorithm or program for controlling a process rather than a novel apparatus itself. The applicant contended, however, that the claims were directed to a method and an apparatus for controlling a process. The process control system in question required measurements to be taken, control of a process in response to certain signals and the calculation or generation of certain signals. The fact that parts of the process control could be implemented on a computer did not mean that the claims were directed to an algorithm or a computer program. The applicant claimed further that the novelty lay in the predictive manner in which a process was controlled rather than in a specific computer program.

The Patent Appeal Board agreed with the applicant's contention and found that while the method and means of the application may have used calculations to arrive at altered signals, the end result of the signals was an altered process and not merely a display of information. Something more than a mere algorithm was advanced by the applicant, and the

19 (1984), 6 C.P.R. (3d) 9 (Pat. Appeal Bd.).
20 (1984), 3 C.I.P.R. 105 (Pat. Appeal Bd.).

application was found to be directed to patentable subject-matter.

In the case of *Re Application for Patent of Rockwell International Corporation (Now Patent No. 1,185,691)*,[21] the applicant sought a patent for a method of monitoring industrial plants such as nuclear power plants. Specifically, vibration and pressure signal sensors were located throughout the plant to monitor various operating components. Signals from the sensors were processed by a computer and compared to signals that were accepted as normal. If the difference between the signals as read by the sensors and the predetermined normal signals exceeded an acceptable level, a warning light or alarm was activated so that the plant could be shut down.

The Board considered the relationship between the subject-matter of the application and that of various other patents that had been issued in the United States relating to apparatus for monitoring computer control processes. The Board concluded that the applicant had made an advance in the art by determining acceptable vibration levels along with the power spectral density analysis at selected points for predicting potential breakdown. Referring to the Federal Court decision in the *Schlumberger Case*, the Board held that the use of a computer to implement a discovery does not change the nature of that discovery.

In the case of *Re Application for Patent of General Electric Co., (Now Patent No. 1,188,775)*,[22] the application related to a failure correction control system including electrical sensors to measure engine control parameters (such as temperatures and pressures) and engine controlled parameters (such as fuel flow). The sensor signals were used to control the position of actuators which varied the engine controlled parameters. A computer was used to compute output signals modifying the control parameters for maintaining a desired level of engine performance. The improvement resided in the modification or replacement of erroneous input signals when the difference between the sensed signals and the computed estimate of the signals exceeded a certain tolerance value.

The Board found that the apparatus consisting of a failure detection and correction system viewed as a whole was directed to a combination of elements and no claims were directed to known elements per se. The Board found further that the discovery of replacing erroneous input signals with model generated signals was performed by means of a computer which continuously compared each control parameter sensor input signal with

21 (1983), 6 C.P.R. (3d) 88 (Pat. Appeal Bd.).

22 (1984), 6 C.P.R. (3d) 191 (Pat. Appeal Bd.). For a similar case, see *Re Application for Patent of Measurex Corp.* (Now Patent No. 1,170,750) 1983, 10 C.P.R. (3d) 93 (Pat. App. Bd.), in which the Board upheld the patentability of an invention for optimizing the performance of a multi-unit power plant. The invention consisted of a process for determining changes in the efficiency of boilers based on the quantities of fuel used, and for calculating the most effective allocation of fuels for a particular combination of desired uses.

the model estimate of the signal. The calculated numbers, that is, the controlled parameters in the application, were not the product or end result of the operation but rather were parameters to be used within a system of controlling an engine. The application was to be distinguished from that of *Schlumberger* in which the measured data were re-computed and plotted for interpretation by an operator. The combination performed a function that the patent laws were designed to protect, that is, a system for controlling an engine, and thus represented statutory subject-matter.

The Patent Appeal Board has also accepted the patentability of various inventions involving information storage and retrieval systems or telecommunications switching systems. In *Re Application for Patent of International Business Machines Corp. (Now Patent No. 1,187,197)*,[23] the subject-matter related to a method for the physical storage of text data in such a way that the log or index of the stored text was continually updated to prevent potential loss such as in the case of a power failure. The method also reduced accessing time and minimized the wear on the electro-magnetic components of the computer. These improvements were made by the tangible operation of recording current data on one dedicated portion physically closest to the storage reading and a recording transducer when storing the data.

The Board found that the applicant's discovery was for a method of storing data in a text processing system and that the disclosed method was directed to more than calculations or a mere scientific principle or abstract theorem. The fact that a person could take the inventive concept and derive a computer program from it to operate a general purpose computer did not make it unpatentable.

In *Re Application for Patent of Janssens (Now Patent No. 1,187,157)*,[24] the applicant sought a patent for a telecommunications switching network controlled by computer processors via peripheral interfacing circuits. Each processor included self-contained memory for storing computer programs and each had access to a further common memory for storing data representing conditions of the peripheral circuits.

The Board noted that the application involved the division of programs into series which were executed by different processors. This provided a means of regulating access to the processors, thereby eliminating the need of queuing the series for processing as required by prior art systems. The improvement lay in an orderly re-distribution of programs between the processors. The system distributed the programs selectively among the processors so that some programs were distributed to a plurality of processors and others were handled by one processor only. The Board concluded that the claims in question defined more than algorithms or

---

23 (1984), 6 C.P.R. (3d) 99 (Pat. Appeal Bd.).

24 (1984), 6 C.P.R. (3d) 213 (Pat. Appeal Bd.).

calculations and were more than the mere execution of programs. As a result, the refusal of the claims was withdrawn.

The case of *Re Application of Patent of Dissly Research Corp. (Now Patent No. 1,188,811)*,[25] related to an information storage and retrieval system for identifying information from a set of records located in a base data file. Each record in the file contained certain predetermined identifiable characteristics ("PIC"). A retrieval file included and maintained arrays of information in the form of binary coded elements which were produced from the information content of the base data file. Each array corresponded to a particular PIC which could be presented in a set of stored records, and each element in the array was representative of the location of a particular item in the base data file. To locate a record, the retrieval file was searched by selecting the arrays representing desired PIC and comparing the binary values of the elements in the selected arrays. The identified record was then extracted and made available such as in display form.

The applicant asserted that he had discovered and claimed a novel method of information storage and retrieval. The method involved an apparatus which included various comparing and selecting components in a properly programmed computer which interacted in an automated manner. The Board found that the application consisted of inter-working elements to store and retrieve predetermined identifiable characteristics and the means to extract certain arrays of information to achieve a useful storage and retrieval system. The system consisted of more than mere calculations or a computer program and therefore was acceptable subject-matter under section 2 of the Patent Act.

In *Re Application No. 268,804 for Patent by Dialog Systems Inc. (Patent No. 1,180,813)*,[26] the Patent Appeal Board was asked to reconsider the guidelines that were adopted by the Board in the *Schlumberger Case*. The application in that case related to a system for recognizing a speech signal. The system involved the conversion of analog voice signals into digital signals that were processed to bring about improved accuracy in matching an unknown speech signal with a known signal.

The Board found that the application in that case went far beyond making certain calculations according to certain formulae. The applicant had used his ingenuity to invent a method of producing an improved, frequency-compensated audio signal which only made use of certain mathematical representations and which was patentable under the Patent Act. The Board found that the applicant had disclosed a system which was directed to a useful end result and not merely directed to making calculations nor to the presentation of an algorithm and its solution.

---

25 (1984), 6 C.P.R. (3d) 420 (Pat. Appeal Bd.).
26 (1984), 6 C.I.P.R. 90 (Pat. Appeal Bd.).

In *Re Application for Patent of Westinghouse Electric Corporation (No. 1), (Now Patent No. 1,199,133)*,[27] the application related to an elevator system and a strategy to provide priority service to a special designated floor in a building. The strategy could be connected to an existing elevator system and could respond to calls from switches located on the main floor and the special floor. It could also select the closest car to service the special floor and determine whether up or down calls should be given priority according to whether the special floor was in the upper or lower half of the building.

The application was initially rejected by the examiner on the grounds that it failed to disclose any novel apparatus or new electronic circuitry to enable a person skilled in the art to carry out the strategies defined in the claim. The examiner stated that the novelty of the applicant's computer system laid solely in the programs implemented by the known programmable system processor to operate the known computer system in a novel manner.

On appeal, the Board held that the subject-matter of the application was directed toward a new strategy that could be placed in an elevator system, and the method and means of obtaining an elevator system operating with the new strategy. Incorporation of the new strategy produced a kind of elevator service that could not be derived from the previous systems, and represented more than an algorithm or program. The Board concluded that the applicant's enhanced elevator service was patentable for purposes of section 2 of the Patent Act.

4. CONCLUSIONS

It would appear from the *Schlumberger Case* that a patent is not issuable in Canada in respect of an invention which involves merely the use of a computer program in order to expedite a machine-operated calculation process. However, a patent may be issued in respect of a claim involving a novel computing apparatus programmed in a certain way, or a method or process based on such an apparatus where the novelty lies in the apparatus itself. Decisions subsequent to the *Schlumberger Case* also suggest that if an invention consists of an improved method or process leading to a useful end result, that method or process may be entitled to protection notwithstanding that a computer program may be used in its implementation. The effect of such decisions would appear to be that patent protection can be obtained for a comptuer-related invention which involves a novel method or strategy for producing a useful end result, even if the method utilizes a computer program in achieving that result.

It has been contended that as a result of decisions subsequent to the

27 (1985), 9 C.P.R. (3d) 202 (Pat. App. Bd.).

Schlumberger Case, operating system programs may be entitled to patent protection. It has been further suggested that the fact that application software is an essential part of a claim should not render it unpatentable, provided that there is sufficient post-solution activity to render the claim as a whole patentable.[28] It would appear that by characterizing a computer-related invention as a method or strategy, and by looking at such an invention in terms of the end results as an entirety rather than in terms of the hardware-related elements, the Patent Appeal Board has shown greater flexibility than is indicated by the principles set out in the *Schlumberger Case*.

28 See G.L. Fisk & A.N. Sadek, "Protecting computer programs by patent; Decisions of the Patent Appeal Board after Schlumberger" (1985), 2 C.C.L.R. 201.

Part III

COMPUTERS, CRIMINAL LAW AND EVIDENCE

7

Computers and the Criminal Law

1. INTRODUCTION

The growing proliferation of computers for business, educational, governmental and personal applications has led to a substantial increase in the potential for abuse or misuse of computer technology. In many cases such conduct will be sufficiently harmful to society as a whole or to individual members of society to require the sanction of criminal law.

The term "computer crime" may have one of two meanings. First, the term may refer to any crime such as fraud or theft committed by or through the use of a computer, that is, where a computer is the instrument or vehicle for the commission of the offence. Second, the term may refer to certain acts involving interference with a computer system or the destruction, alteration or theft of computerized information, that is, where a computer system or computer data or programs are the object of the illegal activity.

While conceptually distinct, the two types of criminal offences may be closely related in practice. In this respect, the unauthorized input or alteration of data for a purpose such as the creation of a fictitious account may itself represent the essential act in the commission of fraud or theft. Moreover, while the unauthorized use of a computer system may itself be an offence, it may also be a means of committing some other criminal act. By way of analogy, the unauthorized entry to another person's physical premises with an intent to commit an offence is itself an offence whether or not any further crime is in fact committed.[1]

There are several practical difficulties with relying on criminal law as a means of preventing the abuse or misuse of computer technology. First, the provisions of the Criminal Code were generally enacted well before the advent of computers, and accordingly may be ill-suited to deal with conduct of this nature. Second, it is often not clear whether as a matter of public policy, various acts involving the misuse of computers should be treated as truly criminal rather than being subject to civil remedies or sanctions under intellectual property legislation. Third, many computer-related crimes may go undetected because of a lack of sufficient audit procedures and security measures established by the owner of a computer

1 Criminal Code, R.S.C. 1970, c. C-34, s. 306.

system.[2] Finally, even if computer crimes are detected, the victims may be unwilling to report such offences because of a fear of public embarrassment, uncertainty whether any loss may be proved or fear of potential liability to shareholders or third parties because of the vulnerability of the computer system.[3]

2. THE NATURE AND SCOPE OF COMPUTER CRIME

There do not appear to be any firm statistics as to the extent of computer crime either in Canada or in other countries. Indeed, such statistics may be either difficult or impossible to compile in view of a lack of clear definition as to what constitutes a computer crime and because of difficulties of detection. In its final report, the Canadian House of Commons Sub-Committee on Computer Crime[4] noted that any offence remotely associated with a computer such as the entry into a computer of falsely altered financial reports, is sometimes described as a computer crime. The Sub-Committee expressed skepticism about reports of estimated annual losses in the billions of dollars. The Report noted that representatives of the Canadian Bankers Association had testified that to their knowledge none of their members had experienced a "pure" computer crime, that is, one in which the computer was instrumental in, rather than incidental to, the offence. The Committee also referred to a survey conducted by the Ontario Provincial Police in 1980-1981 in which of 321 responses received from 648 companies canvassed, only 13 reported a loss through computer crimes. Two-thirds of these acts involved theft of computer time or damage to files or equipment. Only five of these incidents were reported to the police and only three prosecutions were undertaken.

The difficulty of detection and lack of knowledge about computer crimes is demonstrated by a survey conducted by a Task Force on Computer Crime established by the American Bar Association Section of Criminal Justice.[5] The survey was distributed to 1,000 large corporations and government agencies of which 283 responded. Of the respondents, 25 per cent acknowledged a known and verifiable loss due to criminal crime in the preceding 12 months. The total losses reported by the respondents fell between $145 million and $730 million, with individual losses ranging

2 On the other hand, computers may also be used to detect criminal activity, such as by computer monitoring of stock trading to detect illegal trades: see "Case came to light because of computer tracking", *The Globe and Mail*, May 15, 1986, p. B10, col. 2-5 (reprinted from New York Times Service).

3 See Sakolik, "Computer Crime — The Need For Deterrent Legislation" (1980), 2 Computer L.J. 353; Soma, *Computer Technology and the Law* (McGraw-Hill, 1983) at 269-270.

4 *Proceedings*, Issue No. 18 (Ottawa: Supply and Services Canada, 1983).

5 Task Force on Computer Crime, *Report on Computer Crime* (American Bar Association, June 1984).

from $2 million to $10 million. Of those who had experienced a computer crime, about one-third had reported the crime to law enforcement authorities. Other statistics presented before a United States Senate Sub-Committee on Criminal Law and Procedures suggested that only 1 in 100 crimes relating to the computer is detected, only 1 in 5 detected crimes is reported and only 1 in 100 is prosecuted.[6]

In spite of the lack of definitive statistics, it would appear from reported cases and media reports that the incidents of computer abuse have increased substantially in recent years. Moreover, the kinds of criminal acts which have been reported indicate the ability of criminal elements to exploit and abuse computer technology in a wide range of ways.[7]

One of the most common crimes committed through the use of computers involves theft of goods, money or services. In one case, a computer security consultant for a Los Angeles bank gained access to the bank's wire transfer room and programmed the computer to transfer $10.2 million to his personal account.[8] In another case, criminals gained unauthorized access to a railway company's computer system in order to re-route railway boxcars for unloading of their contents.[9] To conceal theft, a computer may be used to alter computerized records by showing the misappropriated inventory as damaged or obsolete.

A computer may also be used to commit theft of services. For example, employees may use a company's computer to run their personal businesses. In one case, organized crime members used a computer illegally to keep track of investments.[10]

Criminals may also gain illegal access to telephone facilities for use of long-distance services. Special programs enable a user to "patch in" to long-distance telephone companies such as AT & T, Sprint and MCI to simulate the tones used by telephone companies' service companies or the beeps made by a pay phone when coins are inserted.[11] In one case, four college students were arrested in Montreal for making $20,000 worth of long-distance telephone calls by using a program that dialed toll-free numbers and then took control of the phone company's electronic switch

6 See L. Bernstein,"Computer Crime Laws Opening Pandora's Box of Problems", *The Globe and Mail,* May 13, 1985, p. B11, col. 1-6.

7 See C. Webber, "Jay Walking on the Electronic Highway" (1983), C.L.Q. 237; D.B. Parker, "How Much Computer Abuse is There?", *International Commercial Crime Conference* (Continuing Legal Education Society of British Columbia, 1981).

8 "System May Fool Computer Bandits", *The Toronto Star,* April 3, 1983, p. H5, col. 1 & 2.

9 A. Bequai, "Organized Crime in the Computer Arena", *Police Chief,* October, 1978, at 24.

10 *Ibid.*

11 See "Computer hackers are turning criminal", *The Toronto Star,* July 24, 1985, col. 1-2 (reprinted from Newsday).

so that the calls could be re-routed to any part of the world.[12]

The use of illegal access devices for phone services proliferates by the exchange of information through electronic bulletin boards. In many cases, illegal use of long-distance telephone services fuels other criminal activities.[13]

One of the most common crimes committed by the use of computers is fraud. For example, a criminal may interfere with the computer system of a financial institution or credit bureau to create fictitious credit histories or references. Similarly, he may attempt to alter the financial data of a company to mislead creditors into believing a company is financially sound.[14] Credit card fraud may be committed by gaining access to a credit bureau's computer system to obtain credit card numbers which may then be shared through electronic bulletin boards.[15]

In contrast to the foregoing situations, in which fraud or theft is committed through the use of computers, other computer abuses may involve physical destruction of data or other property. In one case, a "logic bomb", that is, a program designed to destroy, remove, replace or re-arrange information stored in a computer, was inserted in the software installed in the central computer belonging to the Los Angeles Department of Water and Power, resulting in the computer system being shut down for five days.[16] In another case, a programmer added a program to his employer's computer to provide for a destruction of all records in the computer two years later.[17]

Cases of this kind involve the wilful destruction of data or programs intended to interfere with a company's computer system and the operation of its business. As such, they may be distinguished from the "accidental" deletion of data by a computer "hacker", whose only motive may be to determine as a matter of intellectual curiosity whether he can breach the access code for a company's computer system.

Still other forms of interference with computers may result in possible physical injury to individuals or a community. In one case, a computer hacker gained unauthorized entry to the computer system of New York's Memorial Sloan-Kettering Cancer Centre.[18] In another case, the computer records of the intensive care department of a Los Angeles hospital were

---

12 *Edpacs*, November, 1984, at 9; see also "Phone customers cheat Ma Bell of Thousands", *The Toronto Star*, July 11, 1986, p. A8, col. 5-6.

13 See "Computer hackers are turning criminal", *op. cit.*, footnote 11.

14 See Bequai, *op. cit.*, footnote 9.

15 See "Computer hackers are turning criminal", *op. cit.*, footnote 11.

16 *Computer Law Reporter*, Vol. 3, No. 6 (May, 1985) at 894.

17 D.B. Parker, "How Much Computer Abuse is There?", *International Commercial Crime Conference* (Continuing Legal Education Society of B.C., 1981).

18 *The Globe and Mail*, August 19, 1983, p. 1, col. 1-2.

altered by doubling the doses of medication for all patients.[19] Although no actual harm was done in either case, such breaches of security of a hospital's computer system could result in lethal doses of medication being administered.

A computer could also be used to commit acts of terrorism such as the gaining of unauthorized access to a computerized weapons system to change the instructions for their use. Terrorist groups could conceivably interfere with the computer system of a public transit network, an air system traffic control tower or water supply, any of which could have disastrous consequences. In one reported case, a group of students from Milwaukee gained access to the computer at the United States Nuclear Weapons Research laboratory in Los Alamos, California although the computer system in question did not appear to involve the control or deployment of nuclear weapons.[20]

Even if a person who gains unauthorized access to a computer system does not alter or destroy any data or otherwise interfere with the operation of the system, he may nevertheless cause extensive economic loss. For example, when the computer system of the Sloan-Kettering Cancer Centre was broken into, dozens of medical personnel were required to review the records of some 6,000 patients to ensure that no changes had been made.

3. INTERFERENCE WITH COMPUTER SYSTEMS AND DATA: THE APPLICATION OF THE CRIMINAL CODE

The use of a computer system to commit a "traditional" crime does not by itself give rise to any unique legal issues, although a number of practical problems of detection and enforcement may arise. Accordingly, the emphasis in this chapter will be a review of the provisions of the Criminal Code applicable to various acts involving the use or misuse of computer systems or interference with computer programs or data. Among such acts which may entail criminal sanctions are the following:

(a) theft, vandalism, or destruction of computer hardware or the physical media embodying computer programs or data;

(b) obtaining unauthorized access to a computer system for one or more of the following purposes;

  (i) to obtain various types of computer services (that is, theft of computer time);

19 W. Lowther, "Computer freak pranks turn vicious", *The Toronto Star*, August 21, 1983, p. B3, col. 1-4.

20 *Ibid.*

(ii) to interfere with or to interrupt the processes carried on by a computer (computer vandalism);
(iii) to interfere with or to destroy information or programs stored in the computer or the input or output of a computer system (alteration or destruction of data);
(iv) to appropriate for one's own purposes programs or data stored in the computer or to intercept information being communicated to or from the computer or between a central processing unit and a remote terminal (theft of information);

(c) the unauthorized copying for one's own purposes of confidential computer programs or computer-stored information;
(d) the unauthorized distribution in the course of trade of counterfeit or pirated copies (that is, copies made without the consent of the rightful owner) of computer programs or computerized information such as mailing lists.

Among the provisions of the Criminal Code that may be relevant to the foregoing acts are the following: section 283 (theft); section 287 (theft of telecommunication service); section 338 (fraud); subsection 301.2 (unauthorized use of a computer); section 361 (personation); sections 365 and 368 (offences in relation to trade marks); section 387 (mischief); and subsection 387(1.1) (mischief in relation to data).

Of the foregoing provisions, sections 301.2 and 387(1.1) were enacted in 1985[21] to deal specifically with the abuse of computer technology. The other provisions have been or may be part of the Criminal Code for a longer period of time but have been or may be interpreted to apply to various computer-related offences. The following is a summary of how these provisions may apply to various acts in which computer technology is the object of the illegal activity.

(a) Theft: Criminal Code — Section 283

Subsection 283(1) of the Criminal Code reads as follows:

283. (1) Every one commits theft who fraudulently and without colour of right takes, or fraudulently and without colour of right converts to his use or to the use of another person, anything whether animate or inanimate, with intent,

(*a*) to deprive, temporarily or absolutely, the owner of it or a person who has a special property or interest in it, of the thing or of his property or interest in it,
(*b*) to pledge it or deposit it as security,

---

21 Criminal Law Amendment Act, S.C. 1985, c. 19, proclaimed in force, December 4, 1985.

(*c*) to part with it under a condition with respect to its return that the person who parts with it may be unable to perform, or

(*d*) to deal with it in such a manner that it cannot be restored in the condition in which it was at the time it was taken or converted.

The words "anything whether animate or inanimate" clearly include any tangible or corporeal property. Accordingly, the taking or conversion of any tangible component of a computer system, including tapes, disks, print-outs or other physical media embodying programs or other computerized information, would give rise to an offence under this section.

It is less clear whether section 283 applies where computer programs and information are taken or converted but there is no theft of the physical media on which such information is embodied. To show that section 283 applies to such conduct, it would have to be shown that information *per se* is capable of being stolen. A leading Canadian case which has considered this issue is *R. v. Stewart*.[22]

In the *Stewart* case, the accused was charged with counselling the commission of mischief, theft and fraud under the Criminal Code. The accused was retained as a consultant to assist in forming a hotel bargaining unit and approached an hotel employee to obtain a copy of confidential information containing the names, addresses and telephone numbers of employees. Such information was to be obtained without removing or changing the original records.

At trial, Mr. Justice Krever acquitted the accused on all three counts. He found that information as such did not come within the words "anything whether animate or inanimate" as used in subsection 283(1), nor was it "property" for purposes of subsection 338(1). The acquittal from the counts of counselling fraud and theft was appealed to the Court of Appeal.

A majority of the Court of Appeal held that information which is treated by the holder as confidential is capable of being stolen. Houlden J.A. found that "anything" is a word of everyday use having a wide meaning. Although not all information was property, he could see no reason

> why confidential information that has been gathered through the expenditure of time, effort and money by a commercial enterprise for the purposes of

---

22 (1982), 38 O.R. (2d) 84 (Ont. H.C.); reversed (1983), 42 O.R. (2d) 225 (Ont. C.A.) (currently under appeal to the Supreme Court of Canada). The Alberta Court of Appeal refused to follow the *Stewart* decision in the case of *R. v. Offley* (1986), 45 Alta. L.R. (2d) 23. Referring to the minority judgment of Lacourciere J.A. in the *Stewart* case and the authorities cited therein, the court held that information is intrinsically incapable of being "an inanimate thing" for purposes of s. 283, and its quality as confidential could not make it so. Belzil J.A. also expressed doubt about applying the concept of "returning" something to the owner which had never been truly taken. The decision in the *Offley* case is also under appeal to the Supreme Court of Canada.

> its business should not be regarded as property and hence entitled to the protection of the criminal law.[23]

In reaching this decision, Houlden J.A. relied on two English cases which suggested, at least for purposes of civil law, that information compiled through the expenditure of time and effort for the purposes of sale was valuable property.[24] Reliance was also placed on American cases in which the United States Supreme Court reached a similar conclusion.[25]

While taking the view that confidential information could be regarded as "property", Houlden J.A. recognized the division that existed among jurists with respect to this question. In the case of *Boardman v. Phipps*,[26] two judges of the House of Lords were of the view that confidential information could properly be regarded as trust property, two judges took the contrary view and the remaining judge held that in the particular circumstances of the case at bar, the particular information was not trust property.

On the question of whether the accused had the necessary intent set out in any one of paragraphs (*a*) to (*d*) of subsection 283(1), Houlden J.A. noted that if the information had been copied, its confidential character would have been lost even if it continued to be available to the owner. Since the information could not be restored to the condition in which it was prior to its taking, the accused had the intent prescribed by paragraph (*d*) of subsection 283(1).

Finally, on the question of whether the accused could properly be convicted of theft over $200.00, it was held that if a copy of the list had been prepared by the hotel, the cost of preparation would have exceeded $200.00. Moreover, the information had value over and above the cost of preparation.

Cory J.A. concurred in the views of Houlden J.A. and found that confidential lists which were securely kept fell within "that very wide concept of property set out in s. 283(1)."[27] While noting that a compilation of information such as a business list was a "literary work" for purposes of the Copyright Act, Cory J.A. held that only such a compilation that was confidential could be stolen. He also adopted the following tests set out by Megarry, V.C. in *Thomas Marshall (Exports) Ltd. v. Guinle*,[28] as to when information is to be treated as confidential:

---

23 *Ibid.*, 42 O.R. (2d) at 236-37.

24 *Exchange Telegraph Co. v. Gregory & Co.*, [1896] 1 Q.B. 147; *Exchange Telegraph Co. v. Central News Ltd.*, [1897] 2 Ch. 48.

25 *Bd. of Trade of City of Chicago v. Christie Crain & Stock Co.*, 198 U.S. 236 (1905); *Hunt v. New York Cotton Exchange*, 205 U.S. 322 (1907).

26 [1967] 2 A.C. 46 (H.L.).

27 *Supra*, footnote 22, 42 O.R. (2d) 225 at 241.

28 [1978] 3 All E.R. 193 at 209-10.

> First, I think that the information must be information the release of which the owner believes would be injurious to him or of advantage to his rivals or others. Second, I think the owner must believe that the information is confidential or secret, i.e., that it is not already in the public domain. It may be that some or all of his rivals already have the information: but as long as the owner believes it to be confidential I think he is entitled to try and protect it. Third, I think that the owner's belief under the two previous heads must be reasonable. Fourth, I think that the information must be judged in the light of the usage and practices of the particular industry or trade concerned. It may be that information which does not satisfy all these requirements may be entitled to protection as confidential information or trade secrets; but I think that any information which does satisfy them must be of a type which is entitled to protection.[29]

In a dissenting judgment, Lacourciere J.A. stated that a restricted property concept limited the widest possible sense of the word "anything" in subsection 283(1). The cases relied on by the majority suggesting there was a property right in confidential information insofar as a court of equity would restrain its improper transmission were not helpful in considering whether confidential information could be stolen for purposes of the Criminal Code. Reference was also made to the case of *Oxford v. Moss*[30] in which it was held that confidential information consisting of university examinations did not fall within the English Theft Act, 1968. In addition, in *Rank Film Distributors Ltd. v. Video Information Centre*,[31] the House of Lords held that the pirating of copyrighted films and the selling of unauthorized video cassettes did not constitute theft since the property concept in the Theft Act did not include copyright. Based on these cases, Lacourciere, J.A. concluded that only Parliament could broaden the criminal definition of property to include confidential information.

The applicability of subsection 283(1) to the theft of programs was also considered by the Alberta Queen's Bench in the case of *R. v. Tannas*.[32] The accused in the *Tannas* case was charged with theft under paragraph 283(1)(*d*) on the grounds that he fraudulently and without colour of right converted to his own use confidential computer programs belonging to his former employer, Dome Petroleum. Specifically, the Crown contended that the accused used Dome's confidential computer programs in his

29 *R. v. Stewart, supra*, footnote 22, 42 O.R. (2d) at 245.

30 (1979), 68 Cr. App. R. 183.

31 [1981] 2 All E.R. 76 (H.L.).

32 Alta. Q.B., November 28, 1984; for a report of the indictment and the charge by Mr. Justice Bracco to the jury, see M.P.J. Kratz, Case Comment "*R. v. Tannas*" (1985), 2 C.C.L.R. 205; see also L. Milrad, "Practical Problems of Using the Criminal Law to Protect Computer Software" (1985), Business and the Law, Volume 2, Number 8, August, 57. In view of the subsequent decision of the Alberta Court of Appeal in *R. v. Offley*, the charge to the jury by Bracco J. would not appear to be a correct statement of the current law in Alberta: see footnote 22, *supra*.

possession for sales demonstration purposes and that he intended to deal with such programs in such a way as to destroy their confidential character.

The defence contended that Dome had failed to treat the programs as truly confidential. It argued that Dome permitted information concerning the programs to be freely exchanged, its policy statement regarding confidentiality was not proved to be within the accused's knowledge, such statement indicated only that the accused might be asked to return all materials upon termination of his employment and the accused was in fact not asked for such materials upon leaving the company. It was further contended that the accused intended to use the programs for reference purposes only rather than for purposes of sale and that he did not intend to destroy their confidentiality.

In his charge, Bracco J. instructed the jury that computer software programs and the information contained in them were included in the term "anything" which could be "taken" or "converted" for purposes of subsection 283(1). He noted further that the jury had to be satisfied that the accused did not have an honest belief that he was entitled to make copies of the programs. With respect to the intent prescribed by paragraph (*d*), the judge advised that the jury had to be satisfied that the programs were in fact treated as confidential by Dome, and that the accused intended to and did in fact convert or use the programs in such a manner as to destroy or permanently affect their confidential character.

After six and one half hours of deliberation, the jury returned a verdict of not guilty. Although the basis of the decision is unknown, the jury was presumably not satisfied that all elements of the charge were proved beyond a reasonable doubt. It is possible in this connection that the jury was not satisfied that the company had taken all necessary measures to treat the programs as confidential or that the accused had the necessary intent to destroy the confidentiality of the programs.

Although both the *Stewart* and *Tannas* cases proceeded on the basis that the Crown had to demonstrate the confidential character of the information in question, the rationale underlying the need to prove such confidentiality differed. In the *Stewart* case, proof of the confidential character of the information was necessary in order for the information to come within the word "anything" as used in subsection 283(1). In the *Tannas* case, on the other hand, the need to show such confidentiality arose primarily as an element of proof of the accused's intent to deal with the information in such a manner that it could not be returned to the condition in which it was prior to its taking, that is, its confidential status could not be restored.

The *Stewart* and *Tannas* cases, when contrasted with the decision in the *Offley* case, indicate the uncertainty that remains as to the application of subsection 283(1) of the Criminal Code to certain acts in relation to computer programs, data and services. It is hoped that this uncertainty will be resolved by a decision of the Supreme Court of Canada in either the *Stewart* or *Offley* case. If the majority judgment of the Ontario Court of Appeal is correct, it means that information *per se* can be stolen, provided that it is valuable information collected by the expenditure of time and money and is treated as confidential by its holder. Information of this character should be protected whether it is embodied in a printed medium or held on tapes, disks or other machine readable media. Accordingly, computer programs and computerized information held in data banks would be capable of being stolen provided that the holder takes the necessary steps to preserve their confidentiality. However, as indicated by the *Tannas* case, it may be difficult in any case to prove that the information was truly confidential and that the accused intended to destroy its confidential character.

The *Stewart* case also suggests the appropriation of "computer time" may also give rise to a charge of theft. If the word "anything" includes intangible information, it should also include computer services which have value to the person who owns or controls access to the computer. Accordingly, a person who uses the services of a computer without authorization would be appropriating or converting for his own use valuable computer time for which the expenditure of money would otherwise be required.[33] However, in the same way as a person who compiles confidential information must show that he took the necessary steps to preserve its confidential character, the owner of a computer would have to show that he has taken necessary measures to prevent unauthorized access.

(b) Theft of Telecommunications Service: Criminal Code — Subsection 287(1)

Section 287 of the Criminal Code provides as follows:

> 287. (1) Every one commits theft who fraudulently, maliciously, or without colour of right,
>
> (*a*) abstracts, consumes or uses electricity or gas or causes it to be wasted or diverted, or,
>
> (*b*) uses any telecommunication facility or obtains any telecommunication service.
>
> (2) In this section and in section 287.1, "telecommunication" means any transmission, emission or reception of signs, signals, writing, images, sounds or intelligence of any nature by radio, visual, electronic or other electromagnetic system.

---

33 See Law Reform Commission, Working Paper 19, *Criminal Law, Theft and Fraud* (1977) that would have included computer services within the definition of "property".

The question of whether this section applies to the unauthorized use of a computer facility was considered by the Supreme Court of Canada in the case of *R. v. McLaughlin*.[34] Laskin C.J. on behalf of the Court held that a computer system, consisting of a central processing unit, memory, printers and connected terminals, was a data processing facility rather than a "telecommunication facility" for purposes of section 287. This followed from the fact that the function of a computer was not to channel information to outside recipients, but rather was to permit the making of complex calculations and to process, correlate, store and retrieve information. The fact that the accused was able to make electronic connection with the computer by the use of a remote terminal did not bring his conduct within paragraph 287(1)(*b*).

Estey J. also held that in view of the reference in the definition of telecommunication to a "transmission", it was essential that there be both a sender and a receiver. The operation of a computer involves the participation of only the operator, who does not transmit information by the use of the terminal or console of the computer.

Conduct involving use of a computer may give rise to an offence under section 287 insofar as the computer is used to gain access to a telecommunication facility. For example, a computer modem may be used to gain access to a telephone company's switching system where calls may be re-routed to any location in the world.[35] Alternatively, telecommunication services may be "stolen" by the use of a "red box" to simulate the tones made when coins are inserted into a pay telephone.[36]

(c) Fraud: Criminal Code — Section 338

Section 338. (1) of the *Criminal Code* provides that everyone commits an offence

> who, by deceit, falsehood or other fraudulent means, whether or not it is a false pretence within the meaning of this Act, defrauds the public or any person, whether ascertained or not, of any property, money or valuable security . . .

In the case of *Re London & Globe Finance Corp.*,[37] it was held that to defraud is

> to deprive by deceit: it is by deceit to induce a man to act to his injury. More tersely it may be put, that to deceive is by falsehood to induce a state of

---

34 [1980] 2 S.C.R. 331 (S.C.C.).

35 See *supra*, footnote 12.

36 See L. Surtees, "Using pay-phone fraud box means change for the worse", *The Globe and Mail*, January 10, 1986, p. B14, col. 5-6.

37 [1903] 1 Ch. 728.

mind; to defraud is by deceit to induce a course of action.[38]

A broader definition of fraud was adopted in the case of *Scott v. Metropolitan Police Commissioner*.[39] It was held that to "defraud" was "to deprive a person dishonestly of something which is his or of something to which he is or would or might but for the perpetration of the fraud, be entitled."[40]

The leading Canadian authority on the meaning of fraud is the decision of the Supreme Court of Canada in the case of *R. v. Olan*.[41] Dickson J. on behalf of the court held that the two essential elements to be proved by the Crown were first, dishonesty, and second, deprivation. It was held in this regard that the words "other fraudulent means", as used in subsection 338(1), included not only deceit or falsehood but any means which could properly be stigmatized as "dishonest". The element of deprivation required proof of detriment, prejudice or risk of prejudice to the economic interests of the victim but it was not necessary to prove actual economic loss as the outcome of the fraud.[42]

Based on the tests adopted in the *Olan* case, it has been held that a person may be defrauded notwithstanding that he does not have any nexus or relationship with the accused. In the case of *R. v. Kirkwood*,[43] the accused distributed, either by sale or rental, "counterfeit" video cassettes, that is, tapes duplicated from legitimate tapes without the consent of the copyright owners and packaged in the same way as the originals. The Court of Appeal adopted the finding of the trial judge that the accused knew that he was depriving the copyright owners of revenues they otherwise would have been likely to earn. Moreover, the accused did not attempt to obtain the consent of the rights holders nor did he intend to contribute any revenues to them for the use of the films.

The Court of Appeal held that where the fraud was committed not by deceit or falsehood, but rather by "other fraudulent means", it was necessary to show only dishonesty and deprivation. The first element relates only to the accused's intent and does not require any false representation to operate on the mind of the victim. The second element requires only an awareness of the risk of prejudice to the economic interests of the accused. The fact that there was no relationship between the accused and the victim was irrelevant.

That the same conduct may lead to a conviction on a charge of either

---

38 *Ibid.*, at 732-33.

39 [1975] A.C. 819 (H.L.).

40 *Ibid.*, at 839.

41 [1978] 2 S.C.R. 1175 (S.C.C.).

42 For a full discussion of the implications of the *Olan* decision and the current Canadian law relating to fraud, see D. Ewart, *Criminal Fraud* (Carswell, 1986).

43 (1983), 42 O.R. (2d) 65 (Ont. C.A.).

fraud or theft is supported by the decision in *R. v. Stewart*.[44] In that case, Houlden J.A. found that confidential information was property not only for purposes of subsection 283(1), but also for purposes of subsection 383(1). Houlden J.A. found further that fraud required proof of dishonest deprivation, and it was not essential to prove actual economic loss as the outcome of the fraud. Accordingly, since the hotel could have profited from providing lists of its employees to promotional groups, there was a risk of prejudice to the hotel's economic interests which was sufficient to prove the element of deprivation.

The foregoing cases indicate that subsection 338(1) may apply to any type of interference with a computer system or the taking of computer programs or data where the necessary elements of dishonesty and deprivation can be proved. For example, the unauthorized sale or other distribution of pirated copies of computer programs would appear to constitute fraud in the same way as the unauthorized distribution of pirated video cassettes was found to be fraudulent in the *Kirkwood* case. The only distinction may be that in the absence of an express provision dealing with the protection of computer programs in the Copyright Act, a distributor of unauthorized copies of programs may argue that he honestly believed he had the right to make such copies. However, it is unlikely such an argument would succeed in view of recent case law indicating that programs even in object code form are entitled to protection as literary works under the Copyright Act.

The *Kirkwood* case involved the reproduction and distribution for trade of multiple unauthorized copies of copyright protected material. The court did not consider whether the making of a single unauthorized copy of a copyright work for one's own use could constitute fraud. It is arguable that the making of just one copy would come within subsection 338(1) since it creates at least the risk of prejudice to the owner's economic interests by precluding the purchase of a legitimate copy. However, it is questionable whether as a matter of policy, the making of a single copy of a computer program would be treated by the Crown as justifying the time and expense involved in a prosecution under the Criminal Code.

Finally, subsection 338(1) should also apply to the unauthorized use of a computer facility for the purpose of processing data or obtaining other services, provided that the owner has restricted access to such facility to employees or persons who pay valuable consideration and the user is aware that he is not authorized to use the facility. In such circumstances, the user would appear to have the necessary dishonest intent and the owner of the computer facility would suffer deprivation as a result of the taking of computer services for which payment would otherwise be required.

---

44 *Supra*, footnote 22.

(d) Personation: Criminal Code — Section 361

Section 361 of the Criminal Code provides as follows:

361. Every one who fraudulently personates any person, living or dead,

(*a*) with intent to gain advantage for himself or another person,
(*b*) with intent to obtain any property or an interest in any property, or
(*c*) with intent to cause disadvantage to the person whom he personates or another person,

is guilty of an indictable offence and is liable to imprisonment for fourteen years.

This section may be applicable to a person who gains unauthorized access to a computer system by the use of a password or access code belonging to another person. It would have to be shown in such a case, first, that the accused had the fraudulent or dishonest intent, second, that he "personated" an actual person, and third, that he intended to gain advantage for himself or another person, to obtain property or an interest in property or to cause disadvantage to either the person personated or another person.

As in the case of an offence under section 338, the element of fraud as part of the offence created by section 361 would appear to require some form of dishonesty, which may not amount to deceit or falsehood operating in the mind of the person defrauded.[45] Accordingly, a person who uses an access code or pass word to gain access to a computer knowing that he has no right to do so would have the necessary fraudulent intent for purposes of section 361.

With respect to the second element of the offence prescribed by section 361, it must be shown that the "person", living or dead, who is impersonated actually exists or has existed. Thus it is not an offence under section 361 for a person to assume the identity of a non-existent person, such as by giving a false name when arrested.[46]

With respect to the final element of the offence of personation, it is clear from paragraphs (*a*) and (*c*) that section 361 applies in circumstances which do not involve an intent to gain or to deprive another person of a proprietary interest. It has been held that the word "advantage" referred to in paragraph 361(*a*) is not confined to an economic, proprietary or monetary advantage, but includes only "better position, precedence, superiority or favourable circumstances."[47]

---

45 *R. v. Olan, supra*, footnote 41.

46 *R. v. Cole* (1982), 37 O.R. (2d) 144 (Ont. Dist. Ct.); *R. v. Northrup* (1982), 41 N.B.R. (2d) 610 (N.B.C.A.).

47 *Rozon v. R.* (1974), 28 C.R. (N.S.) 232 (Que. C.A.); *R. v. Hetsberger* (1980), 51 C.C.C. (2d) 257 (Ont. C.A.).

It would appear that a person who gains access to a computer by the unauthorized use of another person's pass word or access code would have the intent to gain advantage for himself or to cause disadvantage to another person within the broad meaning of such terms adopted by the courts. It would not be necessary for the Crown to show that the accused has "stolen" or defrauded the owner of the computer facility of any property, whether consisting of computer programs or information, or the use of computer time. Accordingly, section 361 may be relied upon in circumstances where the application of the fraud and theft provisions of the Criminal Code may be in dispute.[48]

(e) Offences in relation to Trade Marks: Criminal Code — Sections 365 and 368

Sections 365 and 368 of the Criminal Code create two offences in relation to trade marks. Section 365 provides that "everyone commits an offence who, with intent to deceive or defraud the public or any person, whether ascertained or not, forges a trade mark." Section 368 provides that every one "commits an offence who, with intent to deceive or defraud, (*a*) defaces, conceals or removes a trade mark or the name of another person from anything without the consent of that other person. . . ."

In *R. v. Loquet*,[49] the accused was found guilty under section 365 of unlawfully reproducing the trade mark "Magiciel" in a computer program without the consent of the trade mark owner and with the intent of deceiving the purchaser. The accused was also found guilty on two counts under subsection 368(*a*) of concealing and removing the name of the copyright owner from user manuals supplied with the computer programs without the owner's consent.

The court found that in the absence of a written agreement, the accused did not have the authority of the owner to sell the computer programs. Moreover, the accused was expressly prohibited by the terms of a written distribution agreement made with the owner of the programs from copying the user manuals or program disks.

The *Loquet* case indicates that the distribution of unauthorized copies of computer programs may give rise to an offence if a trade mark is forged, or if the name of any person having proprietary rights in computer programs or user manuals is removed without the owner's consent. Such acts will give rise to an offence, however, only in circumstances where there is an intent to deceive or defraud.

---

48 See J. Finlay, "Another Alternative: Personation" (1984), 1 C.C.L.R. 70.

49 (1985), 5 C.P.R. (3d) 173 (Que. Sessions Ct.).

(f) Mischief: Criminal Code — Section 387

Subsection 387(1) of the Criminal Code sets out the kinds of conduct which are deemed to constitute mischief:

> 387. (1) Every one commits mischief who wilfully
>
> (*a*) destroys or damages property,
> (*b*) renders property dangerous, useless, inoperative or ineffective
> (*c*) obstructs, interrupts or interferes with the lawful use, enjoyment or operation of property, or
> (*d*) obstructs, interrupts or interferes with any person in the lawful use, enjoyment or operation of property.

Subsection 387(3) provides that everyone who commits mischief in relation to property of which the value exceeds $1,000 is guilty of an indictable offence and is liable to imprisonment for ten years or an offence punishable on summary conviction. Under subsection 387(4), the maximum term of imprisonment for mischief in relation to property having a value of less than $1,000 is two years.

The offences prescribed by subsection 387(1) all involve various acts relating to "property". Specifically, it must be shown that the accused has destroyed or damaged property, has rendered property useless, inoperative or ineffective, or has obstructed, interrupted or interfered with the lawful use, enjoyment or operation of property or with any person in such lawful use, enjoyment or operation. For purposes of section 387, the word "property" is defined in section 385 to mean "real or personal *corporeal* property". This definition is more restrictive than that set out in section 2 of the Criminal Code which is applicable to the offences of fraud and theft.

The application of the "mischief" provision of the Criminal Code to conduct involving interference with computer tapes was considered by the Ontario Supreme Court in the case of *R. v. Turner*.[50] In that case, the applicants had used telephone connections from Toronto to obtain access to computer tapes installed on a computer facility in Milwaukee. The tapes contained programs used by American companies to process work on their master files. During the course of the telephone connections, the applicants caused the tapes to be mounted on a computer and to be encoded or encrypted so that they could not be properly used by the American owners. The process of encryption did not result in the physical alteration of the tapes nor the destruction of any data. Instead, a locking mechanism was inserted to prevent the American companies from gaining access to the data stored on the tapes.

The applicants relied on the definition of "property" in section 385

---

50 (1984), 27 B.L.R. 207 (Ont. H.C.); affirmed Ont. C.A. (February 14, 1985).

as meaning "real or personal corporeal property", and argued that the alteration of electronic data on a computer tape did not interfere with the tape or its use in its corporeal aspect.

Gray J. held that sections 385 and 387(1) should be given their ordinary meaning. The applicants had made it impossible for the American companies to process their data or to retrieve the information contained on the tapes. The gist of the offence in paragraph 387(1)(*d*) was the interference with the enjoyment of property rather than its physical alteration. Accordingly, the accused's conduct came within such provision insofar as the interference with the intangible information contained on the computer tape affected the right of enjoyment of the property in the tape itself.

The decision in *R. v. Turner* was rendered prior to the enactment of the new section 387(1.1) which created a new offence of mischief in relation to data. The decision may be interpreted as representing a liberal interpretation of subsection 387(1) insofar as there was no actual obstruction or interference with any physical medium. In addition, insofar as the medium on which computer programs are stored represents a different proprietary object than the programs themselves, it is questionable whether there was any actual interference with any corporeal property as such. In any event, in view of the enactment of paragraph 387(1.1) as discussed in greater detail below, it is questionable whether a prosecution would now be brought under section 387 in circumstances similar to those set forth in the *Turner* case.

(g) Unauthorized use of a computer: Criminal Code — Section 301.2

Of the two new computer-related offences created by Bill C-19, the first is that created by section 301.2, which provides as follows:

301.2 (1) Every one who, fraudulently and without colour of right,

(*a*) obtains, directly or indirectly, any computer service,

(*b*) by means of an electromagnetic, acoustic, mechanical or other device, intercepts or causes to be intercepted, directly or indirectly, any function of a computer system, or

(*c*) uses or causes to be used, directly or indirectly, a computer system with intent to commit an offence under paragraph (*a*) or (*b*) or an offence under section 387 in relation to data or a computer system

is guilty of an indictable offence and is liable to imprisonment for a term not exceeding ten years, or is guilty of an offence punishable on summary conviction.

(2) In this section,

"computer program" means data representing instructions or statements that, when executed in a computer system, causes the computer system to perform a function;

"computer service" includes data processing and the storage or retrieval of data;

"computer system" means a device that, or a group of interconnected or related devices one or more of which,

(*a*) contains computer programs or other data, and
(*b*) pursuant to computer programs,
  (i) performs logic and control, and
  (ii) may perform any other functions;

"data" means representations of information or of concepts that are being prepared or have been prepared in a form suitable for use in a computer system;

"electromagnetic, acoustic, mechanical or other device" means any device or apparatus that is used or is capable of being used to intercept any function of a computer system, but does not include a hearing aid used to correct subnormal hearing of the user to not better than normal hearing;

"function" includes logic, control, arithmetic, deletion, storage and retrieval and communication or telecommunication to, from or within a computer system;

"intercept", includes listen to or record a function of a computer system, or acquire the substance, meaning or purport thereof.

At the time of writing, no reported cases had been decided dealing with this new section. However, several comments may be made as to how this provision is likely to be interpreted.

It must first be shown that the conduct of the accused was committed fraudulently *and* without colour of right. As noted above, proof that the accused acted fraudulently will require evidence to show that he acted dishonestly and with an intent to cause deprivation to another person. Proof that the accused acted "without colour of right" will require proof that the accused did not have an honest belief in a state of facts which, if they existed, would be a legal justification or excuse.[51]

The foregoing elements mean that a person who honestly believes he has the right to use a computer system will not be guilty of an offence. For example, the section would not be applicable to a person who uses the services of a computer service bureau, or obtains information from a computer data base, if such person has a *bona fide* belief that he is entitled

51 *R. v. Creaghan* (1982), 1 C.C.C. (3d) 449 (Ont. C.A.).

to do so under the terms of a contract even if such right is found not to exist.

It will be further noted that not every unauthorized use of a computer system gives rise to an offence under section 301.2 but only such uses which involve the "obtaining" of a "computer service", the "interception" of a "function" of a computer system, or the use of a computer system with an intent to obtain any such service or intercept any such function or to cause mischief in relation to data.

The definitions of the italicized expressions give rise to some question as to what conduct is intended to be prohibited by the section. For example, although the term "computer service" is defined to *include* data processing and the storage or retrieval of data, it is not clear whether it may also include other activities which come within the definition of a "function" of a computer system, for example, any communication to, from or within the system.

It will also be noted that what is prohibited under paragraph 301.2(1)(*a*) is only the obtaining of a computer service (for example, data processing or the storage or retrieval of data) rather than the actual taking of computer data or programs. Thus, while the wrongful use of another person's computer system to process one's own data would come within paragraph 301.2(1)(*a*), the retrieval of data belonging to that other person by itself would not appear to come under this provision.

There would also appear to be some difficulties of interpretation with respect to the new offence in paragraph 301.2(1)(*b*) of "intercepting" the "function" of a computer system. The definition of "intercept", as set out in subsection 301.2(2) includes listening to or recording the function of a computer system or acquiring the substance, meaning or purport thereof. This definition is taken from that set forth in section 178.1 of the Criminal Code in respect of the offence created by section 178.11, that is, the interception of a "private communication". The words "private communication" are defined in section 178.1 to mean "any oral communication or any telecommunication . . ."; the word "telecommunication", in turn, is defined in the Interpretation Act to mean "any transmission, emission or reception of signs, signals, writing, images or sounds or intelligence of any nature by wire, radio, visual, or other electromagnetic system."[52]

It will be seen from the foregoing that the word "intercept", as originally set out in section 178.1 and continued in subsection 301.2(2), was intended to deal with the recording, listening to or acquiring the substance of some form of communication of information made either verbally or by some form of electromagnetic system. The term "function", on the other hand, is defined in subsection 301.2(2) to include not only any communication or telecommunication to, from or within a computer

52 R.S.C. 1970, c. I-23, s. 28.

system, but also other functions of the system such as logic, control or arithmetic.

The use of a device for the purpose of acquiring any data being communicated to, from or within a computer system, including from one part of a computer system to another (for example, as between a central processing unit and a remote terminal) would clearly be an offence within the meaning of paragraph 301.2(1)(*b*). This would be the case whether or not any information is in fact "taken". It is not clear, however, how one could "intercept" other functions of a computer system such as "logic" or "arithmetic". In this connection, it would have to be shown that the accused has recorded, listened to or acquired the substance of the logic or arithmetic function of the computer, and not just the information on which such function may operate. While it would be possible to "interfere" with any such function, such as by making it impossible for the functions to be performed, such interference would not necessarily constitute an "interception" of the function. At the same time, if any such act results in an interference with the "use of data", it would be an offence for purposes of subsection 387(1.1) as discussed below.

Reference should also be made to the definitions of "computer program" and "data" set forth in subsection 301.2(2). Specifically, "data" is defined to mean "representations" of information or of concepts that are being prepared or have been prepared for use in a computer system. Under this definition, what is protected is not information *per se* but rather the manner in which it is expressed. The "information" may subsist independently of the "data" insofar as it may remain in the mind of a reader. So too the data must also be distinguished from the medium on which the data may be stored or recorded.[53]

It will be noted that under subsection 301.2(2), data is defined very broadly to include any representations of information or concepts. However, the representation must be prepared in a form suitable for use in a computer system. This would appear to require that the information be embodied in media which can be used in conjunction with a computer. The section clearly contemplates a different status being accorded to data expressed in a form for use on a computer system from other representations of information.

The term "computer program" is defined in subsection 301.2(2) so as to be subsumed under the definition of "data". Specifically, a program means data representing instructions or statements which will cause a computer system to perform a function when executed in a computer system. The subsuming of a "program" within the term "data" represents a departure from the normal usage in the computer business of using the

53 See D.K. Piragoff, "Canadian Legislative Proposals for the Protection of the Integrity of Computer Systems", *Computer Law Update Conference* (Insight, June, 1984).

terms "data" or "data base" to refer to compilations of information stored in or produced by a computer by the use of computer programs.

(h) Mischief in relation to data: Criminal Code — Sections 387(1.1) and 387(5)

The second offence created by Bill C-19 is that of mischief in relation to data, as set forth in a new subsection 387(1.1) which reads as follows:

> (1.1) Every one commits mischief who wilfully
>
> (*a*) destroys or alters data;
> (*b*) renders data meaningless, useless or ineffective;
> (*c*) obstructs, interrupts or interferes with the lawful use of data; or
> (*d*) obstructs, interrupts or interferes with any person in the lawful use of data or denies access to data to any person who is entitled to access thereto.

A new subsection 387(5) was also enacted by Bill C-19 to provide that everyone who commits mischief in relation to data is guilty of an indictable offence and is liable to imprisonment for a term not exceeding ten years or is guilty of an offence punishable on summary conviction. In addition, subsection 387(5.1) provides that every one who wilfully does an act or wilfully omits to do an act that is his duty to do, if that act or omission is likely to constitute mischief in relation to property or data, is guilty of an indictable offence and is liable to imprisonment for a term not exceeding five years, or is guilty of an offence punishable on summary conviction. Finally, subsection 387(8) provides that data has the same meaning as in section 301.2, that is, representations of information or of concepts that are being prepared or have been prepared in a form suitable for use in a computer system.

In order to commit an offence under section 387(1.1), it must be shown that the impugned act was committed "wilfully". Subsection 386(1) provides that a person is deemed to have acted wilfully if he does any act or omits to do an act causing an event to occur if he knows the act or omission will probably cause the event to occur and he is reckless as to whether the event occurs or not.

The new subsection 387(1.1) would clearly apply to the kind of conduct which was the subject-matter of the *Turner* case. Under the new section, it will be sufficient to show interference with the use of the data as expressed, for example, in electromagnetic impulses without the necessity of showing that any person is obstructed in his use or enjoyment of any physical medium in which the information is embodied. At the same time, the section makes it an offence to commit certain acts in relation to representations of information held or prepared in computerized form whereas the same acts in relation to information expressed in printed form may not constitute an offence.

Subsection 386(2) provides further that no person shall be convicted of an offence under section 387 where he proves that he acted "with legal justification or excuse and with colour of right". In *R. v. Creaghan*,[54] it was held that the word "and", as used in this section, should be read as "or". Accordingly, the accused must show that he acted either with legal justification or with colour of right. The words "colour of right" were interpreted to mean an honest belief in a state of fact which, if it existed, would be a legal justification or excuse.

The qualifications set out in subsection 386(2) mean that even if a person wilfully commits any act coming within subsection 387(1.1), he could not be convicted of mischief if he honestly believed he was legally entitled to do the act complained of. For example, the section would not apply to a person who destroys data under an honest but mistaken apprehension that he was authorized to do so by the owner. Similarly, if an owner of a computer service bureau denies access to data stored on the system to a licensed user based on a *bona fide* contractual dispute, such denial of access should not give rise to an offence under section 387(1.1) provided that the owner honestly believed he was entitled to deny such access.

4. CONCLUSIONS

It is likely that the new offences created by sections 301.2 and 387(1.1) of the Criminal Code will be relied on in a number of circumstances where there is an unauthorized use of a computer system or obstruction or interference with computer data or programs. In addition, various other provisions of the Criminal Code will continue to be relied on to deal with other activities of a criminal nature, such as the unauthorized distribution of pirated computer programs, the conversion of confidential programs or data for one's own use, the removal or concealment of trade marks with an intent to defraud, and various other acts that may come within the wide meaning of fraud adopted by Canadian courts. Accordingly, the owners of computer systems and computer programs and data should be aware of how the criminal law may be used to protect their proprietary rights and interests.

54 (1982), 1 C.C.C. (2d) 449 (Ont. C.A.).

8

Computer Documents and the Law of Evidence

1. INTRODUCTION

The establishment of sophisticated methods of computerized record-keeping and data processing has raised important new issues in the law of evidence. Computers have greatly facilitated the conduct of business by reducing the volume of documents and the amount of time required to process, transmit and retrieve information. At the same time, they have raised issues as to the reliability and accuracy of information adduced in evidence. For example, there may be human error in the inputting of information, technical malfunction in the operation of computers, or deliberate interference with or manipulation of computer systems or the information contained therein.

At the time of writing, there are no special statutory rules in the Canadian law of evidence dealing with the admissibility and weight of computer-produced records. Instead, it is necessary to determine the admissibility and weight of such documents based on the common law and statutory rules applicable to other documents.

This chapter will review certain principles of the law of evidence that may restrict the admissibility of computer-produced documents. This is followed by a review of certain common law and statutory rules that represent exceptions to the foregoing principles and that permit the admissibility of computer-produced documents. Specific reference will be made to the provisions of the federal and provincial Evidence Acts dealing with the business document and bank record exceptions to the hearsay rule. Finally, some recent proposals for revision of the Canadian law of evidence governing computer-produced documents are discussed.

2. OBJECTIONS TO THE ADMISSIBILITY OF COMPUTER-PRODUCED RECORDS

At common law, three objections may be made to the introduction into evidence of computer records as proof of their contents. The first is the hearsay rule whereby evidence may be adduced only by a witness having personal or first-hand knowledge of the facts attested to. Under this rule, testimony of a statement made by any other person is in general inadmissible as proof of the contents of such statement. The scope of the hearsay rule was expressed as follows by Dickson J. (as he then was) in the case of *R. v. O'Brien*:[1]

> It is settled law that evidence of a statement made to a witness by a person who is not himself called as a witness is hearsay and inadmissible when the object of the evidence is to establish the truth of what is contained in the statement; it is not hearsay and is admissible when it is proposed to establish by the evidence, not the truth of the statement, but the fact that it was made.

Strict application of the hearsay rule would render inadmissible many business records, including those produced by computers. Many business records are compilations of bookkeeping entries made by a number of persons. As such information is entered into and subsequently processed and stored by a computer, the ultimate record will be one or more further steps removed from the person having personal knowledge of the facts recorded. Accordingly, computer records may be characterized as involving "double" or "multiple" hearsay.[2]

A second objection to the admissibility of computer records is the "best evidence" rule. In general, a document is admissible only if produced in its original form unless the original is unaccounted for or is unavailable.[3] Computer records would often fail to meet this criterion since such records are transcribed from written documents. In other cases, the "original" record may take the form of electronic impulses recorded on a tape, disc or other machine-readable medium, with the human-readable print-out being only a transcript of the original.

The admissibility of computer records may also be in doubt at common law insofar as they may constitute "opinion" evidence. At common law, opinion evidence is admissible only if adduced by an expert. Computer records may be regarded as opinion insofar as they consist of compilations of data generated by a computer, rather than reproductions in computerized form of printed data.[4]

3. RULES OF LAW PERMITTING ADMISSIBILITY OF COMPUTER RECORDS

Notwithstanding the objections set forth above, various common law and statutory rules may be relied upon to permit the admissibility of

---

1 (1977), 38 C.R.N.S. 325 at 327 (S.C.C.).

2 See I. Kyer, "Computer records as Courtroom Evidence" (1984), Computer Law, Vol. 1, No. 8, 57.

3 See *McCormick's Handbook on the Law of Evidence*, 2nd ed. (1972) at 721; *R. v. Cotroni; Papalia v. R.* (1979), 45 C.C.C. (2d) 1 (S.C.C.); and *C.P.R. v. Calgary*, [1971] 4 W.W.R. 241 at 257-258 (Alta. C.A.).

4 See K. Chasse, "Evidentiary Issues in Relation to Computer Printouts", Computer Law Update Conference (Insight, June, 1984).

computer-produced documents. The first is the common law rule permitting the admissibility in certain circumstances of documents produced pursuant to a business duty. In addition, both Parliament and provincial legislatures have enacted provisions in their respective Evidence Acts permitting the admissibility of records produced in the course of an organization's business as well as documents produced by financial institutions.

The rationale underlying these exemptions is that because business documents and bank records are produced in the ordinary course of business and as an integral part of an organization's institutional structure, they are presumed to be accurate and reliable. In other words, such documents may be admitted because of "the inherent circumstantial guarantee of accuracy which one would find in a business context from records which are relied upon in the day to day affairs of individual businesses, and which are subject to frequent testing and cross-checking."[5]

If a document is subject to admissibility under more than one exemption, it will be sufficient to demonstrate compliance with the criteria set forth with respect to only one exemption. With respect to the relationship between the federal and provincial Evidence Acts, the federal Act will apply to all proceedings in matters in respect of which Parliament has jurisdiction including criminal proceedings. The provincial Evidence Acts, on the other hand, apply to all proceedings for which the legislatures have jurisdiction before courts administered by the provinces.[6]

(a) Common Law "Business Record" Exemption

At common law, records made in the course of a business duty were admissible as evidence in the following circumstances.

> . . . to be admissible, the record must have been i) an original entry, ii) made contemporaneously with the event recorded, iii) in the routine, iv) of business, v) by a person since deceased, vi) who was under a duty to do the very thing and record it, vii) and who had no motive to misrepresent.[7]

The circumstances in which business records may be admitted in evidence at common law were reconsidered by the House of Lords in the case of *Myers v. Director of Public Prosecutions*.[8] In that case, the House

5 *R. v. Grimba* (1977), 38 C.C.C. (2d) 469 at 471 (Ont. Co. Ct.).

6 See, *e.g.*, s. 2 of the Evidence Act, R.S.O. 1980, c. 145 which provides that the Act applies to all actions and other matters with respect to which the Legislature has jurisdiction; the term "action" is defined to include any proceeding authorized or permitted to be tried by or before a court under the law of Ontario.

7 J. Douglas Ewart, "Documentary Evidence: The Admissibility at Common Law of Records Made Pursuant to a Business Duty" (1981), 59 Can. Bar Rev. 52 at 54-55.

8 [1965] A.C. 1001 (H.L.).

of Lords refused to admit in evidence manufacturers' records showing the serial numbers stamped on the engine blocks of motor vehicles. Although the evidence in question was described as a kind that would be "eminently reasonable to admit," the Court found that "the existing exception to the hearsay rule which admits evidence of declarations in the course of duty is subject . . . to the firmly established condition that the death of the declarant must be shown."[9] Moreover, it was not for the Court to alter the law to change the conditions for the admissibility of business documents.

A different approach was taken by the Supreme Court of Canada in the case of *Ares v. Venner*.[10] The Court held that notes made by nurses in the course of performing their duties in a hospital were admissible as proof of the statements or observations contained therein. Hall J., on behalf of the Court, adopted the reasoning set forth in the minority judgments in the *Myers* case to the effect that the common law rules as to admissibility should be adapted to meet the necessity created by mass production and modern business. In particular, Hall J. approved the following words of Lord Pearce in the *Myers* case as the basis for admitting documents into evidence:[11]

> I prefer the reasoning of Jessel M.R. in *Sugden v. Lord St. Leonards* (1876), 1 P.D. 154, 241: "Now I take it the principle which underlies all these exceptions is the same. In the first place, the case must be one in which it is difficult to obtain other evidence, for no doubt the ground for admitting the exceptions was that very difficulty. In the next place the declarant must be disinterested; that is, disinterested in the sense that the declaration was not made in favour of his interest. And, thirdly, the declaration must be made before dispute or litigation, so that it was made without bias on account of the existence of a dispute or litigation which the declarant might be supposed to favour. Lastly, and this appears to me one of the strongest reasons for admitting it, the declarant must have had peculiar means of knowledge not possessed in ordinary cases."

Hall J. went on to state the need to adapt the principles set forth in the foregoing case to the circumstances of modern business:

> The individuality of persons in a large factory or business may be difficult or impossible to discover. They do many repetitive and almost automatic tasks concerning which no memory exists. Yet their composite efforts make machines and records whose complexity, efficiency, and accuracy are beyond anything imaginable in 1886. In my view the anonymity of the recorder or the impossibility of tracing him create as valid a necessity as does his death for allowing his business records to be admitted. The principles on which

9 *Ibid., per* Lord Morris, at 1027-1028.
10 [1970] S.C.R. 608 (S.C.C.).
11 *Ibid.*, at 624.

the court sets out to discover the truth about these things remain unchanged, but the way in which those principles are applied must change if the principles are to be honoured and observed.[12]

Based on the foregoing principle, it was held that hospital records, including nurses' notes, made "contemporaneously by someone having a personal knowledge of the matters then being recorded and under a duty to make the entry or record should be received in evidence as *prima facie* proof of the facts stated therein."[13] The Court noted, however, that this would not preclude any party wishing to challenge the records or entries from doing so.

Although the *Ares* case dealt specifically with hospital records, the decision expands the common law principles governing the admissibility of any records made pursuant to a business duty. In *Setak Computer Services Corporation v. Burroughs Business Machines Ltd.*,[14] Griffiths J. indicated that although the decision refers only to hospital records, "it may be inferred that this decision also settles the law applicable to records of other businesses made in similar circumstances."[15] His Lordship did not expand on the "similar circumstances" in which records of other businesses would be admissible. However, express reference was made to the statement in *Ares* indicating that the common law exception applied only to writings or records made by a person speaking from personal observation or knowledge of the facts recorded.

The decision in *Ares v. Venner* was relied on as a basis for finding that computer printouts were admissible at common law in the case of *Tecoglas Inc. v. Domglas Inc.*[16] Rosenberg J. found that the printouts met the conditions for admissibility set forth in *Ares v. Venner*, namely, that there was a necessity for such documents to be produced since there was no other practical way of adducing the evidence before the court and the documents were made in the ordinary course of business in circumstances where there was no motive to misrepresent the facts.

The *Ares* decision clearly expanded the common law doctrine as to the admissibility of business records insofar as it need not be shown that the person who made the records is deceased. However, the scope of the extension of the doctrine remains unclear, particularly insofar as the decision indicates that any business record must be made contemporaneously by someone having a *personal knowledge* of the matters recorded. This would preclude the admissibility of a record as proof of any act not within the personal knowledge of the recorder. On the other hand, the

---

12 *Ibid.*, at 625.
13 *Ibid.*, at 626.
14 (1977), 15 O.R. (2d) 750 (Ont. H.C.).
15 *Ibid.*, at 755.
16 (1985), 51 O.R. (2d) 196 (Ont. H.C.).

Ares decision may be interpreted to mean that the admissibility of any record should be determined based on its presumed reliability or trustworthiness according to the circumstances in which it is prepared and the necessity of modern business. This interpretation would appear to have been adopted by Rosenberg J. in the *Tecoglas* case.

The common law business record exemption would be of limited assistance in permitting the admissibility of computer-produced documents if the doctrine is interpreted to mean that the person making the record must have a personal knowledge of the matters recorded. The computerized record-keeping systems of most businesses are so established that the person entering information into the computer does not have personal knowledge of the facts recorded. Moreover, additional persons may be involved prior to the production of the final output. It could be argued, however, that the computer should be considered as being in the same position as the person recording the information. In other words, as long as a record is compiled or produced as part of the ordinary function of the computer, the output would be deemed to be made by a computer under a duty to produce the result and in the course of the ordinary business in which the computer is used.[17]

In view of the uncertainty as to the scope of the common law business record exemption, it will be preferable in most cases to rely on the statutory business record exemptions to admit computer-produced documents in evidence. Accordingly these exemptions are discussed in detail below.

(b) Statutory "Business Record" Exemption

Section 30 of the Canada Evidence Act[18] provides that where oral evidence in respect of a matter would be admissible in a legal proceeding, a record made "in the usual and ordinary course of business" containing information in respect of that matter is admissible in evidence. The term "business" is defined in subsection 30(12) to mean any business, profession, trade, calling, manufacture or undertaking of any kind carried on in Canada or elsewhere whether for profit or otherwise, including any operation or activity of any government or agency thereof, or any tribunal or court. This definition would appear to include virtually any undertaking or organized activity carried on in any jurisdiction. In *R. v. Grimba*,[19] Callaghan Co. Ct. J. noted that the definition of business was "extremely broad" and that it appeared Parliament intended that any activity or operation carried on in Canada or elsewhere by any government would come within the scope of the section. On this basis, it was held that a

17 See Ewart, *supra*, footnote 7, at 74-75.

18 R.S.C. 1970, c. E-10.

19 *Supra*, footnote 5.

document containing fingerprints prepared in the usual and ordinary course of business by the Federal Bureau of Investigation, being a branch of the United States Department of Justice, was admissible under section 30.

The term "record" is defined in subsection 30(12) to include any book, document, paper, card, tape or other thing on or in which information is written, recorded, stored or reproduced. This definition would include any tape, disc or other machine-readable medium on which information is recorded or stored as well as any document containing information in printed form.

Subsection 30(2) provides that where a record made in the usual and ordinary course of business does not contain information in respect of a matter the occurrence or existence of which might reasonably be expected to be recorded therein, the court may admit the record for the purpose of establishing that fact and may draw the inference that such matter did not occur or exist.

In the event that it is "not possible or practicable" to produce an original record, subsection 30(3) provides that a copy of the record is admissible if accompanied by an affidavit setting out the reasons why the original cannot be produced and a further affidavit by a person who made the copy attesting to its authenticity. The word "copy" is defined in subsection 30(12) to include a print made from a photographic film.

Where the production of a record would not convey the information contained therein because it is kept in a form which requires explanation, a court may admit a transcript of an explanation of the record prepared by a person qualified to do so.[20] The explanation must be accompanied by an affidavit attesting to its accuracy and setting out the qualifications of the person by whom it was prepared.

Subsection 30(6) provides that for purposes of determining the probative value to be given to information contained in a record, the court may examine the record, receive any evidence concerning the record, including the circumstances in which the information therein was written, recorded, stored or reproduced, and draw any inference from the form or content of the record. Subsection 30(9) provides that any person who may reasonably be expected to have knowledge of the making or contents of any record may, with leave of the court, be examined or cross-examined on the record by any party.

In order for a record to be admitted in evidence under section 30, subsection 30(7) provides that at least seven days' notice of intention to produce the document must be given to every other party to the proceeding, and such other party is then entitled to inspect such document on five days' notice to the party intending to produce it.

Subsection 30(10) provides that section 30 is deemed not to render

20 Canada Evidence Act, *supra*, footnote 18, subs. 30(4).

as admissible evidence any of the following: any part of a record made in the course of an investigation or inquiry or in the course of obtaining or giving legal advice or in contemplation of a legal proceeding; any privileged record; a record alluding to a statement by a person who would not be a competent and compellable witness; any record if its production would be contrary to public policy; or any transcript or recording of evidence taken in the course of another legal proceeding. Finally, subsection 30(11) provides that section 30 is in addition to and not in derogation from any other provision of any Act of Parliament or any existing rule of law under which any record is admissible.

A similar set of rules governing the admissibility of business documents is set out in the various provincial Evidence Acts. For example, section 35 of the Ontario Evidence Act imposes two conditions for the admissibility of a writing or record made of an act, transaction, occurrence or event. First, the writing or record must be made in the usual and ordinary course of business, and second, it must be in the usual and ordinary course of such business to make the writing or record at the time of such act, transaction, occurrence or event or within a reasonable time thereafter. The term "business" is defined in subsection 35(1) to include every kind of business, profession, occupation, calling, operation or activity, whether carried on for profit or otherwise. The term "record" is defined to include any information that is recorded or stored by means of any device.

Subsection 35(3) of the Ontario Evidence Act imposes requirements for the giving of notice which are similar to those set out in subsection 30(7) of the Canada Evidence Act.[21] Subsection 35(4) provides that the circumstances of the making of such a record, including lack of personal knowledge by the maker, may be shown to affect its weight, but not its admissibility. Finally, subsection 35(5) provides that section 35 does not affect the admissibility of any evidence that would be admissible apart from such section or renders admissible any writing or record that is privileged.[22]

There would appear to be several differences between the provisions of the federal and Ontario Evidence Acts concerning the admissibility of business records. First, the Ontario Act does not contain any provision dealing expressly with the admissibility of either a copy or an explanation of a business record. Second, unlike the federal Act, the Ontario Act imposes an express statutory requirement that the record be contemporaneous, that is, that it be made at the same time or within a reasonable time of the

---

21 In *Tecoglas Inc. v. Domglas Inc., supra*, footnote 16, it was held that s. 35 does not require a formal or written notice and that actual notice to the other party through a lengthy discovery process complied with the requirements of the section.

22 For a comparative review of the business document and bank record provisions of other provincial Evidence Acts, see Kyer, *supra*, footnote 2, at 77-79.

transaction recorded. Third, the Ontario Act also refers expressly to the lack of personal knowledge of the maker as one of the circumstances that may affect the weight but not the admissibility of a business record; the federal Act merely permits the court to examine all circumstances of the making of the record. Finally, the federal Act contains more detailed provisions as to what documents are deemed not to be admissible based on the business record exemption.

In spite of these differences, the two statutes raise similar issues concerning the admissibility of computer-produced documents. Some of these issues are reviewed below.

(i) *Is double or multiple hearsay admissible under the "business record" exemptions?*

Subsection 30(1) of the Canada Evidence Act indicates that a record containing information concerning any matter is admissible only where oral evidence in respect of such matter would be admissible. It has been argued that these words should be interpreted to mean "where the maker of the record could have given admissible evidence." In other words, a record should be admissible only if its maker is shown to have personal knowledge of the facts recorded.[23]

In *R. v. Grimba*,[24] Callaghan Co. Ct. J. held that section 30 has removed the requirement at common law for the admissibility of a business record that the maker have personal knowledge of the facts recorded. Instead, section 30 contemplated the admission of a record "containing such things as . . . double hearsay."[25]

The interpretation of section 30 as extending to double or multiple hearsay would appear necessary to ensure the admissibility of computer printouts. In this respect, inherent in the nature of computerized record-keeping is the re-formatting and reorganization of information so as to produce new data compilations. As information is re-processed by a computer, its original form may be destroyed, and the final form in which it is recorded may be a number of steps removed from the original.[26]

The admissibility of business records containing double or multiple hearsay would appear to be permitted under section 35 of the Ontario Evidence Act. Subsection 35(4) of that Act refers to the lack of personal knowledge of the maker of a writing or record as one of the circumstances which may be shown to affect its weight but not its admissibility. However, a business record is admissible as proof of its contents only when both

23 See Chasse, *supra*, footnote 4, at 21.
24 *Supra*, footnote 5.
25 *Ibid.*, at 473.
26 See *R. v. Bell* (1982), 35 O.R. (2d) 164 (Ont. C.A.).

the maker of the record and the person giving information to the recorder are acting in the usual and ordinary course of business in entering and communicating an account of an act, transaction, occurrence or event.[27]

(ii) *Is opinion evidence admissible under the "business document" exemption?*

To the extent that computer printouts represent compilations or analyses of data, they may be characterized as opinion evidence requiring that they be adduced by an expert witness. Accordingly, it must be considered whether the business record provisions permit the admissibility of opinion evidence.

Subsection 35(2) of the Ontario Evidence Act provides for the admissibility of a writing or record made of "any act, transaction, occurrence or event." It has been held that this section does not permit the admissibility of a business document to prove the validity of an opinion which would not otherwise be admissible unless given by a duly qualified expert.[28] In the *Setak* case, for example, minutes of a meeting were held not to be admissible to prove the validity of an opinion expressed at a meeting. In the case of *Tecoglas Inc. v. Domglas Inc.*,[29] on the other hand, it was held that computer printouts were admissible either at common law or under section 35 of the Ontario Evidence Act notwithstanding that in making the record, the computer operator was expressing an opinion similar to that which would be expressed by a bookkeeper in making handwritten entries. In other words, the printouts were admissible as business records even though they contained opinion evidence.

Subsection 30(1) of the Canada Evidence Act refers to a record containing information in respect of *a matter*, which may be deemed to include an opinion. In *R. v. Vanderberghe*,[30] the flow-out or printout of a mechanical bookkeeping system was found to come within the meaning of a "record" under section 30. However, there was no specific discussion of whether the record could be characterized as opinion evidence. Moreover, oral testimony was addressed at the trial by a person qualified as an expert in the systems and operations of the bank's computer system.

(iii) *When is a record made in the "usual and ordinary course of business"?*

In *Setak Computer Services Corporation v. Burroughs Business Machines*

---

27 *Setak Computer Services Corp. v. Burroughs Business Machines Ltd., supra*, footnote 14, at 763.

28 See *Adderly v. Bremner*, [1968] 1 O.R. 621 (Ont. H.C.); *Setak Computer Services Corp. v. Burroughs Business Machines Ltd., ibid.*, at 761-762.

29 *Supra*, footnote 16.

30 (1976), 6 C.R. (3d) 222 (B.C.C.A.).

Ltd.,[31] it was held that section 35 (formerly section 36) of the Ontario Evidence Act was aimed not just at "routine" commercial records such as ledger accounts, time-cards and payroll records, but any records which related to the operation of a business as opposed to some purely personal or private activity. The court in that case refused to adopt the reasoning set forth in the decision of the United States Supreme Court in the case of *Palmer v. Hoffman*[32] in which it was held that business records were admissible only if they related to the inherent nature of the business in question. Instead, the court interpreted the Ontario Act to allow the admissibility of documents if made in the usual and ordinary course of some phase, whether principal or ancillary, of the company's business.

(iv) *Is a computer printout a "record"?*

Subsection 30(12) of the Canada Evidence Act defines a "record" to include any "thing" on which information is written, recorded, stored or reproduced. Subsection 35(1) of the Ontario Evidence Act, on the other hand, defines a record as including any "information" recorded or stored by means of any device. In other words, the two provisions differ in their emphasis on information as opposed to the medium in which information is recorded or stored. In the case of section 30 of the federal Act, it would have to be shown that any computer printout sought to be admitted in evidence is an original record, or if only a copy of such record, that the conditions prescribed by subsection 30(3) have been complied with. In making such determination, the courts are likely to be guided by the principles set forth by the Ontario Court of Appeal in the cases of *R. v. McMullen* and *R. v. Bell* dealing with the "bank record" exemptions.[33]

In *R. v. Vanderberghe*,[34] it was held that section 30 of the Canada Evidence Act covered mechanical as well as manual bookkeeping records and the keeping of records, and accordingly the "flow-out" or printout of such a bookkeeping system was admissible under section 30. There was no discussion in that case as to whether the printout constituted an original record or only a copy of that record.

There does not appear to be any case in which a court has been asked to consider the admissibility of information contained in a machine-readable medium, as opposed to a human-readable printout of such information. Although such medium would appear to come within the definition of a "record", it may not be practicable for a computer to be brought into court to "read" the information recorded on such a medium.

31 *Supra*, footnote 14.

32 318 U.S. 109 (1943).

33 See 193-195, *infra*.

34 *Supra*, footnote 30.

(v) *To what extent must a business record be "contemporaneous" with the event recorded?*

As noted above, it is an express requirement of the Ontario Evidence Act that a writing or record be made at the time of the occurrence of an act or event or within a reasonable time thereafter. In *Setak Computer Services Corporation v. Burroughs Business Machines Ltd.*,[35] it was held that where there was some evidence of delay in transcribing an act, the Court must decide in each case, as a matter of fact, whether the time span between the transaction and its recording was so great as to suggest the danger of inaccuracy by lapse of memory.

In the case of a computerized record-keeping system, it would have to be shown that the original computerized entry was made at the same time or within a reasonable time following the occurrence of the event recorded. The fact that the form in which the data is expressed subsequently changes should not affect the admissibility of the record based on the requirement of contemporaneity.

(vi) *When is foundation evidence required to be adduced?*

Subsection 30(6) of the Canada Evidence Act and subsection 35(4) of the Ontario Act permit a court to consider the circumstances of the making of a record as a basis for determining its weight. Neither provision indicates who has the onus of adducing evidence as to the circumstances of the making of the record, and this may vary according to the facts of each case.

The question of whether foundation evidence should be adduced to show the nature of the computer system and the manner in which it processes data was considered in the case of *R. v. Rowbotham*.[36] It was held in that case that in the absence of foundation evidence, computer printouts were not admissible. It was noted that there was no evidence before the court as to where the document came from, how it was prepared, what was the origin of the data contained in the document and what procedures were followed to produce it.

In the event that a court requires evidence as to the nature of a computerized record-keeping system, some of the criteria that may be considered in determining the reliability of the system, and the weight to be given to documents produced by it are the following: whether the input procedures conform to established standards in the industry, whether the programs accurately process data, whether the computer system has been kept under the management and control of a proper person, whether

35 *Supra*, footnote 14.
36 (1977), 33 C.C.C. (2d) 411 (Ont. Gen. Sess. of the Peace).

procedures have been established to prevent unauthorized access to the system and the information contained thereon, and whether back-up copies have been maintained of information recorded on the computer.[37] Additional evidence may be required to be adduced where computerized documents are produced as a result of the use of an external service bureau rather than an internal computer system.

The question of whether foundation evidence need be adduced in order to admit computer-produced records has been considered in several cases dealing with the "bank record" provisions of the Evidence Acts. The principles adopted in these cases, which are reviewed in detail below, would also appear to be applicable to the business document provisions of the Evidence Acts.

(c) Bank Record Exemptions of Evidence Acts

Subsection 29(1) of the Canada Evidence Act provides that a copy of any entry in any "book or record kept in any financial institution shall in all legal proceedings be received in evidence as *prima facie* proof of such entry and of the matters, transactions and accounts therein recorded."

Subsection 29(2) sets out four conditions to the reception and evidence of any such entry. Specifically, it must be proved that

(*a*) the book or record was, at the time of the making of the entry, one of the ordinary books or records of the financial institution;
(*b*) the entry was made in the usual and ordinary course of business;
(*c*) the book or record is in the custody or control of the financial institution; and
(*d*) the copy of the entry must be a true copy thereof.

Proof of the foregoing matters may be given by the manager or accountant of the financial institution and may be given orally or by affidavit.

Subsection 29(5) provides that in any legal proceeding to which a financial institution is not a party, the institution or an officer thereof is not compellable to produce any book or record where the contents can be proved under section 29 or to appear as a witness to prove any matters recorded therein unless by order of the court made for special cause.

Subsection 29(6) provides that upon application by a party to a legal proceeding, a court may order that such party be at liberty to inspect and take copies of any of the entries in the books or records of a financial institution. The person whose account is to be inspected is entitled to two clear days' notice of the application, which notice may be given to the financial institution if it is shown to the satisfaction of the court that such person cannot be notified personally.

---

37 See Chasse, *supra*, footnote 4, at 2-3.

Subsection 29(8) defines a "financial institution" to mean the Bank of Canada, the Federal Business Development Bank and any institution incorporated in Canada that accepts deposits of money from its members or the public, and includes a branch, agency or office of any such institution.

Section 33 of the Ontario Evidence Act also provides for the admissibility of a copy of an entry in a book or record kept in a bank. The word "bank" is defined in that section to mean "a bank to which the Bank Act (Canada) applies or the Province of Ontario Savings Office or any branch, agency or office thereof." The conditions as to the admissibility of any bank record under section 33 of the Ontario Act are substantially the same as those set out in section 29 of the federal Act.

The question of whether a computer printout is a copy of a record kept by a financial institution was considered in the case of *R. v. McMullen*.[38] In that case, a computer printout was prepared by the bank's central computer system at the end of each month for a particular customer based on information received from the local branch. Except for the computer printout, no original account record was kept by the local branch.

On appeal from the trial judge, Linden J. found that a computer printout was a copy of a record kept by a financial institution. He noted that by enacting section 29, Parliament had indicated its faith in the reliability of the records of financial institutions in whatever form they were kept, and the language used by Parliament included records kept by computers.

On further appeal, the Court of Appeal confirmed that the basic purpose of section 29 was

> to avoid the inconvenience to financial institutions and their customers of original bank records being removed from banks for production in legal proceedings and to facilitate the proof of matters entered in bank records. The fact that computerized bank records could not have been within the contemplation of Parliament when s. 29 was first enacted in 1927 . . . is no reason to deny the provision's application to computer records today if such evidence reasonably comes within its language and purpose.[39]

With respect to whether a printout was a "copy" of a record, the Court noted that one possible meaning of that word was "transcript", which may involve the copying out in ordinary writing of data recorded in a different form. Accordingly, the word did not necessarily require that the copy be a duplicate, *in form*, of the original. The Court held further that the word "record", as used in section 29, was not to be interpreted restrictively to mean only a visual, written record.

While accepting that computer printouts come within section 29, the

---

38 (1978), 42 C.C.C. (2d) 67 (Ont. H.C.); affirmed (1979), 47 C.C.C. (2d) 499 (Ont. C.A.).
39 *Ibid.*, at 503 (Ont. C.A.).

Court noted that the demonstration of the reliability of computer evidence was more complex than proving the reliability of written records. Moreover, as a matter of principle, a court should carefully scrutinize the foundation put before it to support a finding of reliability as a condition of admissibility. However, the conditions set out in subsection 29(2), particularly the requirement that the copy of the entry be shown to be a true copy, provided sufficient basis for the court to scrutinize the foundation evidence put before it:

> The nature and quality of the evidence put before the Court has to reflect the facts of the complete record keeping process — in the case of computer records, the procedures and processes relating to the input of entries, storage of information, and its retrieval and presentation. . . . If such evidence be beyond the ken of the manager, accountant or the officer responsible for the records . . ., then a failure to comply with s. 29(2) must result and the printout evidence would be inadmissible.[40]

The issue of the admissibility of computer printouts also came before the Ontario Court of Appeal in the case of *R. v. Bell*.[41] In that case, details of transactions in all chequing accounts of the bank's customers were fed into a central computer in Toronto. The computer was programmed to produce a monthly statement for each account, whereupon any record of the original transactions was erased from the computer's memory. Accordingly, the only record of such transactions was the computer printout produced monthly showing the final balance in the account. The issue was whether such printout was admissible under section 29.

The trial judge held on the authority of *R. v. McMullen* that the information stored in the computer was the record, with the monthly statement being a copy of that record. Because the original record no longer existed, the requirements of subsection 29(2) of the Canada Evidence Act could no longer be met. Specifically, it could not be proved that the printout was a true copy of the original record.

On appeal, Weatherston J.A. for the Court of Appeal held that the *McMullen* case was authority for the proposition that information stored in the computer was *capable* of being a "record kept in a financial institution," and that the computer printout was *capable* of being a copy of that record, notwithstanding its change in form. However, it was not authority for the proposition that the stored information was the only record or that the computer printout was only a copy of that record.

It was held further that although it was always a question of fact as to whether any recorded information was a "record kept in any financial

40 *Ibid.*, at 506-507 (Ont. C.A.).

41 (1982), 35 O.R. (2d) 164 (Ont. C.A.); affirmed without reasons [1985] 2 S.C.R. 287 (S.C.C.).

institution," certain general propositions could be stated in determining that question:

(1) A record may be in any, even an illegible, form.
(2) The form in which information is recorded may change from time to time, and the new form is equally a "record" of that kind of information.
(3) A record may be a compilation or collation of other records.
(4) It must have been produced for the bank's purposes as a reference source, or as part of its internal audit system and, at the relevant time must be kept for that purpose.[42]

It was noted further that it makes no difference for the admissibility of a record that the original information changes form, or becomes absorbed in some larger record. The authenticity of the record is sufficiently guaranteed by compliance with subsection 29(2). In the case at bar, the information changed its form so that the bank's copy of the monthly statement became the record. Since such statement was the only source of reference available to the bank as to the state of the customer's account and since it was intended to be kept for 15 years, it was by itself a record kept in a financial institution.

The decision in *Bell* clearly indicates that a computer printout can constitute a record, as opposed to merely a copy of a record, for purposes of section 29 of the Canada Evidence Act. However it is not clear whether *Bell* has entirely over-ruled the *McMullen* decision with respect to any requirement that foundation evidence be adduced to establish the reliability of a computer system as a condition of the admissibility of printouts produced by the system.

The argument that as a general rule no foundation evidence need be adduced to admit printouts produced by a bank's computer system is based on the grounds that section 29 speaks to the function of banking rather than to a particular form of record-keeping. It is contended that section 29 evinces Parliament's faith in the reliability of banking records, regardless of the system used by a particular bank. Bank records are reliable by their nature because they are subject to continuous processes of checking and cross-checking, because a bank's record-keeping system is relied upon by customers who may be involved in very substantial financial transactions and because banks are subject to extensive governmental regulation.[43]

In contrast to these arguments, it has been contended that the *Bell* decision over-ruled the *McMullen* decision only insofar as it decided that a printout could be treated as the original record. It dealt only with matters of authentication and best evidence and not with the necessity for foundation evidence to be adduced to admit a computer printout. It is

---

42 *Ibid.*, at 166-167.

43 Douglas Ewart, "The Admissibility of Computer-Produced Banking Statements: A Perspective on *R. v. Bell*" (1982), 27 C.R. (3d) 294.

argued further that it is inappropriate to accept the usual affidavit evidence of a bank manager or accountant who is not responsible for the bank's accounting system instead of requiring the attendance of a bank official who can testify in accordance with the *McMullen* standard. Such foundation evidence is also claimed to be necessary on the grounds that the printout cannot be verified as a copy by means of a comparison if the original no longer exists, a record may contain computer-generated information and not just computer-stored information and a bank provides only a limited time for an account owner to have errors corrected. For all of these reasons, it is contended that new standards are required to ensure adequate "circumstantial guarantees of reliability" for computer-produced documents beyond those provided by the conditions precedent set out in the existing bank records exemptions in the Evidence Acts.[44]

In considering the merits of the foregoing arguments, it should first be noted the *McMullen* decision clearly stands for the proposition that any requirement for proof of the reliability of the computer system must be found within the provisions of subsection 29(2) itself. To the extent that the decision was based on the need to prove "the facts of the complete record keeping process" in order to show that a printout was a "true copy", such a requirement would not be applicable where, as in the *Bell* case, the printout can be adduced as the original record itself. In appropriate cases, however, a court may still require foundation evidence as to the nature and reliability of a bank's computer system in order to show that the entry in the record was made in the usual and ordinary course of business. This may involve evidence being adduced as to whether the bank's computerized record-keeping system is consistent with the usual manner in which business is conducted within the bank.

The court would also appear to have a general discretion to require detailed foundation evidence in any case where the reliability of a computer system is placed in question based on subsection 29(5) of the Canada Evidence Act. This provision entitles a court, by order made for special cause, to compel an officer to appear as a witness to prove any matters set out in any bank record.

If foundation evidence is required to prove the reliability of computer-producted records, such evidence should be given by a person familiar with and responsible for the bank's computer system, rather than a branch manager or accountant. In addition, section 29 would appear inadequate to deal with the case where a bank's computer operations are carried out at an external service bureau rather than within the bank itself. In this respect, subsection 29(2) provides that as one of the conditions of admissibility of a record, proof must be adduced that the record is in the custody or control of the financial institution.

44 See K. Chasse, *supra*, footnote 4, at 47-51.

4. PROPOSALS FOR REVISION OF THE LAW OF EVIDENCE

The revision of the federal and provincial Evidence Acts to deal with such issues as the admissibility of computer documents has been the subject of a number of reports and proposals for legislation. In 1975, the Law Reform Commission of Canada published its *Report on Evidence*.[45] This was followed by a report published by the Ontario Law Reform Commission in 1976.[46] The Uniform Law Conference of Canada subsequently established a Federal-Provincial Task Force on Uniform Rules of Evidence which published its report in 1982.[47]

The draft Evidence Act included in the Task Force Report contained certain criteria for the admissibility of a business record made by a computer. Specifically section 171 of the draft Act provided as follows:

> 171 (1) A business record made by computer is admissible in evidence in the same manner as any other business record if the proponent proves that
>
> (*a*) the data or information on which the record is based are of a type or types regularly supplied to the computer during the regular activities of the person or organization from which the record originates;
> (*b*) the entries into the data or information bank on which the record is based were made in the usual and ordinary course of business; and
> (*c*) the computer program used in producing the record reliably and accurately processes the data or information in the data or information bank.
>
> (2) The proof required by subsection (1) may be made by producing an authenticating affidavit of the custodian of the record or any other qualified witness, based on his information and belief.

The foregoing provision was rejected by the Uniform Law Conference on the grounds that if a computer print-out were used by a business in its usual and ordinary course of business, that fact would provide a sufficient guarantee of reliability. In addition, such provisions would require the calling of many additional witnesses. In some cases, the satisfaction of the conditions would be impossible even though the evidence may be reliable.[48]

On November 18, 1982, the Canada Evidence Act, 1982 (Bill S-33) was given first reading in the House of Commons. The Bill was subsequently given second reading and referred to the Standing Committee on Legal and Constitutional Affairs but died on the order paper on June 29, 1984.

---

45 Ottawa: Information Canada, 1975, Catalogue Number J31-15/1975.

46 Report on the Law of Evidence (Queen's Printer, 1976).

47 Report of Federal/Provincial Task Force on Uniform Rules of Evidence (Carswell, 1982).

48 See "Decisions of the Uniform Law Conference of Canada on the Recommendations of the Federal/Provincial Task Force on the Uniform Rules of Evidence", Appendix 1 to the Report (Carswell, 1982), at 522.

Bill S-33 contained several provisions designed to deal with the admissibility of computer printouts. First, subsection 130(1) of the Bill would have defined an "original" to mean "in relation to stored or processed data or information, any print-out or intelligible output that reflects accurately the data or information or is the product of a system that does so." Section 153 of the Bill would have provided for the admissibility of a business record whether or not any statement contained therein is hearsay or a statement of opinion, subject, in the case of opinion, to proof that the opinion was given in the usual and ordinary course of business.

Subsection 155(1) of Bill S-33 would have permitted a court to examine a record, and to receive evidence orally or by affidavit, including evidence as to the circumstances in which the information in the record was written, recorded, stored or reproduced, in determining whether to admit any record, or in determining the probative value to be given to the record. Subsection 155(2) of Bill S-33 would have provided that where evidence respecting the authenticity or accuracy of a business record is to be given, the court may require the evidence of the custodian of the record or other qualified witness to be given orally or by affidavit.

The foregoing provisions of Bill S-33 would clarify several aspects of the admissibility of computer produced documents. First, the Bill would clearly indicate that computer printouts can constitute originals of records. This would be subject to the qualification that the printout or intelligible output must be shown to reflect accurately the data or information or is the product of the system that does so. It would also provide for the admissibility of multiple hearsay or a statement of opinion. Finally, it would permit evidence to be adduced by a custodian of a record or other witness as to its authenticity or accuracy.

Subject to the foregoing changes, Bill S-33 would preserve as part of the Canadian law of evidence, the concept of a record being made in the "usual and ordinary course of business". It would also permit a court to examine the circumstances of the making of a record for purposes of determining whether it may be admitted in evidence or the probative value to be given to it.

It has been suggested that if the foregoing provisions had been enacted in law, extensive foundation evidence would still be required as a condition of the admissibility of computer printouts. It has been further suggested that in determining the kind of foundation evidence that may be necessary, Canadian courts should seek guidance from legislation enacted in such countries as the United Kingdom and South Africa which set out detailed criteria for the admissibility of evidence generated by computers.[49] Such criteria may have the effect of shifting the onus of demonstrating the

49 See K. Chasse, "Computer Printouts — Designing Provisions for the New Canada *Evidence Act*: Part 2" (1985), 2 C.C.L.R. 150.

reliability of documents produced by a computerized record keeping system to the proponents of their admissibility.[50]

At the time of writing, a new Evidence Act has not been introduced. It is hoped that a new Bill will be forthcoming shortly which will clarify the conditions for the admissibility of computer records. In particular, it is hoped that this Bill will deal with some of the problems that arise as a result of the definition of an "original" for purposes of a document produced by a computer, and the nature and extent of the foundation evidence that must be adduced in order to admit any such document.

50 *Ibid.*, at 157.

Part IV

FREEDOM OF INFORMATION, PRIVACY, DATA PROTECTION AND ELECTRONIC FUNDS TRANSFERS: THE CANADIAN EXPERIENCE

9

The Regulation of Information Technology in Canada: An Overview

The purpose of this Part of the text is to review several separate but related issues concerning the regulation of information technology in Canada. Specifically, it is proposed to consider Canadian laws governing freedom of information, privacy, data protection, transborder data flow and electronic funds transfers.

What is common to all of these issues is that they relate to the manner in which information may be collected, the uses to which it may be put and the manner in which it may be processed, disseminated and transmitted from one location to another.

The issues relating to the regulation of information technology involve a number of laws and transcend many legal disciplines. For example, the rights of citizens with respect to information collected by governments are to be found in freedom of information and privacy legislation enacted by Parliament and provincial legislatures. Other laws have been enacted to deal with the manner in which information may be collected and used by certain industries within the private sector. Tort law must also be considered as a means of preventing abuses in the manner in which information is collected and used by individuals or corporations in relation to one another. Self-governing privacy codes adopted by corporations represent a further means for the protection of the rights of citizens in whose favour they are adopted.

The issues considered in this Part may also involve various areas of law according to the nature of the undertaking carried on by the business or organization which collects or disseminates information as well as the purposes for which the information is used. For example, the collection and use of information about employees by personnel management departments may have implications for the rights and duties of employers and employees under labour law. Laws governing the regulation and control of the financial industry, including banks, insurance companies, trust companies and investment dealers, must take into account the manner in which information is collected and used by such institutions. Similarly, laws governing the provision of health care by hospitals and the medical profession must take into account the manner in which information about patients may be collected and used.

The issues considered in subsequent chapters of this Part clearly transcend the use of computers. For example, the laws and regulations governing such matters as privacy and data protection are in many cases applicable to information collected or held in either printed or computerized form. The main significance of the growing use of computers is the potential for information abuse that they entail and the consequent need of the development of new laws or the expansion of existing laws. In this regard, the growing use of microcomputers, the establishment of integrated office networks and the convergence of computers and communications have facilitated the collection and processing of large volumes of information and its widespread dissemination. Such developments have raised new concerns about personal privacy and the ability of governments to exercise control over the collection and use of information.

Among the factors to be considered in determining the scope of the application of laws governing information technology in Canada are the following:

(a) What is the subject matter of the information (for example, personal information about private citizens, financial information about businesses, statistical data or other compilations of information)?
(b) What persons or entities have collected the information (for example, governments, quasi-government agencies or businesses)?
(c) What are the purposes for which the information is collected and held and the uses to which the information is to be put (for example, for internal use such as in personnel management or for provision to third parties)?
(d) What is the nature of the business carried on by the collector of the information (for example, banking, insurance or a consumer reporting agency)?
(e) In what form is the information kept (for example, manual records vs. computerized form)?

The foregoing factors must be considered in determining not only the scope of the existing laws governing the collection, processing and dissemination of information, but also the public policy interests underlying the possible expansion of such laws in the future. Each of these factors must also be considered in determining the administrative procedures, enforcement mechanisms and regulatory agencies that should be established to deal with problems of information abuse.

The following chapters will review laws in five areas relevant to privacy, data protection, transborder data flow and electronic funds transfer. Particular emphasis will be placed on laws relevant to the collection and use of information in computerized form. These areas are as follows:

(a) Laws relating to the duties of governments and the rights of citizens with respect to information collected by governments, particularly personal information about private citizens;

(b) Laws relevant to the rights of private citizens with respect to information collected by other persons, institutions or organizations in the private sector;

(c) Laws relevant to the transmission or transfer of information collected in Canada or about Canadian citizens to or from Canada;

(d) The implications for Canada of international legal developments relating to privacy, data protection and transborder data flow;

(e) Some legal issues relating to the establishment of electronic funds transfer systems.

Although not intended to be exhaustive, it is hoped that this review will provide a perspective on existing Canadian approaches with respect to the regulation of information technology, that is, the control of the collection and use of information *per se*.

10

Freedom of Information, Privacy and Data Protection: The Public Sector

1. INTRODUCTION

The purpose of freedom of information legislation is to provide citizens with the right to obtain access to information collected and held by governments. Such legislation is based on the principle that in a democratic society, information collected by governments at public expense should be readily available to the public. It is only by such means that private citizens can properly judge the merits of any actions or decisions taken by governments that may affect all aspects of their lives.

Privacy legislation, on the other hand, is intended to restrict the powers of governments to collect personal information about citizens and how such information is used. The purpose of such legislation is to protect the individual against undue interference with or invasion of his personal life. Specifically, the individual is protected against the improper collection, dissemination or use of personal information by governments.

In some respects, privacy and freedom of information legislation reflect competing public policy interests: to the extent that the individual is given the right to control how information about him is collected and held, there must be corresponding restrictions on the rights of governments to collect and disseminate information of this nature. At the same time, privacy and freedom of information laws are complementary since it would be impossible for individuals to exercise effective control over the collection and use of personal information without a right of access to the documents and records of government bodies. Accordingly, the right of "access" may be regarded as the common denominator of both privacy and freedom of information legislation.

The following sections will review the principal Canadian laws dealing with the rights of citizens to obtain access to information held by governments, and the duties of governments in dealing with personal information collected by them about private citizens. Both federal and provincial legislation in this area will be considered.

2. THE ACCESS TO INFORMATION ACT (CANADA)

The Access to Information Act was enacted together with the Privacy Act as 29-30-31 Elizabeth II, c. 111, and came into force on July 1, 1983. The following sections will review, in turn, the major provisions of the Access to Information Act and the Privacy Act.

(a) Access to Information Act (Canada)

(i) *Purpose of the Act and Basic Rights of Citizens*

Subsection 2(1) of the Access to Information Act provides that the purpose of the Act is to provide a right of access to information in records under the control of any government institution in accordance with three principles: first, that government information should be available to the public; second, that necessary exceptions to the right of access should be limited and specific; and third, that decisions on the disclosure of government information should be reviewed independently of government. Subsection 2(2) provides that the Act is intended to complement existing procedures for access to government information and is not intended to limit access to government information normally available to the public. In accordance with the principles set out in section 2, it would appear that the Act should be interpreted broadly so that any ambiguity would be resolved in favour of providing access to any government record.

Subsection 4(1) of the Act sets out the basic principle that every person who is a Canadian citizen or a permanent resident has a right to, and on request shall be given, access to any record under the control of a government institution. Although only private citizens and not corporations are entitled to make a request for access to any record held by a government institution, subsection 4(2) provides that the Governor in Council may extend such right to include persons not referred to in subsection 4(1).

The word "record" is defined in section 3 of the Act to include

> any correspondence, memorandum, book, plan, map, drawing, diagram, pictorial or graphic work, photograph, film, microfilm, sound recording, videotape, machine-readable record, and any other documentary material, regardless of physical form or characteristics and any copy thereof.

Subsection 4(3) of the Act provides that subject to the regulations, any record that does not exist but can be produced from a machine-readable record under the control of a government institution using normal hardware, software and technical expertise used by the institution shall be deemed to be a record under the control of the institution. However, under section 3 of the Access to Information Regulations,[1] a record that

1 SOR/83-507, P.C. 1983-1667.

does not exist but can be produced from a machine-readable record need not be produced where such production would unreasonably interfere with the operations of the institution affected.

(ii) *Dissemination of Information Concerning Records*

The Act provides for the preparation at least annually of a publication containing a description of the institutions subject to the Act, of the classes of records under the control of each institution, of the manuals used by the employees of each institution in carrying out programs and activities, and of the name and address of the appropriate officer of each institution to whom requests for access to records should be made.[2] Such publication is to be updated at least semi-annually by bulletins and the initial publication and bulletins are to be made available throughout Canada.[3] These materials are now generally available through government offices, post offices and taxation offices.

(iii) *Procedures for Requesting Access and Dealing with Requests*

Any request for access to a record is to be made in writing to the institution having control of the record. Such request must be in sufficient detail to enable an experienced employee to identify the record with a reasonable effort.[4] Upon any request being made, the head of the institution concerned must, within 30 days, notify the person making the request whether access is to be provided and if applicable, provide such access.[5] Section 8 of the Act sets out a procedure for transferring a request for access to a record to another government institution if that institution has a "greater interest" in the record. The time limit for responding may be extended for reasonable periods of time if the request is for, or necessitates a search through, a large number of records and meeting the original time limit would "unreasonably interfere with the operations of the government institution," if consultations are necessary to comply with the request but cannot reasonably be completed within the original time limit, or if notice of the request is to be given under subsection 28(1) to a "third party".[6]

The head of an institution who refuses a request must notify the applicant that the record does not exist, or set out the specific provision of the Act on which the refusal was based. There is no requirement to state whether a record exists, although the notice must state the provision

2 Access to Information Act, S.C. 1980-81-82-83, c. 111, s. 5.
3 *Ibid.*
4 *Ibid.*, s. 6.
5 *Ibid.*, s. 7.
6 *Ibid.*, s. 9.

on which refusal could reasonably be expected to be based if the record existed. The notice must state further that the person who made the request has the right to make a complaint to the Information Commissioner.[7]

(iv) *Payment of Fees*

Section 11 of the Act provides for the payment of various fees by a person who requests access to a record, including an initial application fee, the costs of reproduction, payments on account of the time in excess of five hours required to search for the record or prepare for disclosure, and fees in respect of the preparation of a record produced from a machine-readable record.[8] The actual fees payable by a person who requests access to a record are set forth in section 7 of the Access to Information Regulations.

(v) *Exemptions to Right of Access*

Sections 13-27 of the Act set out a number of cases in which a government institution is either required or entitled to refuse disclosure of any record requested under the Act. Among the records in respect of which refusal of access is *mandatory* under the Act are the following:

(i) records containing information obtained in confidence from a foreign government, an international organization of countries, or a provincial or municipal government, unless the government or organization consents to the disclosure of the information or makes the information public;[9]

(ii) a record containing personal information as defined in the Privacy Act, unless such disclosure is consented to by the individual to whom it relates, the information is publicly available or disclosure is made in accordance with section 8 of the Privacy Act;[10]

(iii) any record
  (a) containing trade secrets of a third party;
  (b) containing financial, commercial, scientific or technical information that is confidential information supplied by a third party and is treated consistently in a confidential manner by the third party;[11]

---

7 *Ibid.*, s. 10.

8 *Ibid.*, s. 12.

9 *Ibid.*, s. 13.

10 *Ibid.*, s. 19.

11 In *Maislin Indust. Ltd. v. Min. for Industry, Trade & Commerce* (1984), 80 C.P.R. (2d) 253 (Fed. T.D.), it was held that in order to be exempted from disclosure under this provision, the information had to meet the objective test of being confidential in character *and* the subjective test of being so treated by the third party.

(c) containing information the disclosure of which could reasonably be expected to result in material financial loss or gain to a third party or to prejudice its competitive position; or

(d) containing information the disclosure of which could reasonably be expected to interfere with the contractual or other negotiations of a third party.[12]

However, a record containing the foregoing information may be disclosed if it contains the results of product or environmental testing carried out by or for a government institution, if the disclosure is consented to by the third party in question, or if the disclosure is in the public interest relating to public health, public safety or the protection of the environment and such public interest clearly outweighs the protected interest of the third party;[13]

(iv) any record containing information, the disclosure of which is restricted by any provision set out in the Privacy Act.[14]

The head of an institution is given a *discretionary* power to refuse to disclose any record if the record contains information concerning any of the following matters:

(i) information the disclosure of which could reasonably be expected to be injurious to the conduct by the Canadian government of federal-provincial affairs or the conduct of international affairs, the defence of Canada or an allied state, or the detection, prevention or suppression of subversive or hostile activities;[15]

(ii) information obtained by an investigative body specified in the regulations in the course of investigations relating to the detection, prevention or suppression of crime or the enforcement of any law; information relating to investigative techniques or plans for investigation; information the disclosure of which could be injurious to the enforcement of any law or be injurious to the security of penal institutions; or information that could reasonably be expected to facilitate the commission of an offence;[16]

(iii) information the disclosure of which could reasonably be expected to threaten the safety of individuals;[17]

(iv) trade secrets or financial, commercial, scientific or technical information belonging to the Government having substantial value; information the disclosure of which may prejudice the competitive position of a government institution; scientific or technical informa-

12 *Op. cit.*, s. 20.
13 *Ibid.*, s. 20.
14 *Ibid.*, s. 24.
15 *Ibid.*, ss. 14-15.
16 *Ibid.*, s. 16.
17 *Ibid.*, s. 17.

tion obtained through research by an officer or employee of a government institution, the disclosure of which could deprive such person of priority of publication; and information the disclosure of which could reasonably be expected to be materially injurious to the financial interests of the Canadian Government or the ability of the Government to manage the economy or could result in undue benefit to any person;[18]

(v) any record containing advice or recommendations developed by or for a government institution or Minister; an account of consultations or deliberations involving officials or employees of a government institution or Minister or a Minister's staff; positions or plans developed for the purpose of negotiations carried out on behalf of the Government; and plans relating to the management of personnel or the administration of a government institution that have not as yet been implemented;[19]

(vi) any record containing information relating to testing or auditing procedures or techniques or details of specific tests or audits to be conducted;[20]

(vii) any record containing information that is subject to solicitor-client privilege.[21]

(vi) *Procedures for Protection of Rights of Third Parties*

Section 28 of the Act sets out certain procedures to be followed where the head of a government institution intends to disclose any record requested under the Act that contains or that the head of an institution believes may contain any protected information of a third party. For purposes of this section, a "third party" is defined as any person, group or organization other than the person who made the request or a government institution.[22] Specifically, the head of the institution must notify the third party within 30 days that the request has been made and of his intent to disclose the record. The notice must describe the contents of the record that belong to, were supplied by or relate to the third party, and

18 *Ibid.*, s. 18.

19 *Ibid.*, subs. 21(1). Subs. 21(2) provides that the above exceptions do not apply to an account of, or a statement of reasons for, a decision made in the exercise of a discretionary power or an adjudicative function that affects the rights of a person, or a report prepared by a consultant or adviser who is not an officer of a government institution or a staff member of a Minister. In *Information Commr. of Can. v. C.R.T.C.* (1986), 9 C.P.R. (3d) 1 (Fed. T.D.), Jerome A.C.J. found that the preparatory notes and communications among Board members leading to a C.R.T.C. decision were subject to the exemption, even though the final reasons for decision were required to be disclosed.

20 *Ibid.*, s. 22.

21 *Ibid.*, s. 23.

22 *Ibid.*, s. 3.

indicate that the third party is entitled, within 20 days, to make representations to the head of the institution, as to why the record should not be disclosed. Within 30 days from the date of the notice from the head of the institution, the third party must be notified whether the record is to be disclosed. The third party, in turn, has an additional 20 days to request a review of the decision by the Court under section 44.

Any third party also has a right to request a review of any decision made by the head of an institution to disclose a record made on the recommendation of the Information Commissioner.[23]

(vii) *Complaints and Investigations*

Section 30 of the Act authorizes the Commissioner to receive and investigate complaints under the Act. The Commissioner may also initiate a complaint if he is satisfied that there are reasonable grounds to investigate a matter relating to requesting or obtaining access to records under the Act.[24] All complaints must be in writing and, where they relate to a request for access to a record, be made within one year from the time of receipt of the request.[25]

Before commencing an investigation of a complaint under the Act, the Commissioner must notify the head of the government institution concerned and any third party affected by the request of his intention to do so.[26] Subject to the Act, the Information Commissioner may determine the procedures to be followed in the performance of his duties.[27] Section 35 requires every investigation to be conducted in private, and although a person affected is entitled to make representations to the Commissioner, no one is entitled to have access to the representations made by any other person. Section 36 of the Act sets out the powers that may be exercised by the Commissioner in the course of investigating any complaint; these include the power to examine a record under the control of any government institution.

Section 37 provides that if the Information Commissioner finds that a complaint is well founded, he must provide the head of the institution concerned with a report of his findings and any recommendations that he considers appropriate. The head of the institution may also be requested to notify the Commissioner within a specified time of any action proposed to be taken with respect to the recommendations. The results of the investigation are also to be provided to the complainant and any third

23 *Ibid.*, s. 29.
24 *Ibid.*, s. 30.
25 *Ibid.*, s. 31.
26 *Ibid.*, ss. 32-33.
27 *Ibid.*, s. 34.

party affected.[28]

If the Commissioner is not provided with the notice requested from the head of the institution as to any action proposed to be taken, or if he considers any action proposed to be taken as inadequate or inappropriate, the Commissioner must so advise the complainant and may include such comments as he thinks fit. If access to a record is refused, the Commissioner must also notify the complainant of his right to apply to the Court for a review of the matter.[29]

(viii) *Judicial Review*

Section 41 provides that any person who has been refused access to a record requested under the Act and who has made a complaint to the Information Commissioner may apply to the Court for a review of the matter within 45 days after being provided with a report of the results of the investigation by the Commissioner, or such longer period of time as is permitted by the Court. Judicial review or the right to appear as a party to any review may also be requested in certain circumstances by the Commissioner himself or by any third party to whom notice of a request for access to a record has been given.[30]

In any hearing under the Act, the Court may examine a record to which the Act applies that is under the control of an institution provided that it must take reasonable precautions against disclosing any information on the basis of which disclosure of any part of a record could be refused under the Act.[31] The head of a government institution has the burden of proving that it is authorized to refuse to disclose a record.[32] Where the Court determines that the head of the institution is not authorized to refuse to disclose a record under the Act, it shall order the head of the institution to disclose the record, subject to such conditions as it deems appropriate, or shall make such other order as it deems appropriate.[33] However, in the case of a refusal to disclose a record based on certain grounds such as an alleged injury to federal-provincial or international affairs, the enforcement of any law or the conduct of investigations, the Court shall order disclosure of the record only if it finds that the head of the institution concerned did not have reasonable grounds for refusing to do so.[34] The

28 *Ibid.*, s. 37.
29 *Ibid.*, s. 37.
30 *Ibid.*, ss. 42-44.
31 *Ibid.*, ss. 46-47
32 *Ibid.*, s. 48.
33 *Ibid.*, s. 49. Unless the Court makes such determination, the applicant's right to a record is subject to the discretion of the head of the institution: *Information Commr. of Can. v. C.R.T.C.*, *supra*, footnote 19.
34 *Ibid.*, s. 50.

Court shall also order the head of the institution not to disclose a record if it finds that he is required not to make such disclosure, subject to such other order as the Court deems appropriate.[35]

(ix) *Exclusions*

Section 69 of the Act provides that the Act does not apply to confidences of the Queen's Privy Council for Canada, including Cabinet discussion papers, draft legislation or records of communications between Ministers of the Crown. The complete exclusion of all records containing information of this nature means that such records are not subject to the provisions of the Act relating to complaints to, or investigations by, the Information Commissioner, or appeals to the Federal Court.

(x) *Other Provisions*

Other provisions of the Act provide for the establishment of the office of the Information Commissioner and that of one or more Assistant Information Commissioners, and set out the duties and powers of each.[36] The Act also provides for the appointment of officers and employees and for the delegation by the Information Commissioner of certain of its powers to assistants.

The Act also provides for the Information Commissioner to submit an annual report and special reports to Parliament.[37] Annual reports are also to be submitted by the head of every government institution.[38]

Administration of the Act is to be reviewed on a permanent basis by such committee as is designated by Parliament for such purpose. The committee which is designated must undertake a comprehensive review of the Act within three years of its coming into force and submit a report of such review within a further one-year period.[39]

(b) Privacy Act (Canada)

(i) *Background and Definitions*

Section 2 of the Privacy Act[40] provides that the purpose of the Act is to extend the present laws of Canada that protect the privacy of individuals with respect to personal information about themselves held by government institutions and that provide individuals with a right of

35 *Ibid.*, s. 51.
36 *Ibid.*, ss. 54-57.
37 *Ibid.*, ss. 38-40.
38 *Ibid.*, s. 72.
39 *Ibid.*, s. 75.
40 S.C. 1980-81-82-83, c. 111, Sch. II.

access to such information. The term "personal information" is defined in section 3 of the Act to mean "information about an identifiable individual that is recorded in any form"; some ten categories of information coming within such definition as well as certain exclusions are also listed. Section 3 also defines a government institution to mean any department or Ministry of state listed in the schedule to the Act; the departments and Ministries are substantially the same as those which are subject to the Access to Information Act.

(ii) *Restrictions on Collection, Retention and Disposal of Personal Information*

Section 4 of the Act provides that no personal information is to be collected by a government institution unless it relates directly to an operating program or activity of the institution. Personal information is to be collected "whenever possible" directly from the individual to whom it relates and the individual is to be informed of the purpose for which the information is collected; however, the foregoing provisions do not apply where compliance might result in the collection of inaccurate information or defeat the purpose or prejudice the use for which information is collected.[41] Section 6 of the Act imposes certain other duties on a government institution, including the duty to retain personal information for a period of time prescribed by regulation, to ensure that information is accurate, up-to-date and complete, and to dispose of the information in the manner prescribed by regulation. Under section 4 of the Privacy Regulations,[42] personal information must be retained for a period of at least two years from the time the information is collected for an administrative purpose, or where a request for access has been received, until the individual has had an opportunity to exercise all rights under the Act.

(iii) *Protection of Personal Information*

Section 7 of the Privacy Act provides that as a general rule, personal information under the control of a government institution is not to be used except for the purpose for which it was obtained or for a use consistent with that purpose or for a purpose for which the information may be disclosed to that institution under subsection 8(2). Subsection 8(1) provides that personal information under the control of a government institution shall not, without the consent of the individual to whom it relates, be disclosed by the institution except in accordance with section 8. Subsection

41 *Ibid.*, s. 5.
42 SOR/DORS/83-508.

8(2) sets out the circumstances in which personal information under the control of an institution may be released, including the following:

(a) for the purpose for which the information was collected or for a use consistent with that purpose;
(b) for any purpose for which any act of Parliament or any regulation thereunder authorizes disclosure;
(c) for the purpose of complying with any subpoena or warrant or any rule of court relating to the production of information;
(d) to the Attorney General for use in legal proceedings involving the Crown in right of Canada;
(e) to an investigative body prescribed by regulations for the purpose of enforcing any law or carrying out an investigation;
(f) under any agreement or arrangement with any province, other state, or international organization of states for the purpose of administering or enforcing any law or carrying out an investigation;
(g) to a Member of Parliament to enable him to assist the individual to whom it relates;
(h) to employees of the institution or to the office of the Comptroller General for audit purposes;
(i) to the Public Archives for archival purposes;
(j) to any person or body for research or statistical purposes, subject to certain qualifications including an undertaking from such person or body not to disclose the information in a form that would permit identification of the individual to whom it relates;
(k) to any association of aboriginal people, Indian band or government institution for the purpose of researching or validating the claims of any aboriginal peoples;
(l) to any government institution for the purpose of locating an individual to collect a debt owing by that individual to Her Majesty;
(m) for any purpose where in the opinion of the head of the institution concerned, the public interest in disclosure clearly outweighs any invasion of privacy or the disclosure would clearly benefit the individual to whom the information relates.

In the case of any disclosure under sub-paragraph (e) above, the head of any institution is required to keep for such period of time as is prescribed by regulation a record of any request received and of any information disclosed.[43] Section 7 of the Privacy Regulations[44] provides that such records must be kept for two years from the time of any request.

In the case of disclosure under sub-paragraph (m), the head of the institution must notify the Privacy Commissioner prior to the disclosure

43 Privacy Act, *op. cit.*, subs. 8(4).

44 *Op. cit.*

if practicable, or otherwise forthwith on disclosure, and the Commissioner in turn may notify the individual concerned.[45]

Section 9 of the Act provides that the head of a government institution is required to retain a record of any use by the institution of any personal information or any use or purpose for which the information is disclosed.

(iv) *Personal Information Banks and Indexes*

Section 10 requires the head of an institution to include in a "personal information bank" all personal information under the control of the institution that has been used or is being used for an administrative purpose or that is organized or is intended to be retrieved by the name of an individual or an identifying number or symbol assigned to an individual.

Section 11 requires the Minister responsible for the administration of the Act to publish not less than annually an index of all personal information banks setting out certain information relating to each bank, including a description of the class of individuals included in the bank, the name of the institution having control of the bank, the title and address of the person to whom requests relating to information in the bank should be sent, a statement of the purposes for which information in the bank was collected, a statement of the retention and disposal standards applicable to information in the bank, and if applicable, an indication that the bank was designated as an exempt bank. The index must also set out a description of all classes of personal information under the control of the institution not included in personal information banks.[46] The index is to be made available throughout Canada in conformity with the principle that every person is entitled to reasonable access to the index.[47]

(v) *Right of Access to Personal Information*

Section 12 of the Privacy Act sets out the general principle that every Canadian citizen or permanent resident is entitled to access to any personal information about him contained in a personal information bank, and any other personal information with respect to which the individual is able to provide sufficiently specific information as to its location to enable the institution to retrieve the information. In addition, an individual is entitled to request correction of any personal information, require a notation to be made reflecting any corrections requested but not made, and require that any person or body to whom the information was disclosed within

45 Privacy Act, S.C. 1980-81-82-83, c. 111, subs. 8(5).
46 *Ibid.*, para. 11(1)(*b*).
47 *Ibid.*, subs. 11(3).

the previous two years be notified of the correction or request for correction.[48]

The procedures to be followed by an individual in requesting access to personal information and the manner in which such requests are to be dealt with are set forth in sections 13-17 of the Act and in sections 8-11 of the Regulations. Such procedures are similar to those applicable to requests for access to information under the Access to Information Act.

(vi) *Exemptions*

Section 18 of the Privacy Act provides that the Governor in Council may designate certain information banks containing files consisting of certain types of personal information as exempt banks, and that the head of an institution may refuse to disclose any personal information contained in an exempt bank. The Act also sets out various classes of personal information for which the head of an institution is required to refuse a request for access; specifically, the head of an institution must refuse access to personal information obtained in confidence from the government of another state or a provincial or municipal government,[49] as well as information obtained by the RCMP while performing policing services for a province or municipality.[50]

The Act also sets out certain classes of personal information for which disclosure is a matter of discretion on the part of the head of the institution concerned. Such classes include: information that may be injurious to the conduct of federal-provincial or international affairs, the defence of Canada or an allied state or the detection, suppression or prevention of hostile or subversive activities;[51] information relating to law enforcement and investigation;[52] information collected for the purposes of determining whether to grant security clearances;[53] information collected by certain agencies relating to an individual under sentence for an offence under any Act of Parliament;[54] information the disclosure of which may threaten the safety of individuals;[55] information about an individual other than the person making the request;[56] information that is subject to solicitor-client privilege;[57] and information relating to an individual's physical or mental

48 *Ibid.*
49 *Ibid.*, s. 19.
50 *Ibid.*, subs. 22(2).
51 *Ibid.*, ss. 20-21.
52 *Ibid.*, subs. 22(1).
53 *Ibid.*, s. 23.
54 *Ibid.*, s. 24.
55 *Ibid.*, s. 25.
56 *Ibid.*, s. 26.
57 *Ibid.*, s. 27.

health where disclosure would be contrary to the individual's best interests.[58]

(vii) *Complaints and Investigations*

Section 29 of the Privacy Act authorizes the Privacy Commissioner to receive and investigate complaints under the Act, or to initiate a complaint on his own accord. The procedures for investigation of complaints, as set forth in sections 31-35 of the Act are substantially the same as those applicable to investigation of complaints under the Access to Information Act. The Privacy Commissioner may also carry out investigations of the files kept in personal information banks designated as exempt banks under the regulations.[59] If he considers that any file should not be contained in an exempt bank, he must provide a report of his findings and recommendations to the head of the institution concerned. If no action is taken on such recommendations, or if any action taken is considered inadequate or inappropriate, the Commissioner may apply for a court order under section 43.[60] The Information Commissioner is also entitled to carry out investigations to determine whether an institution has complied with the provisions of sections 4-8 of the Act, dealing with the duties of an institution with respect to the collection, retention, disposal and disclosure of personal information.[61]

(viii) *Review by Federal Court*

Section 41 permits any individual who has been refused access to personal information to apply to the Court for a review of the matter, provided that a complaint is first made to the Privacy Commissioner. The Privacy Commissioner may also apply to the Court in certain circumstances. The procedures relating to judicial review, as well as the powers of the Court on such an application, as set forth in sections 44-52 of the Act, are substantially the same as those set out in the Access to Information Act.

(ix) *Other Provisions*

Other sections of the Privacy Act provide for the Privacy Commissioner to submit an annual report and special reports to Parliament;[62]

---

58 *Ibid.*, s. 28.
59 *Ibid.*, s. 36.
60 *Ibid.*
61 *Ibid.*, s. 37.
62 *Ibid.*, ss. 38-40.

provide for the Office of the Privacy Commissioner, his staff, and the delegation of certain of his duties to an Assistant Privacy Commissioner;[63] provide for the Commissioner to carry out special studies on the privacy of individuals;[64] and set out the general powers and duties of the Commissioner.[65] Section 70 of the Act provides that the Act does not apply to "confidences of the Queen's Privy Council for Canada", including Cabinet memoranda, or discussion papers, records of Cabinet deliberations, communications among Ministers or draft legislation. Section 71 of the Act imposes certain duties on the head of each government institution with respect to the manner in which personal information banks are to be maintained and kept under review. Other sections of the Act provide for periodic reports to be submitted to Parliament by the heads of institutions subject to the Act and for a permanent review of the administration of the Act by a Parliamentary committee.[66] The Cabinet is also authorized to make various regulations relating to the enforcement and administration of the Act.[67]

3. AN ACT TO PROVIDE FOR FREEDOM OF INFORMATION AND PROTECTION OF INDIVIDUAL PRIVACY (ONTARIO) (BILL 34)[68]

Bill 34 incorporates in one Bill many of the provisions and principles set forth in the federal Access to Information Act and Privacy Act. Administration of the Bill is to be under the responsibility of a single official known as the Information and Privacy Commissioner. Although the purposes of the Bill, as set forth in section 1, are virtually identical to those of the federal Acts, there are some important differences in the manner in which the purposes of the legislation are implemented.

(a) Right of Access

Unlike the federal Act, which grants a right of access to records only to private citizens and permanent residents, section 10 of the Ontario Bill grants such right to any "person". This would appear to include corpo-

63 *Ibid.*, ss. 53-59.

64 *Ibid.*, s. 60.

65 *Ibid.*, ss. 61-67.

66 *Ibid.*, ss. 72-75.

67 *Ibid.*, s. 77.

68 The following summary is based on the provisions of Bill 34 as given first and second reading by the Ontario Legislature on April 22, 1986 and as reflecting changes proposed in the summer of 1986 by the Attorney General for Ontario; as at the end of 1986, the Bill is being examined by the Standing Committee on the Legislature and was expected to be enacted in substantially the form proposed in 1987.

rations and other artificial persons. Section 11 of the Ontario Bill also imposes on the head of an institution an overriding obligation, which supersedes other provisions of the Act, to disclose any record to the public or the persons affected if the head has reasonable and probable grounds to believe that it is in the public interest to do so and the record reveals a grave environmental, health or safety hazard to the public.

(b) Exemptions

Like the federal legislation, Bill 34 sets out two classes of exemptions to the general right of access to public documents, that is, those exemptions where the head of an institution has an obligation not to permit access and those where refusal of access is discretionary. The head of an institution is required not to permit access to records which would reveal the substance of deliberations of the Executive Council or its committees[69] and records containing personal information which may not, except in specified circumstances, be disclosed to any person other than the individual to whom the information relates.[70] In addition, the head of an institution is required to obtain the consent of Cabinet prior to disclosing any record if the disclosure might prejudice the conduct of intergovernmental relations by the Ontario Government or reveal information received in confidence from another government or an organization of states.[71]

Bill 34 as originally introduced provided for a discretionary rather than mandatory exemption for the disclosure of certain information supplied to the government by any "third party". However, under changes to the Bill proposed by the Attorney General, subsection 17(1) provides that a head of an institution *shall* refuse to disclose a record that reveals a trade secret, or scientific, technical, commercial, financial or labour relations information supplied in confidence, implicitly or explicitly, where the disclosure could reasonably be expected to: (a) prejudice significantly the competitive position or interfere significantly with the contractual negotiations of a person or organization; (b) result in information of the same kind no longer being supplied to the institution where the information was supplied in confidence and it is in the public interest that similar information continue to be provided to the institution; or (c) result in undue loss or gain to any person, group, committee or financial institution or agency. However, subsection 17(1) does not apply where the public interest in the disclosure of a record outweighs the interest of any person, group

69 Bill 34, s. 12.

70 *Ibid.*, s. 21; the circumstances in which personal information may be disclosed are discussed in greater detail below.

71 *Ibid.*, s. 15.

or organization in its continued confidentiality.[72] Any record described in subsection 17(1) may also be disclosed with the consent of the person concerned.[73]

Under the foregoing provisions, the criteria to be met before the head of an institution may refuse to disclose a record containing information supplied by a third party are somewhat more onerous than under the federal Act (for example, it must be shown that disclosure will prejudice *significantly* a person's competitive position or interfere *significantly* with contractual negotiations). In addition, the section would appear to give the head of an institution a wider discretion than under the federal Act to disclose a record based on the public interest, which test is not subject to any qualification of reasonableness. As noted in further detail below, the Ontario Bill also differs from the federal legislation in that the right of a party to appeal any decision by the head of an institution lies only to the information and Privacy Commissioner and not to any court of law.

Like the federal Access to Information Act, Bill 34 permits the head of an institution not to disclose a record that would reveal information with respect to advice or recommendations to the government by public servants or consultants. However, such exemption does not extend to records containing information with respect to a number of matters, including factual material, a statistical survey, a report by a valuator, an environmental impact statement, certain test results, a report or study on the performance or efficiency of an institution, feasibility or technical studies relating to a government policy, a report of results of field research, certain plans, proposals or reports (but not those which are to be submitted to Cabinet) and the reasons for any final decision, order or ruling of an officer of an institution in the exercise of any discretionary authority.[74]

Among the classes of records for which the head of an institution is given a discretionary power to refuse to disclose are the following: various records containing information where the disclosure could reasonably be expected to cause injury or prejudice relating to law enforcement or investigation;[75] records the disclosure of which may be prejudicial to the defence of Canada or any allied state or be injurious to the detection, suppression or prevention of espionage, sabotage or terrorism;[76] records containing trade secrets or other information belonging to the Government of Ontario, information the disclosure of which could be prejudicial to the economic interests of an institution or injurious to the financial interests

72 *Ibid.*, subs. 17(2).
73 *Ibid.*, subs. 17(3).
74 *Ibid.*, s. 13.
75 *Ibid.*, s. 14.
76 *Ibid.*, s. 16.

of the Government of Ontario or its ability to manage the economy; information relating to negotiations carried on by an institution, its administration or its proposed plans, policies or projects, provided that records containing any such information may be disclosed if the public interest in its disclosure outweighs the interest of the Government in its continued confidentiality;[77] recoreds subject to solicitor-client privilege;[78] records the disclosure of which may threaten the safety or health of an individual;[79] and records containing information that is publicly available or that is reasonably expected to be published within 90 days.[80]

The Ontario Bill provides that the head of an institution shall not disclose personal information to any person other than the person to whom the information relates except in the following circumstances: upon the written consent of the individual affected, in compelling circumstances affecting the health or safety of an individual, where the information was collected specifically for the purpose of creating a record available to the general public, where the disclosure is authorized under any Act of Ontario or Canada for a research purpose in some circumstances, or if the disclosure does not constitute an invasion of personal privacy.[81] With respect to the last-mentioned exemption, subsection 21(2) requires the head to consider all the relevant circumstances including certain specified criteria. Subsections 21(3) and 21(4) also set out respectively some circumstances where disclosure is presumed to constitute an invasion of personal privacy and other circumstances where it is presumed not to be such an invasion.

(c) Access Procedure

The procedures to be followed in making a request for access to a record under the Ontario Act are substantially the same as those applicable under the federal Access to Information Act.[82] Where the request is for a record that the head of an institution has reason to believe might contain information about a third party referred to in subsection 17(1), or personal information intended to be disclosed on the grounds that it does not constitute an unjustified invasion of personal privacy, notice of the request is to be given to the person "to whom the information relates".[83] Any such person is entitled to make representations to the head of the institution concerned as to why the record in question should not be disclosed. If the head of the institution subsequently decides to disclose the record, the

77 *Ibid.*, s. 18.
78 *Ibid.*, s. 19.
79 *Ibid.*, s. 20.
80 *Ibid.*, s. 22.
81 *Ibid.*, s. 21.
82 *Ibid.*, ss. 24-27.
83 *Ibid.*, s. 28.

person affected is to be given notice of his right to ask the Commissioner to review the decision within twenty days after the record is given.[84] Where the head of an institution refuses to give access to a record, he must set out in the notice given to the person who made the request the specific provision of the Bill under which access is refused, the reason the provision applies to the record, the name and office of the person responsible for making the decision to refuse access and the right of the person who made the request to appeal to the Commissioner for a review of the decision.[85] Section 30 of the Bill provides that as a general rule a person who is given access to a record is entitled to receive a copy unless it is not practicable to reproduce the record in which event he is entitled to examine the record; where reasonably practicable, such person is also entitled to examine a record in accordance with the regulations.[86]

(d) Protection of Personal Information

Part III of Bill 34 contains detailed provisions for the protection of individual privacy. Section 35 sets out the general rule that no person shall collect personal information on behalf of an institution unless the collection is expressly authorized by statute, used for the purposes of law enforcement or necessary to the proper administration of a lawfully authorized activity. Section 36 provides that except in specified circumstances, personal information shall be collected only from the individual to whom the information relates. Any person about whom information is collected must be notified of the legal authority for the collection, the purposes for which such information is collected, and the identity of a public official who can answer the individual's questions about the collection.[87] Other provisions in Part III set out the manner in which personal information may be used and disclosed, provide for personal information collected by an institution to be kept in data banks for which an index is to be prepared annually, provide for records to be kept of any use by an institution of personal information, provide for the rights of individuals to obtain access to and to correct any personal information held by an institution and set out the procedures whereby an individual may obtain access to records containing personal information about himself.[88] Section 45 of the Act sets out certain circumstances in which a head of an institution may refuse to disclose personal information.

---

84 *Ibid.*, s. 28.
85 *Ibid.*, s. 29.
86 *Ibid.*, subs. 30(2).
87 *Ibid.*, subs. 36(2).
88 *Ibid.*, ss. 37-44.

(e) Appeal

The rights of appeal with respect to requests for disclosure of records are set forth in Part IV of the Bill. Section 46 provides that any person who has made a request for access to a record, for access to personal information or for correction of personal information, as well as any third person who may be affected by disclosure of a record and to whom notice of a request is provided under subsection 28(1), may appeal any decision of the head of an institution to the Commissioner. Any appeal is to be made within a period of 30 days from the date of the decision appealed from.[89] Upon receiving a notice of appeal, the Commissioner must notify the head of the institution concerned and any other affected person.[90] The Commissioner is entitled to authorize a mediator to investigate the circumstances of an appeal and to try to effect a settlement of the matter.[91] Where no such settlement is effected, the Commissioner is bound to conduct an inquiry to review the head's decision, which inquiry is to be conducted in accordance with the provisions of section 48 of the Act. In any such inquiry, the onus is on the head to prove that a report falls within one of the specified exemptions under the Act.[92] After all evidence is adduced, the Commissioner is bound to make an order disposing of the issues raised by the appeal.[93] Where the Commission upholds the decision of a head that the head may refuse to disclose a record or part of a record, the Commission shall not order the head to disclose the record or part.[94] The Commissioner or any person acting under his authority is not compellable to give evidence in any legal proceeding, nor does any proceeding lie against the Commissioner or such other person for anything done in good faith in the performance of any power, duty or function under the Act.[95]

Unlike the federal Act, the Ontario Bill does not provide for any procedure for judicial review of decisions made by the Commissioner.

(f) Other Provisions

Other provisions of the Bill provide for the preparation and dissemination of publications showing the records kept by institutions;[96] for the preparation of guidelines by institutions as to the interpretation of provisions

89 *Ibid.*, subs. 46(2).
90 *Ibid.*, subs. 46(3).
91 *Ibid.*, s. 47.
92 *Ibid.*, s. 49.
93 *Ibid.*, subs. 50(1).
94 *Ibid.*, subs. 50(1a).
95 *Ibid.*, s. 51.
96 *Ibid.*, ss. 31-32.

of enactments for which they are responsible;[97] for the preparation of an annual report by the head of an institution to the Commissioner;[98] for the levying of fees in respect of the costs of searches;[99] and for the preparation of an annual report by the Commissioner to the Speaker of the Legislature.[100]

4. NOVA SCOTIA FREEDOM OF INFORMATION ACT

(a) Background and Definitions

The recitals to the Nova Scotia Freedom of Information Act[101] set out the background on which the Act is based. Specifically, the recitals state that the province has had a system of responsible government since 1848, that the people of the province should be protected from secrecy in the conduct of public business by officials, and that these principles can be better maintained by assuring the people of the province that the Government is operating openly and by providing to the people access to as much information in the hands of the Government as possible without impeding the operation of Government or disclosing personal information about other persons.

The Nova Scotia Act defines the term "access" to mean the opportunity to examine an original record or the provision of a copy at the option of the Government.[102] The term "information" is defined to mean information in any form, including information that is written, photographed, recorded or stored in any manner and on file or in the possession or under the control of a department, including personal information.[103] The term "personal information" is defined to mean information respecting a person's identity, residence, dependents, marital status, employment, borrowing and repayment history, income, assets and liabilities, credit worthiness, education, character, reputation, health, physical or personal characteristics or mode of living.[104]

(b) Right of Access

Section 3 of the Act sets out certain kinds of information to which any person is entitled to obtain access. The list includes information about

---

97 *Ibid.*, s. 33.
98 *Ibid.*, s. 34.
99 *Ibid.*, s. 53.
100 *Ibid.*, s. 54.
101 S.N.S., 1977, c. 10, proclaimed in force November 1, 1977.
102 *Ibid.*, subs. 2(*b*).
103 *Ibid.*, subs. 2(*f*).
104 *Ibid.*, subs. 2(*g*).

the organization of a department, administrative staff manuals and instructions to staff affecting the public, rules of procedure, descriptions of forms available, statements of general policy, final decisions of administrative tribunals, personal information pertaining to the person making the request, the annual report and regulations of a department, programs and policies of a department and any amendments to the foregoing.

Section 4 of the Act provides that a person shall not be permitted access to information that might reveal personal information about another person; might result in financial gain or loss to a department or influence negotiations in progress; would jeopardize the ability of a department to function on a competitive commercial basis; might be injurious to relations with another government; would be likely to reveal information obtained during an investigation concerning the alleged violation of any statute; would be detrimental to the custody, control or supervision of persons under sentence; would be likely to disclose legal opinions provided to a department or opinions or recommendations by public servants in matters to be decided by a Minister or the Cabinet; would be likely to disclose draft legislation or regulations; or would be likely to disclose information the confidentiality of which is protected by any enactment.

(c) Provisions respecting Personal Information

Subsection 6(1) of the Act sets out certain rights of any person with respect to personal information about him contained in a file by a department; specifically any such person may request that the information be corrected and amended, may request that it not be used for any purpose other than that for which it was provided without his consent, and may seek injunctive relief to correct or amend such information. Corresponding to these rights, each department is subject to certain duties with respect to personal information files maintained by it, including the duty not to make the information available to another person or department for "another purpose" without the person's consent; to maintain all necessary and lawful records (which are to be current and accurate); to disclose the existence of all data banks and files containing personal information; to refrain from making a person's name available for mailing list use without his consent; and to permit a person to have access at all reasonable times to personal information kept about him.[105] The Act does not set out the purposes for which personal information may be collected or the circumstances in which, or the persons to whom, personal information may be disclosed.

105 *Ibid.*, subs. 6(2).

(d) Procedures for Obtaining Access to Information

Any person seeking access to information kept by a department may apply to the department where the information is kept either in person, by telephone or in writing.[106] If access is not forthcoming as a result of such an application, a formal request may be made under section 10. Such a request must be made in writing to the Deputy Head of the Department who must respond within fifteen days. Any denial of a request by the Deputy Head must be in writing and must specify the reasons for the denial and set out the appeal procedures available to the person making the request.[107]

(e) Appeal Procedures

If a request for access to information is denied, the person making the request may, within a period of 15 days, appeal the denial to the Minister, who must, within a further 30 days, review the request and either affirm, vary, or overrule the denial.[108] If any denial of access to information has been affirmed by the Minister, the person to whom the information is denied may make a further appeal to the House of Assembly.[109] Such an appeal is to be presented by a member and dealt with in accordance with the Rules and Forms of Procedure of the House.[110] Unlike the federal freedom of information and privacy legislation, the Nova Scotia Act does not provide for the establishment of the office of a commissioner to investigate complaints under the Act, nor does it contain provisions for the appeal of any denial of access to the courts.

5. AN ACT RESPECTING FREEDOM OF INFORMATION (NEWFOUNDLAND)

(a) Purpose of Act and Definitions

Section 3 of the Newfoundland Act Respecting Freedom of Information[111] provides that the purpose of the Act is to provide a right of access by the public to information in records of departments, and to subject that right only to specific and limited exceptions necessary for the operation of such departments and for the protection of personal privacy. The term "department" is defined to mean any department of the government of the province, or any board, commission, or body listed in the schedule

---

106 *Ibid.*, s. 9.
107 *Ibid.*, s. 11.
108 *Ibid.*, s. 12.
109 *Ibid.*, s. 13.
110 *Ibid.*, s. 13.
111 S.N. 1981, c. 5 (assented to June 16, 1981).

to the Act. The term "information" is defined substantially as set out in the Nova Scotia Act.

(b) Right of Access

Section 4 indicates that any Canadian citizen, any permanent resident of Canada or any corporation incorporated under a law of Canada or a province and carrying on business in the province has a right to and shall, on request, be permitted access to information under the control of a department. The Lieutenant Governor in Council is required to publish a list of all departments and a list of the general class of information prepared by or under the control of each department.[112] A request for information under the Act must be in writing and must be addressed to the head of the department in which the information is kept or filed.[113] The request must be in sufficient detail as to time, place and event and if the information is unable to be identified, the head must so advise the person requesting the information and invite him to supply additional details.[114] The head of a department who receives a request for information must grant, deny or refer the request to another department within 30 days.[115] Where the request is granted, the head must, upon payment of the prescribed fees, allow the information to be inspected and reproduced.[116]

(c) Exceptions

The Newfoundland Act sets out three classes of exemptions, that is, those which are non-discretionary, those relating to personal information, and certain discretionary exceptions. The non-discretionary exceptions are set forth in section 9 and include information in records obtained in confidence under an agreement with another government, and information that relate to proposals, recommendations, deliberations, consultations, briefings or background papers prepared by or for a Minister of the Crown or the Executive Council. There is also a general non-discretionary exception with respect to any records required to be kept confidential under the provisions of any Act of the province. Such exceptions do not apply in respect of information where disclosure is authorized by the Premier of the province.[117]

Section 10 of the Act prohibits access to personal information

---

112 *Ibid.*, s. 5.
113 *Ibid.*, s. 6.
114 *Ibid.*
115 *Ibid.*, s. 7.
116 *Ibid.*, s. 8.
117 *Ibid.*, s. 9.

respecting an identifiable individual other than certain classes of information, such as information relating to an individual who was an officer or employee of a department and that relates to his position or functions. Section 11 of the Act also sets forth certain discretionary exceptions, including information relating to federal-provincial negotiations; information respecting the enforcement of any law; information the disclosure of which would have a substantial adverse effect on the economic interests of the province; legal opinions or advice provided by a law officer of the Crown or privileged communications; proposed legislation or regulations; and certain financial, commercial, scientific or technical information, the disclosure of which would affect either the continued access to such information or the competitive position of a person or result in undue financial loss or gain to a person.

(d) Appeal Procedure

A person who is not satisfied with the decision of the head of a department, or who does not receive a response to a request for access to information within the prescribed time, may appeal the decision to the Ombudsman. The Ombudsman must investigate the matter within 30 days of receiving the referral.[118] The Ombudsman must provide a report in writing of his opinion and reasons to the head of the institution, the person who referred the matter to him, and to any third party to whom notice of the request was made by the head.[119] Where the Ombudsman makes his report, the head of a department must decide within 30 days whether to follow the recommendation or to make some other decision.[120] Within 30 days of receiving the decision of the head, any person affected, including a third party, may appeal the decision to the Trial Division, which shall determine the matter *de novo* and may examine any information in camera in order to determine on the merits whether such information may be withheld under the Act.[121] The Act also provides that if a request for information includes trade secrets or other prescribed information affecting a third party, that party may intervene to make representations relating to the disclosure of the information or may appeal the decision to the Ombudsman in the same manner as the person who made the request for the information.[122]

In any proceeding under the Act, the burden of establishing that access to the information required under the Act may or must be refused, shall

118 *Ibid.*, s. 12.
119 *Ibid.*, subs. 13(1).
120 *Ibid.*, subs. 13(3).
121 *Ibid.*, s. 14.
122 *Ibid.*, s. 15.

be on the head concerned.[123]

6. RIGHT TO INFORMATION ACT (NEW BRUNSWICK)

(a) Right of Access

Section 2 of the New Brunswick Right to Information Act[124] provides that subject to the Act, every person is entitled to request and receive information relating to the public business of the province. The term "public business" is defined in section 1 of the Act to mean any activity or function carried on or performed by a department. The term "department" is defined to mean any department of the government of the province, any Crown agency or Crown corporation, any other branch of the public service, or any body or office, the operation of which is effected through money appropriated for the purpose and paid out of the Consolidated Funds.

(b) Procedure for Obtaining Access

The procedures set forth in the New Brunswick Act with respect to obtaining access to information kept by a department are substantially the same as set forth in the Newfoundland Act. Specific requests for information must be made with sufficient particularity in writing to the Minister of a Department, who must respond within 30 days.[125] Section 4 of the Act provides for the inspection and/or photocopying of the information upon payment of the prescribed fees. Subsection 4(3) provides that where a request for information is granted, the information shall be provided only in the language or languages in which it was made. In addition, subsection 4(4) provides that when the document containing the information that has been requested has been destroyed or does not exist, the appropriate Minister must advise the applicant of such fact.

(c) Exceptions

Section 5 of the Act provides that an appropriate Minister may only deny a request for information or a part thereof in accordance with subsection 4(4) (that is, where the information does not exist) and section 6. Denial of access for information must be in writing and the reasons for such denial must be stated. In addition, the applicant must be provided with the necessary forms for a review under the Act.[126]

The New Brunswick Act does not distinguish between mandatory and discretionary exemptions. Instead, section 6 of the Act provides that there

---

123 *Ibid.*, s. 16.
124 S.N.B. 1978, chap. R-10.3 (assented to June 28, 1978), as amended S.N.B. 1985, c. 67.
125 *Ibid.*, s. 3.
126 *Ibid.*, s. 5.

is no right to information where its disclosure would reveal information the confidentiality of which is protected by law; would reveal personal information given on a confidential basis concerning another person; would cause financial loss or gain to a person or department or would jeopardize negotiations leading to an agreement or contract; would violate the confidentiality of information obtained from another government; would be detrimental to the proper custody, control or supervision of persons under sentence; would disclose legal opinions provided to a person or department by a Crown officer or privileged communications; would disclose opinions or recommendations by public servants for a Minister or the Executive Council; would disclose proposed legislation or regulations; would provide information gained by the police in the course of an investigation or information reported to the Attorney General with respect to an illegal activity; or would impede an investigation, enquiry or any administration of justice.

(d) Appeal Procedure

Section 7 of the Act provides that where an applicant is not satisfied with the decision of an appropriate Minister, or where a Minister fails to reply to a request within the prescribed time, the applicant may refer the matter either to a judge of the Supreme Court or to the Ombudsman. However, where the matter is referred to the Supreme Court, it may not thereafter be referred to the Ombudsman.[127] In addition, if the matter is referred to the Ombudsman, it may not thereafter be referred to a judge of the Supreme Court, subject to a further appeal procedure if the Minister fails to carry out the recommendation of the Ombudsman.[128]

Any inspection of the information which is the subject-matter of a request made by the Ombudsman or a Supreme Court judge must be in camera without the presence of any person.[129] A Supreme Court judge is authorized to hold a hearing and may order the Minister to grant the request, or where he failed to reply to a request either grant or deny the request, or may make any other order that is appropriate.[130] The Ombudsman, on the other hand, is authorized only to make a recommendation to the Minister who must either carry out the recommendation or make such other decision as he thinks fit.[131] Where the person making the referral is not satisfied with the decision of the Minister based on the recommendation of the Ombudsman, that person may appeal the matter to a judge

---

127 *Ibid.*, subs. 7(2).
128 *Ibid.*, subs. 7(3).
129 *Ibid.*, subs. 7(4).
130 *Ibid.*, s. 8.
131 *Ibid.*, s. 10.

of the Supreme Court.[132]

7. THE FREEDOM OF INFORMATION ACT (MANITOBA)

(a) The Right of Access

Section 3 of the Manitoba Freedom of Information Act[133] provides that, subject to the Act, every person has, upon application, a right of access to any record in the custody or under the control of a department, including any record which discloses information about the applicant. The right of access is defined to mean either the right to examine or to obtain a copy of the record. With respect to a record stored by electronic means, the head of a department is presumed to have given access where the head permits the applicant to examine a transcript of the record, provides the applicant with a copy of such transcript, or permits the applicant to view or hear a record produced for visual or aural reception.

(b) Procedure for Obtaining Access

Section 5 of the Act provides that any officer or employee of a department who receives an application must promptly inform the applicant of his right to file a complaint with the Ombudsman in the event the department fails to respond to the application within 30 days. In addition, section 6 requires the head of a department which receives an application to notify the applicant whether he is to be given access to the record, and if so, to give such access, within 30 days of the application. If such access is refused, the notice must set out the right of the applicant to file a complaint with the Ombudsman, indicate if the record does not exist or cannot be located and set out the provision of the Act upon which any refusal to give access is based.[134]

Other provisions of the Act set out the fees to be payable in connection with the examination of a record or the obtaining of a copy, provide for the forwarding of an application to another department if that other department has "a greater interest" in the record, and provide for the extension of the time limits within which the head of a department must respond to an application.[135] In addition, section 12 of the Act provides for the severability of exempt information, and permits the head of a department to supply the applicant with additional information as may be necessary to explain or interpret the record. Section 13 permits an applicant given access to a record which discloses information about

132 *Ibid.*, s. 11.

133 Bill 5, 34 Eliz. II, 1985 (assented to July 11, 1985).

134 *Ibid.*, subs. 6(2).

135 *Ibid.*, ss. 7-11.

himself to submit to the Department a written objection respecting any error or omission of fact which the applicant alleges is contained in the record. Such objection forms part of the record and may not be destroyed, altered or removed.[136]

(c) Complaints

A complaint may be filed with the Ombudsman by any applicant who is refused access to a record or who receives a notice of extension of the time by which an applicant is to be advised whether he is to receive a record.[137] The Ombudsman may also investigate any complaint by a person about the adequacy or availability of the access guide to be published under the Act.[138] The Ombudsman is also authorized to initiate a complaint about any matter under the Act.[139]

In investigating any complaint, the Ombudsman is given certain powers of investigation, the right to delegate the exercise of such powers and certain rights of entry.[140] Every investigation by the Ombudsman is to be conducted in private although the complainant and the head of the department involved must be given a reasonable opportunity to make representations respecting the complaint.[141] Subject to certain qualifications, the Ombudsman must maintain secrecy in respect of matters which come to his knowledge in the course of an investigation, and he must take every reasonable precaution to avoid the disclosure of any exempt information.[142]

Upon completing an investigation, the Ombudsman must send a report of his findings and recommendations to the applicant and the head of the department concerned.[143] If the Ombudsman finds that the complaint is not justified, he must notify the applicant of his right to appeal to the court for access to the record.[144] The head of the department must respond within 14 days of receiving the report from the Ombudsman indicating any action which the head has taken or proposes to take to implement the recommendations and any refusal to take any action to implement such recommendations.[145] The report indicating the departmental response must be sent by the Ombudsman to the applicant. Where the response

136 *Ibid.*, s. 13.
137 *Ibid.*, s. 14.
138 *Ibid.*, s. 15.
139 *Ibid.*, subs. 15(2).
140 *Ibid.*, ss. 16-17.
141 *Ibid.*, ss. 19-20.
142 *Ibid.*, s. 21.
143 *Ibid.*, subs. 25(1).
144 *Ibid.*, subs. 25(2).
145 *Ibid.*, s. 26.

indicates that the head has rejected a recommendation by the Ombudsman to give the applicant access to a record, the Ombudsman must state his opinion as to whether or not the response of the head is justified, and set out the right of the applicant to appeal to the court and the time limit for filing such an appeal.[146] If the head of the department has decided that the applicant will be given access, such access must be given within ten days.[147]

(d) Appeal Procedure

An applicant who is refused access to a record and who has filed a complaint with the Ombudsman may appeal the refusal of access to the court by filing an originating notice of motion within 30 days of receiving the Ombudsman's report.[148] The head of the department involved is to be the respondent in any such appeal but the Ombudsman is entitled to be a party.[149] In disposing of an appeal under the Act, the court is required to take every reasonable precaution, including receiving representations *ex parte* and conducting hearings in camera, to avoid disclosure of any exempt information.[150] If the court finds that the record in question falls within an exemption, the court shall not order the head to give the applicant access to the record, even if the exemption in question merely authorizes rather than requires the head to refuse to give the applicant access to the record.[151] The decision of the court under section 35 is final and binding and there is no appeal from such decision.[152]

(e) Exceptions

As in the case of the Acts of other Canadian jurisdictions, the Manitoba Act contains both mandatory and discretionary exceptions. Specifically, the head of a department is *required* to refuse access to any record which discloses a confidence of the Cabinet, including proposals prepared by a Minister and certain communications between Ministers;[153] any record the disclosure of which would constitute an unreasonable invasion of the privacy of a third party;[154] commercial information belonging to a third party (subject to certain qualifications if the disclosure of such information

146 *Ibid.*, s. 27.
147 *Ibid.*, s. 28.
148 *Ibid.*, s. 30.
149 *Ibid.*
150 *Ibid.*, s. 33.
151 *Ibid.*, subs. 35(2).
152 *Ibid.*, s. 36.
153 *Ibid.*, subs. 38(1).
154 *Ibid.*, s. 41.

is in the public interest);[155] and information obtained in confidence from another government.[156]

The head of a department is given a discretionary authority to refuse to give access to any record which discloses policy opinions, advice or recommendations to a department or a Minister, or a plan relating to the administration of a department or the contents of a draft enactment. These exemptions are subject to various limitations, such as with respect to environmental impact statements, test results, results of research, instructions or guidelines to officers or employees of a department or a consultant's report.[157]

Other discretionary exceptions set out in the Act include records relating to law enforcement and legal proceedings;[158] any record which discloses a trade secret of a department or the government, or innovative scientific or technical research the disclosure of which may be expected to prejudice the competitive position of a department or the government or injure the financial interests of a department;[159] any record the disclosure of which could be injurious to the conduct of federal-provincial relations;[160] a record relating to testing or auditing procedures;[161] any material which is to be made public;[162] any personal records made before the coming into force of the section;[163] and any record the disclosure of which may lead to physical or serious psychological harm to the applicant or a third party.[164]

(f) Other Provisions

Section 50 of the Act provides for the publication of an access guide containing such information as instructions on the exercise of the right of access, a list of personnel involved in the administration of the Act and appropriate departmental facilities, a description of the records in the custody or control of each department and a subject index of terms describing the functions of departments.[165] The access guide is to be made available to the public and is to be updated at least once every two years.[166] Annual reports are also to be provided by the designated Minister under

155 *Ibid.*, s. 42.
156 *Ibid.*, s. 45.
157 *Ibid.*, s. 39.
158 *Ibid.*, s. 40.
159 *Ibid.*, s. 43.
160 *Ibid.*, s. 44.
161 *Ibid.*, s. 46.
162 *Ibid.*, s. 47.
163 *Ibid.*, s. 48.
164 *Ibid.*, s. 49.
165 *Ibid.*, s. 50.
166 *Ibid.*, ss. 51-52.

the Act as well as by the Ombudsman.[167] Section 63 of the Act preserves any rights of access under any custom or practice, or under any other Act of the legislature or regulation.

8. AN ACT RESPECTING ACCESS TO DOCUMENTS HELD BY PUBLIC BODIES AND THE PROTECTION OF PERSONAL INFORMATION (QUEBEC)

(a) Application of Act

Section 1 of the Quebec Act[168] states that the Act applies to documents kept by a public body in the exercise of its duties whether it keeps them itself or through the agency of a third party. Section 1 provides further that the Act applies whether the documents are recorded in writing or print, on sound tape or film, in computerized form or otherwise. Section 3 of the Act provides that the public bodies which are covered by the Act include the Government, the Conseil exécutif, the Conseil du Trésor, government departments and agencies, municipal and school bodies and health services and social services establishments. The term "public bodies" also includes the Lieutenant Governor, the National Assembly, agencies whose members are appointed by the Assembly and every person designated by the Assembly to an office under its jurisdiction. However, the courts are deemed not to be public bodies and are not included in the Act.

(b) Access to Documents Held by Public Bodies

Chapter II of the Quebec Act sets out the general right of every person to obtain access to the documents held by a public body. Section 14 of the Act provides that no public body may deny access to a document for the sole reason that it contains information that it must or may refuse to release, except where the information forms the substance of the document; in other cases, the public body must give access to the requested document after deleting only the information to which access is not authorized. The right of access under the Act applies only to documents that can be released without requiring computation or comparison of information.[169] Every public body is required to classify its documents in such a manner as to allow their retrieval and maintain a current list of the classification of the documents.[170] An index must also be published

167 *Ibid.*, ss. 54-55.

168 An Act Respecting Access to Documents held by Public Bodies and the Protection of Personal Information, S.Q. 1982, c. 30.

169 *Ibid.*, s. 15.

170 *Ibid.*, s. 16.

and distributed in every region of Quebec.[171]

(c) Restrictions to the Right of Access

Among the classes of information which the government or any department *may* refuse to disclose are the following: information received from another government, or information the disclosure of which would be detrimental to relations with another government;[172] information relating to negotiations in progress between public bodies;[173] information relating to the financial operations of a public body;[174] certain industrial secrets, financial or technical information or management information;[175] information relating to the negotiation of a collective agreement or contract;[176] information concerning the deliberations of the Conseil exécutif;[177] legal opinions concerning the application or validity of a law or a draft of a bill or regulations;[178] any study the disclosure of which may affect the outcome of judicial proceedings;[179] the records of the deliberations of a meeting of the board of directors of a public body;[180] a recommendation or opinion of a civil servant to a public body presented less than ten years earlier;[181] a recommendation or opinion made by an agency under the jurisdiction of a public body to that body until the final decision is made public;[182] a study prepared in connection with a recommendation to a public body;[183] aptitude and similar tests;[184] or information affecting auditing by the Auditor General.[185]

Among the classes of information which a public body is prohibited from releasing are the following: industrial secrets or other confidential information of a third party;[186] other information supplied by a third party that would hamper contract negotiations, result in losses or reduce the third party's competitive margin, without his consent;[187] certain informa-

---

171 *Ibid.*, s. 17.
172 *Ibid.*, ss. 18-19.
173 *Ibid.*, s. 20.
174 *Ibid.*, s. 21.
175 *Ibid.*, s. 22.
176 *Ibid.*, s. 27.
177 *Ibid.*, s. 30.
178 *Ibid.*, s. 31.
179 *Ibid.*, s. 32.
180 *Ibid.*, s. 35.
181 *Ibid.*, s. 37.
182 *Ibid.*, s. 38.
183 *Ibid.*, s. 39.
184 *Ibid.*, s. 40.
185 *Ibid.*, s. 41.
186 *Ibid.*, s. 23.
187 *Ibid.*, s. 24.

tion affecting the administration of justice and public security;[188] and certain internal communications of the Conseil exécutif, the Conseil du Trésor or cabinet committees, which may not be disclosed for a period of 25 years from the date of the information in question.[189]

(d) Procedure for Obtaining Access

The procedure to be followed in order for a person to obtain access to a document is set forth in Chapter II, Division III. Under section 43, a request for access may be made in writing or orally. However, only a decision on a written request may be reviewed under the Act and any person making an oral request must be so advised.[190] The response by the person in charge must be made not later than 20 days from the date on which a request is received.[191] The person in charge must give the reasons for any refusal to disclose the information and indicate the provision of the Act on which the denial is based.[192]

(e) Protection of Personal Information

Section 53 of the Act sets out the general rule that nominative information is confidential unless the person about whom the information relates authorizes its disclosure. In any document, information concerning a natural person which allows the person to be identified is nominative information.[193] However, personal information which by law is public is not nominative information,[194] nor is the name of a person except where it appears with other information.[195] Section 57 of the Act sets out certain information which is deemed to be public information. Section 59 of the Act prohibits the release by a public body of any nominative information without the consent of the person concerned except in certain specified circumstances.

Division II of Chapter III of the Act sets out various restrictions on the collection, keeping and use of nominative information. In particular, section 64 prohibits any person, on behalf of a public body, from collecting nominative information if it is not necessary for the carrying out of the attributions of the body or the implementation of a programme under its management. In addition, the person collecting nominative information

188 *Ibid.*, ss. 28-29.
189 *Ibid.*, s. 33.
190 *Ibid.*, s. 45.
191 *Ibid.*, s. 47.
192 *Ibid.*, s. 50.
193 *Ibid.*, s. 54.
194 *Ibid.*, s. 55.
195 *Ibid.*, s. 56.

must identify himself and inform the person concerned of the body on whose behalf the information is being collected, the use to which it will be put, who will have access to the information, whether a reply is obligatory, the consequences of a refusal to reply and the rights of access and correction provided by law.[196] In any case where a public body proposes to obtain nominative information which has already been assembled from any person or private body, the public body must inform the Commission.[197] Section 67 of the Act also provides for a written agreement to be made where the release of nominative information from one public body to another is authorized by law. Such a disclosure agreement may also be made with other public bodies or with other governments to allow the disclosure of nominative information for the carrying out of an Act or for an investigation, as well as with a person or private body.[198] Any such agreement must provide necessary measures to ensure the confidentialityof the information, must be submitted to the Commission for an opinion and is subject to government approval.

Division III of Chapter III sets out certain requirements with respect to the establishment and maintenance of files of personal information. In particular, the Act sets out procedures to ensure the accuracy of any nominative information. In addition, the Government may authorize the public body to establish a confidential file under conditions set forth in the government's order.[199]

Division IV of Chapter III of the Act sets out certain rights of persons concerned by nominative information. Such rights include the right of access (which right may be restricted where the information is kept in a confidential file) and the right of correction to be exercised in accordance with the procedures set out in the Act.[200]

(f) Powers of Commission d'accès à l'information

Division I of Chapter IV of the Act provides for the establishment and organization of the Commission, including the procedures for the appointment of members of the Commission. Division II of Chapter IV sets forth the duties and powers of the Commission. Section 122 provides that the object of the Commission is to hear, to the exclusion of every other court, a request for review made under the Act. The Commission may prescribe the conditions applicable to any personal information file; may grant a person or an agency the authorization to receive nominative

---

196 *Ibid.*, s. 65.
197 *Ibid.*, s. 66.
198 *Ibid.*, s. 68.
199 *Ibid.*, ss. 80-81.
200 *Ibid.*, ss. 83-102.

information from a personal information file for study, research or statistical purposes under certain conditions to ensure its confidentiality; may investigate whether a personal information file contains information that was entered and used in accordance with any order; may make orders that nominative information be corrected or deleted; may order a public body to take measures it considers necessary to meet the conditions provided for by the Act or the prescriptions of the Commission; may order the destruction of a personal information file or recommend to the government that the order authorizing the establishment of a confidential file be amended or revoked.[201] The Commission is also responsible for publishing, at least once every two years, an index of all files held by public bodies.[202] In any case where the Commission considers that appropriate measures have not been taken to implement any recommendation made to a public body, it may so notify the Government or submit a special report to the National Assembly or set out the situation in its annual report.[203] The Committee on the National Assembly must designate the Committee to study the special report, which study must take place within 60 days of its tabling in the National Assembly.[204]

(g) Review Procedure

The review and appeal procedures applicable to requests for access to information are set out in Chapter V. Any person whose request has been denied by the person in charge of access to documents or of protection of personal information may apply to the Commission for a review of the decision.[205] Such a review may also be requested for any decision as to the time prescribed for processing a request, the mode of access, the application of section 9 (granting the general right of access) and the fee payable. Any such application must be made within 30 days of the date of the decision, although the Commission may for serious cause release the applicant from a failure to observe the time limits.[206] A third party who has been given notice of an intention to release information and has submitted written observations in connection therewith may also apply to the Commission for a review of any decision to grant access.[207] A hearing of the matter may be held by any member of the Commission acting alone.[208] On an application for review, the Commission must give the

201 *Ibid.*, ss. 124-128.
202 *Ibid.*, s. 132.
203 *Ibid.*, s. 133.
204 *Ibid.*, s. 134.
205 *Ibid.*, s. 135.
206 *Ibid.*
207 *Ibid.*, s. 136.
208 *Ibid.*, s. 139.

parties an opportunity to submit their observations.[209] The Commission is given all powers necessary for the exercise of its jurisdiction and it may make every order it considers appropriate to protect the rights of the parties and decide on every matter of fact or of law. In particular, it may order a public body to release a document or part of a document, to refrain from doing so, to correct, complete, clarify, update or delete any nominative information, or to discontinue the use or release of nominative information.[210]

Any action prescribed to be taken by the Commission is executory within 15 days after its receipt by the parties unless an appeal is brought, while any prohibited action is executory from the time of delivery to the public body.[211] However, section 145 provides that where the Government considers it necessary in the public interest, it may require a public body to postpone the execution of a decision of the Commission ordering the release of a document or information for such period of time as it indicates. Such an order, which results in a stay in a request for access to the document or of any proceedings in the appeal from the decision of the Commission, must be tabled in the National Assembly within 15 days. Section 146 of the Act provides that every decision of the Commission on a question of fact within its competence is final.

(h) Appeal Procedure

Section 147 of the Act provides that a person who is "directly interested" may bring an appeal from a decision of the Commission before three judges of the Provincial Court on any question of law or jurisdiction. However, no such appeal may be brought except with leave of a judge of the Provincial Court who may do so if in his opinion the question ought to be examined in appeal.[212] A motion for leave to appeal must be filed in the office of the Provincial Court, at Montreal or at Quebec, within 30 days from the decision after notice to the parties and the Commission.[213] The appeal is brought by filing with the Commission a notice served on the parties within ten days of the date of the decision giving leave for the appeal.[214] The procedures for such appeal are set forth in sections 151-153 of the Act. The decision of the three judges of the Provincial Court is final.[215]

---

209 *Ibid.*, s. 140.
210 *Ibid.*, s. 141.
211 *Ibid.*, s. 144.
212 *Ibid.*, s. 147.
213 *Ibid.*, s. 149.
214 *Ibid.*, s. 150.
215 *Ibid.*, s. 154.

(i) Other Provisions

Other provisions of the Act provide for the making of regulations with respect to various matters by the Government;[216] for the imposition of penalties on persons who commit certain offences under the Act;[217] for the grant of a right of redress to a natural person who is wronged by a decision of a public body in respect of nominative information which is inaccurate, or which has been collected, kept or released in contravention of the Act;[218] and for the obligation of a public body to compensate a person for prejudice resulting from the unlawful infringement of the person's right in respect of personal information.[219] The Act provides that in general its provisions prevail over any contrary provision, with any inconsistent law respecting the protection of personal information ceasing to have effect on October 1, 1986.[220] Other transitional provisions provide for delays in the period before a public body is bound by certain provisions of the Act.[221]

9. CONCLUSIONS

The foregoing review indicates that at the federal level and in many provincial jurisdictions, Canadians have extensive rights to gain access to information held by public bodies, and to restrict the collection, disclosure and use of personal information about themselves by governments. In all cases, the legislation sets out a number of circumstances where access to public information may be denied. Such denial may be based either on mandatory exemptions, where the responsible official is required not to permit access, or discretionary exemptions, where the head in question is entitled but not required to refuse access. The various statutes differ significantly in the nature and scope of such exemptions, with the statutes of Canada, Manitoba and Quebec and Bill 34 in Ontario containing very detailed lists of exemptions. The various statutes also provide procedures for the protection of information supplied to governments by "third parties", that is, persons other than those who make a request for access. All statutes also contain procedures for the protection of personal information, that is, information about identifiable individuals.

The various Canadian statutes contain very different procedures relating to the review of, or appeal from, the decisions of the heads of institutions as to whether or not to release information. These procedures

---

216 *Ibid.*, ss. 155-157.
217 *Ibid.*, ss. 158-165.
218 *Ibid.*, s. 166.
219 *Ibid.*, s. 167.
220 *Ibid.*, ss. 168-169.
221 *Ibid.*, see generally ss. 175-182.

may involve the submission of complaints to a Commissioner who is empowered either to make recommendations or to render a final decision, a review of such decisions by the provincial Ombudsman, appeals to the courts (either before or after review by a Commissioner or other public official) and appeals to the Legislature.

It will be noted from the foregoing review that the existing statutes have not specifically addressed some of the problems of privacy protection that arise as a result of the widespread computerization of personal and other information held by government bodies. Such computerization makes it more difficult for citizens to be fully informed and to exercise effective control over the collection of personal information about them by governments. In particular, existing legislation does not deal with issues arising from the potential cross-linking of a number of data banks held by various institutions and governments.

11

The Right of Privacy and the Private Sector

1. THE RIGHT OF PRIVACY IN CANADIAN LAW: AN OVERVIEW

Information privacy, which may be regarded as one aspect of the right of privacy, has been defined as "the claim of individuals, groups or institutions to determine for themselves when, how and to what extent information about them is communicated to others."[1] Canadian law has not generally recognized a specific cause of action for invasion of privacy although recent authorities suggest that such an action may be asserted in some circumstances.

Even in the absence of a specific cause of action, various forms of legal protection for the privacy of individuals exist. For example, many provisions of the Criminal Code directed to the protection of private property or the right of the individual to be free from bodily injury and the threat of such injury deal at least indirectly with the protection of personal privacy. Moreover, many civil rights recognized at common law, such as those relating to trespass, nuisance, defamation and breach of confidence may also be relied upon to prevent undue interference with an individual's personal life.

Although various causes of action at common law may be relied upon to protect rights to privacy, the protection granted may be limited in scope. For example, in the case of an action for defamation, it must be shown not only that the information which is disseminated is untrue, but also that damage has been caused to an individual's reputation as a result. In general, such a cause of action does not enable an individual to prevent the collection and/or the widespread dissemination of information which is accurate, nor does it enable the individual to control the uses that may be made of such information.

The law relating to breach of confidence is also of limited assistance in protecting the individual against the collection, dissemination or use of personal information by third parties. As discussed in Part II, Chapter 5, *supra*, the courts will impose an obligation to respect confidential information based on general principles of equity according to the nature

---

1 A.F. Westin, *Privacy and Freedom* (Atheneum, 1967).

of the relationship between a disclosing party and the recipient or the circumstances in which the information is obtained. The information that may be protected includes not just trade secrets or other confidential business information, but also personal information disclosed in confidence or obtained as a result of a particular relationship. For example, in *Argyll v. Argyll*,[2] an injunction was granted to restrain the plaintiff's former husband from publishing confidential information concerning their married life. The court noted that the communications in question were ones which "the policy of the law recognises should be protected" and it was not to be deterred from exercising its jurisdiction "merely because it is not already provided with fully developed principles, guides, tests, definitions and the full armament for judicial decision."[3]

Some formulations of the principle of breach of confidence suggest that the principle may apply where the information is used by a person who is not himself in a confidential relationship with the plaintiff. Thus, in *Saltman Engineering Co. v. Campbell Engineering Co.*,[4] Lord Greene formulated the principle as follows:

> If a defendant is proved to have used confidential information, directly or indirectly obtained from a plaintiff, without the consent, express or implied, of the plaintiff, he will be guilty of an infringement of the plaintiff's rights.

Most authorities indicate, however, that the nature of the relationship or the circumstances of disclosure must be considered in determining whether equity will restrict the use of any information obtained by the recipient. Accordingly, such cause of action appears to have only limited, if indeed any, application to the collection, dissemination or use of information about one person by another person who does not have any contractual or other relationship with the subject.

Another cause of action relating to privacy that has found some judicial favour is that of appropriation of personality. In *Krouse v. Chrysler Canada Ltd.*,[5] the plaintiff was a professional football player whose likeness was used without his consent in the promotion of a product. At trial, the plaintiff succeeded in his claim for damages for the appropriation of his likeness for commercial purposes. Although the trial court's decision was reversed on appeal, the Court of Appeal recognized that the appropriation of personality was a tort actionable in damages.[6]

The decision in the *Krouse* case was applied in *Saccone v. Orr*[7] as a basis for awarding damages against the defendant for playing in public

2 [1965] 1 All E.R. 611.
3 *Ibid.*, at 625.
4 [1963] 3 All E.R. 413 at 414 (C.A.).
5 (1972), 25 D.L.R. (3d) 49 (Ont. H.C.); reversed (1974), 1 O.R. (2d) 225 (Ont. C.A.).
6 See also *Athans v. Cdn. Adventure Camps Ltd.* (1977), 17 O.R. (2d) 425 (Ont. H.C.).
7 (1981), 34 O.R. (2d) 317 (Ont. Co. Ct.).

a tape of the plaintiff's telephone conversation made without his consent. An account of the conversation was subsequently published. For want of a better description, Jacob Co. Ct. J. described the defendant's act as an invasion of privacy. In spite of the contention of defendant's counsel that no such action existed, he held that the plaintiff had a right of recovery and was entitled to damages.

It would appear that a cause of action for the violation of the right of privacy may also exist in the Province of Quebec. In *Robbins v. Canadian Broadcasting Corp.*,[8] the plaintiff had written to criticize a television program broadcast by the defendant. On a subsequent program, the producer displayed the plaintiff's name and address and invited viewers to write or telephone him. As a result, the plaintiff received a number of offensive telephone calls and letters. In awarding the plaintiff damages, the court found that the defendant had committed a fault under Article 1053 of the Civil Code, which provides that every person capable of discerning right from wrong is responsible for the damage by his fault of another, whether by positive act, imprudence, neglect or want of skill.[9]

A provision of the Criminal Code directly relevant to the protection of privacy is section 178.11, which was enacted pursuant to the Protection of Privacy Act.[10] Section 178.11 provides that any person who wilfully intercepts a private communication by electromagnetic, mechanical or other device without the consent of the originator or receiver commits an offence. A "private communication" is defined in section 178.1 to mean any oral communication or any telecommunication made under circumstances in which it is reasonable for the originator to expect that it will not be intercepted by any person other than the person intended to receive it. The term "intercept" is defined in section 178.1 as including to "listen to, record or acquire a communication or acquire the substance, meaning or purport thereof."

Section 178.18 of the Criminal Code provides that it is a further offence for any person to possess, sell or purchase any electromagnetic, acoustic, mechanical or other device knowing that its design renders it primarily useful for the surreptitious interception of private communications. Still another offence is created by section 178.20 which prohibits any person from using, disclosing the contents of, or disclosing the existence of any private communication without the consent of the originator or receiver.

Other provisions of the Criminal Code set forth the procedures to be followed to obtain authorization to intercept a private communication, provide for emergency authorizations to be granted in certain circumstan-

8 (1957), 12 D.L.R. (2d) 35 (Que. S.C.).
9 See Glenn, "Civil Responsibility — Right to Privacy in Quebec" (1974), 52 Can. Bar Rev. 297.
10 S.C. 1973-74, c. 50.

ces, set out certain exemptions to the offences relating to the possession of devices and the disclosure of the contents of private communications, provide for the admissibility of intercepted communications in evidence and provide for the awarding of damages in favour of a person whose private communication has been wrongfully intercepted.[11]

2. PROVINCIAL PRIVACY ACTS

Four Canadian provinces have enacted legislation making it a tort for any person to violate the privacy of another person, and giving a cause of action to any person whose right of privacy is violated. Such legislation has been enacted in Manitoba,[12] Saskatchewan,[13] British Columbia[14] and Newfoundland.[15]

Section 2 of the Manitoba Act provides that a person commits a tort, actionable without proof of damages, if he "substantially, unreasonably, and without claim of right, violates the privacy of another person." Section 2 of the Saskatchewan Act, section 1 of the British Columbia Act and section 3 of the Newfoundland Act provide that it is a tort for a person to violate the privacy of another person "wilfully and without claim of right."

None of the Privacy Acts contains a definition *per se* of the concept of privacy. However, section 3 of the Manitoba Act sets out certain acts by which the privacy of a person may be violated. These acts include: surveillance of a person or his home by any means, including eavesdropping, watching or following; listening to or recording a person's conversation; the unauthorized use of a person's likeness, name or voice for the purposes of advertising, promotion or any other gain of the user; or the use of a person's letters, diaries or other personal documents without his consent. Section 3 of the Saskatchewan Act and section 4 of the Newfoundland Act provide that any of the foregoing acts, if committed without the consent of the individual involved, is *prima facie* proof of a violation of the privacy of such individual.

Subsection 1(4) of the British Columbia Act provides that without restricting the generality of the offence, privacy may be violated by surveillance or eavesdropping. Section 3 of the British Columbia Act further provides that it is a tort for a person to use the name or portrait of another person for the purpose of advertising or promoting the sale of, or other trading in, any property or service, without that other person's consent.

---

11 Criminal Code, ss. 178.12-178.23.
12 S.M. 1970, c. 74/P. 125.
13 R.S.S. 1979, c. P. 24.
14 R.S.B.C. 1979, c. 336.
15 S. N. 1981, c. 6.

Section 4 of the Manitoba Act sets out certain remedies to which a plaintiff is entitled if the conduct of a defendant is found to constitute an invasion of privacy. These remedies include damages, an injunction, an accounting for profits and an order for delivery up of all articles or documents that have come into the plaintiff's possession as a result of the wrongful act.[16]

Subsection 4(2) of the Manitoba Act provides that in awarding damages for a violation of the privacy of a person, the court shall have regard to the circumstances of the case, including the nature, incidence and occasion of the act; the effect of the violation on the health, welfare, social, business or financial position of the individual; any relationship between the parties; any distress, annoyance or embarrassment suffered by the individual; and the conduct of the individual and the defendant, including any apology or other amends made by the defendant. Factors similar to the foregoing are to be taken into account in determining whether any act is an invasion of privacy of an individual under section 6 of the Saskatchewan Act and subsection 3(2) of the Newfoundland Act. Section 1 of the British Columbia Act provides that the nature and degree of privacy to which a person is entitled in any situation or in relation to any matter is that which is reasonable in the circumstances, having due regard to the lawful interests of others, and that in determining whether the act or conduct of a person is a violation of another's privacy, regard must be given to the nature, incidence and occasion of the act and to any relationship between the parties.

The right of action created by the statutes in Manitoba, Saskatchewan and Newfoundland and the remedies therein are in addition to and not in derogation of any other right of action or remedy that may be available. However, any damages awarded may be taken into account in assessing damages under any other proceeding.[17]

Section 5 of the Manitoba Act sets forth certain defences that may be pleaded by a defendant in an action for invasion of privacy. These defences include that the plaintiff consented to the act; that the defendant did not know or had no reasonable grounds for believing that the act was an invasion of privacy; that the act was reasonable, necessary for, and incidental to the exercise or protection of a lawful right involving the defence of person, property or other interest of the defendant; that the defendant acted under lawful authority; that the act was engaged in by a peace officer or public officer in the course of an investigation and was not disproportionate, was not committed in the course of a trespass, was within the scope of the officer's duties and was reasonably necessary in

16 See also s. 7 of the Saskatchewan Act and s. 6 of the Newfoundland Act.

17 See s. 6 of the Manitoba Act, s. 8 of the Saskatchewan Act, and s. 7 of the Newfoundland Act.

the public interest; and if the violation occurred through a publication of any matter, there were reasonable grounds for believing it was in the public interest, it was privileged or it was fair comment on a matter of public interest. Similar defences are found in section 4 of the Saskatchewan Act, section 2 of the British Columbia Act and section 5 of the Newfoundland Act.

Section 10 of the Manitoba Act provides that an action for an invasion of privacy is to be commenced within two years of the occurrence of the violation, or two years after the person becomes aware or could have become aware through reasonable diligence of its occurrence, but in any event within four years. Section 9 of the Saskatchewan Act provides that any action is to be commenced within two years of the discovery of the alleged violation of privacy. Section 10 of the Newfoundland Act provides that no action lies after the expiration of two years when the violation of privacy first became known or should have become known, and in any case, after the expiration of seven years from the date the violation occurred. The British Columbia Act is silent as to when an action is to be commenced.

Section 10 of the Saskatchewan Privacy Act, section 5 of the British Columbia Act and section 11 of the Newfoundland Act provide that a right of action for invasion of privacy is extinguished by the death of the individual whose privacy is alleged to have been violated. The Manitoba statute is silent as to whether such right of action survives the individual's death.

The foregoing review indicates that in four Canadian provinces the violation of the privacy of any person gives rise to a statutory cause of action. In the absence of any comprehensive definition as to what is meant by privacy, it would appear that the unauthorized collection of personal information which may then be kept in a computerized data bank would not by itself constitute an invasion of an individual's right of privacy. However, if any acts referred to in the various statutes such as eavesdropping or watching were committed as the means of gathering information, such acts would come within the statutes. Similarly, a statutory invasion of privacy may occur if incorrect personal information is disseminated or if such information is used for a purpose other than that for which it is collected without the consent of the person affected.

3. CONSUMER REPORTING ACTS

A number of Canadian provinces have enacted legislation governing the collection and use of information about private individuals which is intended to be provided to third parties for a consideration. Such legislation has been enacted in Ontario,[18] British Columbia,[19] Manitoba,[20] Newfoundland,[21] Nova Scotia,[22] Prince Edward Island[23] and Saskatchewan.[24]

In general, the various Consumer Reporting Acts provide that no person shall conduct or act as a consumer reporting agency unless he is registered or holds a licence under the applicable legislation.[25] However, in the case of Manitoba, a "personal reporting agency" is not required to obtain a special licence or to be registered under the Act.

Subsection 1(*a*) of the Ontario Act defines a "consumer reporting agency" as a person who for gain or profit or on a regular co-operative non-profit basis furnishes "consumer reports". A "consumer report" is defined to mean a written, oral or other communication by a consumer reporting agency of "credit information" or "personal information", or both, pertaining to a consumer for consideration in connection with certain specified purposes. These purposes, which represent the intended uses for which a consumer report may be furnished, are set out in paragraph 8(1)(*d*) of the Act, and include use in connection with the collection of a debt; use in connection with a tenancy agreement; use for employment purposes; use in connection with the underwriting of insurance; use to determine a consumer's eligibility for any matter under a statute or regulation; or other use for which a person has a direct business need for information in connection with a business or credit transaction involving a consumer.

The term "credit information" is defined to mean information about a number of characteristics of a consumer, namely, name, age, occupation, place of residence, previous places of residence, marital status, spouse's name and age, number of dependents, particulars of education or professional qualifications, places of employment, previous places of employment, estimated income, paying habits, outstanding debt obligations, cost of living obligations and assets. The term "personal information" is defined to mean information other than credit information about a consumer's character, reputation, health, physical or personal characteristics or mode of living or about any other matter concerning the consumer.[26] The term "consumer" is defined to mean a natural person but not including a person

18 See Consumer Reporting Act, R.S.O. 1980, c. 89.
19 See Credit Reporting Act, R.S.B.C. 1979, c. 78.
20 See the Personal Investigations Act, S.M. 1971, c. 23/P. 33.
21 See the Consumer Reporting Agencies Act, S. N. 1977, c. 18.
22 See Consumer Reporting Act, S.N.S. 1973, c. 4/C-49.
23 See Consumer Reporting Act, R.S.P.E.I. 1974, c. C-18.
24 See the Credit Reporting Agencies Act, R.S.S. 1978, c. C-44.
25 See s. 3 of the Ontario Act, s. 3 of the British Columbia Act, s. 10 of the Newfoundland Act, s. 4 of the Nova Scotia Act, s. 3 of the Prince Edward Island Act and ss. 3-4 of the Saskatchewan Act.
26 Consumer Protection Act, R.S.O. 1980, c. 89, paras. (1)(1)(*d*) and (*j*).

engaging in a transaction in the course of carrying on a business, trade or profession.[27] Similar definitions are set forth in the legislation of other provinces.

In general, an applicant is entitled to registration or renewal of registration as a consumer reporting agency except on certain specified grounds, such as where the applicant cannot reasonably be expected to be financially responsible in the conduct of his business; the past conduct of the applicant affords reasonable grounds for belief that he will not carry on business in accordance with law and with integrity and honesty; or the applicant is, or if registered would be, in contravention of the Act or regulations.[28]

In the case of Saskatchewan, the Registrar may grant a licence where in his opinion the applicant is suitable to be licensed and the proposed licensing is not for any reason objectionable. However, the Registrar may refuse to grant a licence if after investigation he is for any reason of the opinion that the applicant should not be granted a licence.[29] Section 11 of the Saskatchewan Act also sets out certain additional grounds for suspending or cancelling a licence, such as where the licensee has violated any provision of the Act, had made a material misstatement in any application for the licence or in any information submitted to the Registrar, is guilty of any misrepresentation, fraud or dishonesty, or has demonstrated his incompetency, unfitness, or untrustworthiness to operate or act as a credit reporting agency.[30]

The various provincial Acts impose restrictions on the circumstances in which information in the files of a consumer reporting agency may be disclosed. For example, section 8 of the Ontario Act provides that no such information shall be furnished except: (a) in response to a court order; (b) in accordance with written instructions from the consumer to whom the information relates; (c) in response to any order or direction made under the Act; or (d) in a consumer report given to a person who it has reason to believe will use the information for specified purposes, namely, those for which the agency is entitled to collect any information. Similar provisions are found in section 10 of the British Columbia Act, section 5 of the Manitoba Act, section 9 of the Nova Scotia Act, section 8 of the Prince Edward Island Act and section 19 of the Newfoundland Act.

Section 9 of the Ontario Act requires every consumer reporting agency to adopt all procedures reasonable for ensuring accuracy and fairness in the contents of its consumer reports. Subsection 9(3) also provides that

27 *Ibid.*, para. 1(1)(*a*).

28 See s. 4 of the Ontario Act, ss. 5-6 of the British Columbia Act, s. 11 of the Newfoundland Act, s. 5 of the Nova Scotia Act and s. 4 of the Prince Edward Island Act.

29 The Credit Reporting Agencies Act, R.S.S. 1978, c. C-44.

30 *Ibid.*, s. 11.

certain information may not be included in a consumer report. Specifically, the report may not include any credit information based on evidence that is not the best evidence reasonably available; any unfavourable personal information unless reasonable efforts have been made to corroborate the evidence on which such information is based; information with respect to such matters as judgments, bankruptcy, non-payment of taxes, criminal convictions, writs or other adverse items of information where more than seven years have expired from the date of any such occurrence; any information as to race, creed, colour, sex, ancestry, ethnic origin or political affiliation; or any information given orally in the consumer report unless the content of the oral report is recorded in the file. Any consumer reporting agency is also required to maintain in its file all the material and information which the subject is entitled to have disclosed to it under section 11 of the Act.[31]

Section 10 of the Ontario Act sets out information which a consumer is entitled to obtain in respect of the activities of a consumer reporting agency. Specifically, upon request, a consumer is entitled to be informed whether a consumer report has been or is to be referred to in connection with a specified transaction, and if so, the name and address of the agency supplying the report. In addition, before any person procures or causes an agency to prepare a report containing *personal* information, he must notify the consumer of such fact in writing before the report is requested. A person proposing to extend credit to a consumer and who refers or will be referring to a consumer report containing *credit* information only must so notify the consumer at the time of the application for credit. A person extending credit to a consumer must also not divulge to other credit grantors or a consumer reporting agency any personal information respecting the consumer except with the consent of the consumer or on his referral, unless the person notified the consumer in writing at the time of the application of the credit of his intention to do so. In addition, where any adverse action, such as the denial of a benefit is taken in relation to a consumer because of information received from a consumer reporting agency, the user of such information must deliver notice of that fact to the consumer who is entitled, upon request, to be notified of the nature and source of the information or the name and address of the consumer reporting agency that supplied such information.[32]

Section 11 of the Ontario Act requires every consumer reporting agency to disclose to any consumer who makes a written request for the same, any information pertaining to the consumer held by the agency, the

---

31 For similar restrictions in other provinces see s. 11 of the British Columbia Act, s. 4 of the Manitoba Act, s. 20 of the Newfoundland Act, s. 10 of the Nova Scotia Act, s. 9 of the Prince Edward Island Act, and s. 18 of the Saskatchewan Act.

32 Consumer Reporting Act, R.S.O. 1980, c. 89, s. 10.

sources of any credit information, the names of the recipients of any consumer report and copies or particulars of any such reports provided within a specified time period. The consumer is also entitled to be informed of his right to protest any information contained in the file.[33] Section 12 of the Ontario Act further provides that where a consumer disputes the accuracy or completeness of any item of information, the agency is required within a reasonable time to use its best endeavours to confirm or complete the information and must correct, supplement or delete the information in accordance with good practice. Section 12 also provides for the corrected information to be made available to persons who received a consumer report within specified time periods. The consumer reporting agency may also be ordered by the Registrar to amend or delete any information or restrict or prohibit the use of any information that is inaccurate or incomplete, or that does not comply with the Act.[34]

The various provincial Acts also contain detailed provisions with respect to the investigation of complaints and enforcement. For example, in the case of the Ontario Act, where the Registrar receives a written complaint in respect of a consumer reporting agency and if he so directs in writing, the agency is required to furnish the Registrar with such information respecting the matter complained of as the Registrar requires.[35] The Registrar is authorized to enter upon or to authorize another person to enter upon the business premises of the agency to make an inspection in relation to the complaint.[36] Section 16 of the Ontario Act also provides that the Minister may order an investigation into any matter to which the Act applies, while the Director under the Act may appoint any person to conduct an investigation if he believes on reasonable and probable grounds that any person has contravened the Act or regulations or

33 *Ibid.*, s. 11.

34 *Ibid.*, s. 13. For similar restrictions with respect to notices to be given to the consumer, disclosure of information to be provided to a consumer, and the rights of a consumer to obtain a correction of any incorrect information held by an agency, see ss. 12-17 of the British Columbia Act, ss. 6-11 of the Manitoba Act, ss. 23-24 of the Newfoundland Act, ss. 11-14 of the Nova Scotia Act, ss. 10-13 of the Prince Edward Island Act and ss. 21-26 of the Saskatchewan Act. In the case of the Saskatchewan Act, there is a further requirement that where a consumer disputes the completeness or accuracy of any information contained in the file of a credit reporting agency but the agency is of the opinion that the information in the file respecting the consumer is reasonably accurate and should not be deleted, the consumer is entitled to file a brief statement setting forth the nature of his dispute; a copy of such statement is to be provided in any subsequent credit report containing the information in respect of which the dispute was filed. In addition, where any information held by a credit reporting agency respecting a consumer is deleted or a statement of dispute is filed, every person to whom a credit report was furnished during the immediately preceding twelve months must be so notified: subss. 25(4) and 25(5) of the Saskatchewan Act.

35 Consumer Reporting Act (Ont.), subs. 13(1).

36 *Ibid.*, subs. 15(3).

committed an offence under the Criminal Code or any other relevant law. The Director may also apply to a judge for an order directing compliance with the Act.[37] Section 22 of the Act also provides that any person who furnishes false information, fails to comply with any order, direction or requirement under the Act or contravenes any provision of the Act, and every director or officer of a corporation who knowingly concurs in such act, is guilty of an offence and is liable to a fine of not more than $2,000, or in the case of a corporation $25,000, or to imprisonment of a term not more than one year.

Similar provisions to those set forth in the Ontario Act with respect to the investigation of complaints and enforcement of the Act are found in sections 18-25 of the British Columbia Act, sections 12-19 of the Manitoba Act, sections 17-23 of the Nova Scotia Act and sections 15-22 of the Prince Edward Island Act. Section 26 of the British Columbia Act also creates a civil remedy in favour of any consumer who has suffered any loss, damage or inconvenience as a result of a contravention of the Act. Such a person may obtain judgment for damages suffered or $100, whichever is greater. In the case of Manitoba, on the other hand, section 16 provides for an exemption from civil liability on the part of any user, personal reporter or personal reporting agency unless such person is or ought to be reasonably aware that part or all of the information in the report is false or misleading or was obtained negligently. Section 25 of the Newfoundland Act authorizes the Registrar to apply to a judge of the Supreme Court requiring any person to comply with the Act or regulations, or any order or direction thereunder. Sections 29-31 of the Saskatchewan Act contain provisions prescribing offences and penalties that may be imposed under the Act.

4. CONCLUSIONS

The foregoing review indicates that various causes of action may be asserted to protect the rights of privacy of Canadian citizens. However, many of these causes of action provide only limited protection, and may have little, if any, application to claims for information privacy, that is, the right of an individual to control the collection, dissemination and use of information about him by others.

In several Canadian provinces, a specific statutory right to privacy has been created. The legislation in these provinces sets out certain acts by which an individual's right of privacy may be violated or which are *prima facie* proof of such violation. They also indicate some factors to guide the court in awarding damages, set out certain defences, and indicate when a cause of action must be brought. Such laws are significant not

37 *Ibid.*, s. 20.

only for the provinces in which they have been enacted, but also for other provinces, since the principles reflected in such legislation may be taken into account in what appears to be a trend toward the recognition of a right of privacy at common law.

This chapter also reviewed consumer reporting legislation enacted by a number of Canadian provinces governing the collection of information about individuals and its provision to third parties for a consideration. This legislation is significant because it represents the only comprehensive set of controls governing the maintenance of data banks, either in computerized or printed form, by the private sector in Canada.

12

Laws Regulating the Transfer of Data and Technology Outside Canada

1. INTRODUCTION

No laws of general application have been enacted in Canada to regulate the transfer or transmission of computerized data outside Canada. However, certain laws of this nature have been enacted to regulate specific industries, such as consumer reporting agencies and banks. In addition, other industries such as insurance companies are subject to certain statutory requirements with respect to the retention of records in Canada. The Foreign Extraterritorial Measures Act enacted by the Canadian Parliament in 1984 authorizes restrictions or prohibitions to be imposed against the disclosure or production of various documents, including computer records, to foreign tribunals. Finally, the Export and Imports Permits Act may also apply in some circumstances to the export of computer technology. The following is a summary of these laws.

2. CONSUMER REPORTING ACTS

As noted above, Consumer Reporting Acts have been enacted by several provinces to regulate businesses that collect personal or credit information concerning individuals and that provide this information for a fee to third parties. Although these Acts do not prohibit the transfer or transmission of information outside Canada, they impose certain requirements with respect to the location of files to be kept of information that may be reported by a consumer reporting agency. For example, subsection 9(2) of the Ontario Consumer Reporting Act[1] provides as follows:

> (2) A consumer reporting agency shall not report,
>
> . . .
>
> (b) any information that is not extracted from information appearing in files stored or collected in a repository located in Canada regardless of whether or not the information was obtained from a source outside

1 R.S.O. 1980, c. 89.

> Canada, except where the consumer report is in writing and contains the substance of any prior information orally acquired that conforms to the requirements of this Act.

The foregoing provision means that in general, any information included in a consumer report would have to be provided from files stored in Canada. The word "file" is defined in subsection 1(*g*) of the Act to mean all information pertaining to a consumer that is required to be retained by a consumer reporting agency, regardless of the manner or form in which information is stored. This definition would appear to include any information recorded in machine-readable form so that a consumer report could not include information retrieved from computer data banks maintained outside Canada. However, this would not prevent the maintenance of duplicate files or records in another country.

The exception set forth in paragraph 9(2)(*b*) appears to mean that a written consumer report may contain information extracted from files located outside Canada only if the information was acquired orally by the consumer reporting agency. However, the exception must be read with paragraph 9(3)(*m*) which provides that a consumer reporting agency shall not include in a consumer report any information given orally unless the content of the oral report is recorded in the file. Since the exception in paragraph 9(2)(*b*) applies only to information *acquired* rather than *given* orally, it would appear that wherever an oral report is to be provided, the information would have to be provided from files located in Canada regardless of how the information was obtained.

Provisions similar to paragraph 9(2)(*b*) of the Ontario Consumer Reporting Act, but without the exception with respect to information orally acquired, have been enacted in subsection 11(2) of the British Columbia Credit Reporting Act, subsection 9(2) of the Prince Edward Island Consumer Reporting Act and subsection 10(2) of the Nova Scotia Consumer Reporting Act.

3. BANK ACT

Sections 155 and 157(4) of the Bank Act[2] impose certain duties on banks carrying on business in Canada with respect to the location for the maintenance of records and the processing of information. Specifically, section 155 requires a bank to prepare and maintain at its head office or such other location in Canada designated by the directors various records including "adequate corporate accounting records". In addition, subsection 157(4) provides as follows:

> (4) A bank shall maintain in Canada

---

2 S.C. 1980-81-82-83, c. 40.

(a) a record showing, for each customer of the bank on a daily basis, particulars of the transactions between the bank and that customer and the balance owing to or by the bank in respect of that customer; and

(2) all registers and other records referred to in subsection (1);

and shall maintain and process in Canada any information or data relating to the preparation and maintenance of such records.

The registers and other records referred to in subsection 157(1) are those required or authorized by the Act to be prepared and maintained by a bank, all of which may be in a bound or loose-leaf form or in a photographic film form, or may be entered or recorded by any system of mechanical or electronic data processing, or any other information storage device that is capable of reproducing any required information in intelligible form within a reasonable time. Registers and records maintained in one form may be converted to another form, whereupon the originals may be destroyed.[3]

The effect of subsection 157(4) is that all registers and records required or authorized to be kept by any bank must be maintained in Canada, and information or data relating to the preparation and maintenance of such records must be both maintained and processed in Canada. However, certain exceptions to these requirements are set forth in subsections 157(5) and (6) of the Bank Act. Specifically, subsection 157(5) provides that subject to subsection (6), a bank may maintain copies of or take extracts from any records or registers referred to in subsection (4) and may further process information or data *relating to such copies or extracts* inside or outside Canada. Subsection 157(6) provides that where copies or extracts of any records or registers are maintained, or information or data relating to such copies or extracts are further processed, outside Canada, the bank must inform the Inspector and provide him with a description of the copies or extracts maintained outside Canada, of the further processing carried on outside Canada and such other information as the Inspector may require. Moreover, if the Inspector at any time is of the opinion that the further processing outside Canada of information relating to copies or extracts is incompatible with the fulfilment of his responsibilities, or he is advised by the Minister that in the Minister's opinion such further processing is not in the national interest, the Inspector must direct the bank to further process information or data relating to such copies or extracts in Canada.

Subsection 157(7) provides that subsection (4) does not apply in respect of records of a bank branch outside Canada or of clients of such a branch.

The scope of subsection 157(4) is somewhat unclear with respect to the right of a bank to *further* process information or data relating to *copies*

3 Bank Act, subs. 157(2).

or extracts of records and registers outside Canada. It would appear that if the form in which information is recorded is changed (for example, from printed to computerized form), and the originals are destroyed, subsequent forms of expression of those records may become originals.[4] Such subsequent forms would not be considered as "copies" processed outside Canada under subsection 157(5) of the Bank Act.

Questions also arise as to the rationale underlying the restrictions against processing data outside Canada and the circumstances in which either the Inspector or the Minister may direct all processing to be carried on in Canada. It would appear that such restrictions are intended to enable the Inspector to exercise his responsibilities for the protection of the interests of depositors, shareholders and creditors. However, such restrictions may reflect "extraneous" considerations such as the protection of a domestic data processing industry. It is also unclear whether such an extraneous factor would be considered by the Minister in determining whether the further processing outside Canada of information relating to copies or extracts would not be in the "national interest" for purposes of subsection 157(6).[5]

4. LEGISLATION PROVIDING FOR RETENTION OF RECORDS IN CANADA

Various Canadian laws require original records and documents kept by certain industries to be maintained in Canada. For example, section 24 of the Foreign Insurance Companies Act,[6] provides that "Every company shall keep at the chief agency in Canada original records and documents sufficient to enable the chief agent to prepare and furnish the required statement of Canadian business. . . ." Although the Act does not define the term "original records and documents" it would appear that copies of any records and documents, including copies in computerized form, could be maintained outside Canada. In addition, information and data with respect to such records and documents could be processed outside Canada. Accordingly, the transformation of records from printed to machine-readable form should not affect the ability to transfer computerized information outside Canada as long as the printed records are maintained in this country.

4 See decision of the Ontario Court of Appeal in *R. v. Bell* (1982), 35 O.R. (2d) 164 (Ont. C.A.) in which it was held that a computer printout became an original record for purposes of the Evidence Act (Canada) once the original materials on which the printout was based were destroyed.

5 For a further discussion of issues arising from section 157, see Pat Learmonth and Yves Desjardins-Siciliano, "Transborder Data Flow and the Canadian Bank Act" (1984), 1 C.C.L.R. 73.

6 R.S.C. 1970, c. I-16.

5. FOREIGN EXTRATERRITORIAL MEASURES ACT

The Foreign Extraterritorial Measures Act[7] empowers the Canadian Government to take certain steps to counteract measures taken by tribunals or governments of other states that may affect international trade and commerce or infringe Canadian sovereignty. In particular, the Act authorizes restrictions to be imposed on the production of records kept in Canada to a foreign tribunal. For purposes of the Act, a "record" is defined to include various types of documents and records, including a machine-readable record, and any other documentary material, regardless of physical form or characteristics, and any copy or portion thereof.

Subsection 3(1) of the Act provides that where in the opinion of the Attorney General of Canada, a foreign tribunal has exercised, is exercising or is proposing to exercise jurisdiction or powers of a kind or in a manner that has affected or is likely to affect adversely significant Canadian interests in relation to international trade or commerce involving a business carried on in Canada, or that otherwise has infringed or is likely to infringe Canadian sovereignty, he may impose certain orders. Specifically, he may prohibit or restrict the production or disclosure to the tribunal of any records in Canada or under the control of a Canadian citizen or person resident in Canada, the doing of any act in Canada that would result in the records being produced, disclosed or identified to a foreign tribunal, or the giving of any information relating to the records to or for the purposes of the tribunal by a Canadian citizen or resident.

Subsection 3(2) of the Act provides that where an order is made under subsection 3(1), a tribunal in Canada shall not, for purposes of proceedings before the foreign tribunal, receive a record whose production has been prohibited, or a record whose production has been restricted so as to result in a contravention of the order. The Attorney General may also apply to a superior court for a warrant to require the seizure of any records and their delivery to the court or some other person for safekeeping.

Section 5 of the Act also authorizes the Attorney General, with the concurrence of the Secretary of State for External Affairs, to require a notice to be given of certain measures taken by a foreign state or tribunal, or to impose an order prohibiting any person from complying with such measures. Section 8 of the Act also provides that the Attorney General may declare certain foreign judgments in proceedings under any antitrust law not to be enforced or recognized in Canada.

It is likely that the Foreign Extraterritorial Measures Act will be invoked only in extraordinary circumstances to prohibit the disclosure of records held in Canada to foreign tribunals. However, the Act is broad enough to apply to records held in computerized form and must be taken

---

7 S.C. 1984, c. 49.

into account where any such records are to be adduced in evidence in proceedings before foreign tribunals. Questions may arise as to the enforceability of orders made under the Act where duplicate copies of records originating in Canada are kept in another country, or where a non-Canadian has access by data transmission facilities to records stored in a computer facility in Canada.

6. EXPORT AND IMPORT PERMITS ACT

The Export and Import Permits Act[8] authorizes the Governor in Council to establish a list of goods, to be called an Export Control List, including any article the export of which he deems it necessary to control for any one of several purposes. These purposes include ensuring that goods of a military nature will not be made available to any destination wherein their use might be detrimental to the security of Canada, and ensuring that there is an adequate supply of the article in Canada for defence or other needs. Section 4 authorizes the Governor in Council to establish a list of countries, to be called an Area Control List, including therein any country to which he deems it necessary to control the export of any goods. Section 5 of the Act also authorizes the establishment of an Import Control List including any goods the import of which the Governor in Council deems it necessary to control for any one of several specified purposes.

The items which are subject to export control under the Export and Import Permits Act are set forth in the Export Control List.[9] The list includes various types of electronic computers and related equipment which come within Group 3, entitled "General Purposes Industrial Machinery and Electronic Devices". In addition, Group 10, "Miscellaneous Goods and Materials" includes item 10003, being "technical data in material form", including technical drawings, photographic negatives and prints, recordings, design data and technical and operating manuals that can be used in the design, production, operation, or testing of equipment and materials described in Groups 3 to 9 unless the same are publicly available. The procedures to be followed to obtain an application for permits under the Act are set forth in the Export Permit Regulations.[10]

The Area Control List to which the export of any goods is controlled under section 4 of the Act includes Albania, Bulgaria, Czechoslovakia, Democratic People's Republic of Korea, German Democratic Republic and East Berlin, Hungary, Libya, Mongolia, Poland, Romania, Socialist Republic

8 R.S.C. 1970, c. E-17, as amended.

9 C.R.C. 1978, c. 601.

10 C.R.C. 1978, c. 602. For a more detailed analysis of the procedures for obtaining export permits, see J. Hefferman, "Export Control of Computer Goods" (1986), 3 C.C.L.R. 170.

of Vietnam and Union of Soviet Socialist Republics.[11] Although exports of any goods to these countries requires an export permit, some products may be exported under a General Export Permit.

The Export and Import Permits Act must be reviewed to determine whether any permits are required for the export of computer equipment and related items, including disks with computer programs or data.[12] However, the technical data referred to in item 10003 of the Export Control List must be in material form and accordingly, the Act and regulations do not appear to be applicable to the transmission of data from Canada to other countries by electronic means.

---

11 SOR 81-543, as amended.

12 The Export Control List also prohibits the export of goods originating in the United States unless they have been further processed or manufactured outside the United States resulting in a substantial change in the value, form and use of those goods: see Hefferman, *op. cit.*, at 173. The export from Canada of computer products that originate in or that contain parts that originate in the United States, or that are the "direct product" of technical data obtained from the United States, may also be subject to U.S. export control laws and regulations: see Joseph H. Price, "U.S. Export Controls — Their Effect on International Transactions (with focus on Computers and Software)", paper presented at Canadian Computer Law Association Conference, Toronto, October 5-6, 1986.

13

International Developments in Privacy, Data Protection and Transborder Data Flow: The Implications for Canada

1. INTRODUCTION

As discussed in earlier chapters of this Part, no laws of general application have been enacted in Canada setting forth the standards to be met in the establishment of data banks within the private sector. In addition, there are no such laws governing the transmission of data to or from Canada, although laws regulating the transfer, processing and/or retention of data outside Canada within particular industries have been enacted.

Recent international developments in the area of data protection and transborder flows directly affect Canada in several respects. First, various international agreements and covenants, as well as legislation enacted by other countries, set out certain standards to be met in the collection, use and transmission of personal information which over a period of time may acquire the status of customary international law in their application to governments. Second, insofar as the Canadian private sector fails to comply with international standards governing the collection, use or dissemination of personal information, either based on legislation or as a result of voluntary compliance, such failure may affect the free flow of information into Canada from other countries. Finally, the imposition of unduly restrictive controls not related to privacy protection on the transfer of data from Canada to other countries may result in reciprocal controls being imposed on the transfer of data to Canada from such other countries. Accordingly, this section will review some recent international developments in data protection and transborder data flows and consider their implications for Canadian law and practice.

2. INTERNATIONAL AGREEMENTS GOVERNING DATA PROTECTION AND TRANSBORDER DATA FLOW

Several international conventions and agreements have been adopted dealing with the protection of personal data and transborder data flows.

First, in September, 1980, the Organization for Economic Co-operation and Development (the "OECD"), consisting of some 19 European countries together with Canada, United States, Japan, New Zealand and Australia, adopted the Guidelines on the Protection of Privacy and Transborder Flows of Personal Data (hereinafter called the "OECD Guidelines").[1] The OECD Guidelines were adopted by 19 member countries on September 23, 1980 and subsequently by the remaining five member countries. A further Declaration on Transborder Data Flows was adopted by the governments of the OECD Member countries on April 11, 1985.

The Council of Europe Convention for the Protection of Individuals with regard to the Automatic Processing of Personal Data (the "Council of Europe Convention")[2] was opened for ratification in January, 1981 and came into force on October 1, 1985, being three months after the date on which it was ratified by five signatory states. The first five states to sign the Convention were Sweden (September, 1982), France (May, 1983), Spain (January, 1984), Norway (February, 1984) and West Germany (June, 1985). Ratification by any state can take place only when a signatory country has adopted domestic legislation to reflect the Convention. The Council of Europe Convention will be ratified by the United Kingdom when the provisions of the United Kingdom Data Protection Act come into full force.[3]

Unlike the Council of Europe Convention, the OECD Guidelines do not have any binding legal effect, but instead represent a voluntary code with which member states are expected to comply. Both the Convention and the Guidelines provide for the principle of free flow of data between countries, subject to certain basic standards being adopted to provide for the protection of personal data in relation to both the public and the private sectors. The principle that member countries should facilitate transborder data flows was also set forth in the OECD Declaration adopted in April, 1985. While the Council of Europe Convention applies to personal data (that is, data about identified or identifiable individuals) held in computerized form, the OECD Guidelines and Declaration deal with such data held in any form, whether printed or computerized.

In addition to the OECD and the Council of Europe, a number of other international organizations have been concerned with legal, technical and economic aspects of international data transmissions, particularly by electronic means. Some of these organizations have issued policy state-

1 Guidelines on the Protection of Privacy and Transborder Flows of Personal Data, Annex to Recommendation of the Council, September 23, 1980, Organization for Economic Co-operation and Development, Paris, 1981.

2 Convention for the Protection of Individuals with regard to Automatic Processing of Personal Data (January 28, 1981), Eur. T.S. No. 108.

3 See N. Frome, "Data Protection Laws in the EEC", Paper presented at Pacific Rim Symposium on Technology and the Law, Singapore, October, 1985.

ments on transborder data flows while other organizations have been concerned with the circumstances in which documents transmitted by electronic means should be accepted, the means by which they may be authenticated and their use in court proceedings. Some of the international organizations which have been active in this area are the United Nations Commission on International Trade Law ("UNCITRAL"), the Hague Conference on Private International Law, the International Chamber of Commerce, the International Maritime Committee, the International Law Association, the Economic Commission for Europe, the Inland Transport Committee, the International Maritime Organization, the International Civil Aviation Organization, the International Rail Transport Committee and the Customs Co-operation Council.[4]

It is beyond the scope of this text to review the work of all of these international organizations in the area of transborder data flow and the use of automated data. However, the following sections will review the conventions and agreements of general application which have been adopted by the OECD and the Council of Europe.

(a) OECD Guidelines on the Protection of Privacy and Transborder Flows of Personal Data

The preamble to the Recommendation of the Council of Europe, to which the OECD Guidelines are annexed, notes the common interest of member countries in protecting privacy and individual liberties and in reconciling fundamental but competing values such as privacy and the free flow of information. The preamble recognizes further that transborder flows of personal data contribute to economic and social development and such flows may be hindered by domestic legislation concerning privacy protection. The Recommendation calls upon member countries to take four steps to advance the free flow of information and to avoid the creation of unjustified obstacles to the development of economic and social relations among member countries, namely:

(i) to take into account in their domestic legislation the principles concerning the protection of privacy and individual liberties set forth in the Guidelines;
(ii) to endeavour to remove or to avoid creating, in the name of privacy protection, unjustified obstacles to transborder flows of personal data;
(iii) to co-operate in the implementation of the Guidelines; and
(iv) to agree on specific procedures of consultation and co-operation for the application of the Guidelines.

---

4 For a review of the activities of these organizations, see "Legal Implications for Automatic Data Processing," Report of the Secretary-General, United Nations Commission on International Trade Law, 19th Session, A/CN.9/279, April 7, 1986.

The Guidelines themselves state that they apply to personal data, whether in the public or private sector, which, because of the manner in which they are processed, or because of their nature or the context in which they are used, pose a danger to privacy and individual liberties. The term "personal data" is defined to mean any information relating to an identified or identifiable individual (data subject). Although the Guidelines apply to personal data generally, paragraph 3 provides that they should not be interpreted as preventing: (a) different protective measures being taken to different categories of personal data according to their nature and the context in which they are collected, stored, processed or disseminated; (b) the exclusion from the application of the Guidelines of personal data which do not obviously contain any risk of privacy and individual liberties; or (c) their application only to automatic processing of personal data.

The Principles of National and International Application set forth in the Guidelines are intended to represent minimum standards, which are capable of being supplemented by additional measures for the protection of privacy and individual liberties. In addition, the Guidelines provide that any exceptions to such Principles, including those relating to national sovereignty, national security and public policy, should be as few as possible and made known to the public.

The eight Basic Principles of National Application, as set forth in Paragraphs 7-14 of the Guidelines, are as follows:

> <u>Collection Limitation Principle</u>
> 7. There should be limits to the collection of personal data and any such data should be obtained by lawful and fair means and, where appropriate, with the knowledge or consent of the data subject.
>
> <u>Data Quality Principle</u>
> 8. Personal data should be relevant to the purposes for which they are to be used, and, to the extent necessary for those purposes, should be accurate, complete and kept up-to-date.
>
> <u>Purpose Specification Principle</u>
> 9. The purposes for which personal data are collected should be specified not later than at the time of data collection and the subsequent use limited to the fulfilment of those purposes or such others as are not incompatible with those purposes and as are specified on each occasion of change of purpose.
>
> <u>Use Limitation Principle</u>
> 10. Personal data should not be disclosed, made available or otherwise used for purposes other than those specified in accordance with Paragraph 9 except:
> (a) with the consent of the data subject; or
> (b) by the authority of law.

Security Safeguards Principle
11. Personal data should be protected by reasonable security safeguards against such risks as loss or unauthorized access, destruction, use, modification or disclosure of data.

Openness Principle
12. There should be a general policy of openness about developments, practices and policies with respect to personal data. Means should be readily available of establishing the existence and nature of personal data, and the main purposes of their use, as well as the identity and usual residence of the data controller.

Individual Participation Principle
13. An individual should have the right:
- (a) to obtain from a data controller, or otherwise, confirmation of whether or not the data controller has data relating to him;
- (b) to have communicated to him, data relating to him
  - (i) within a reasonable time;
  - (ii) at a charge, if any, that is not excessive;
  - (iii) in a reasonable manner; and
  - (iv) in a form that is readily intelligible to him;
- (c) to be given reasons if a request made under subparagraphs (a) and (b) is denied, and to be able to challenge such denial; and
- (d) to challenge data relating to him and, if the challenge is successful, to have the data erased, rectified, completed or amended.

Accountability Principle
14. A data controller should be accountable for complying with measures which give effect to the principles stated above.

The Basic Principles of International Application, as set forth in Part 3 of the Guidelines, are first, that member countries should take into consideration the implications for other Member countries of domestic processing and re-export of personal data; second, Member countries should take all reasonable and appropriate steps to ensure that the transborder flows of personal data, including transit through a Member country are "uninterrupted and secure"; third, Member countries should refrain from restricting transborder flows of personal data to another Member country except where the latter does not substantially observe the Guidelines or where the re-export of such data would circumvent its domestic privacy legislation; and fourth, that Member countries should avoid developing laws, policies and practices in the name of the protection of privacy and individual liberties, which would create obstacles to transborder flows of personal data exceeding requirements for such protection.

Part 4 of the Guidelines, entitled "National Implementation", calls on Member countries to establish procedures and institutions for the protection of privacy and individual liberties in respect of personal data. In particular, Members are called upon to endeavour to adopt appropriate domestic legislation, encourage self-regulation, such as by codes of conduct, provide for reasonable means for individuals to exercise their rights, provide for appropriate sanctions and remedies in respect of the failure to comply with the principles set out in the Guidelines, and ensure that there is no unfair discrimination against data subjects. Finally, the Guidelines provide for various measures of international co-operation, including that Member countries should advise other countries of the details for the observance of the principles, ensure that procedures adopted for transborder flows of personal data and the protection of privacy and individual liberties are simple and compatible with those of other Member countries, establish procedures for the exchange of information related to the Guidelines and mutual assistance in procedural and investigative matters, and work towards the development of principles to govern the applicable law in the case of transborder flows of personal data.

The OECD Guidelines were adhered to by Canada at the OECD Council meeting held in July, 1984. However, the only legislation enacted by the federal Parliament and by several provinces deals with the protection of personal data in the public sector and no corresponding legislation dealing with the private sector has been adopted. In accordance with the principles set forth in Part 4 of the Guidelines with respect to the procedures for National Implementation, Canada has encouraged and supported self-regulatory mechanisms for the enforcement of the principles in the Guidelines, including the adoption of voluntary codes of conduct. In this regard, the Canadian Government has requested some 135 Canadian companies to commit themselves voluntarily to the OECD Guidelines, and has indicated that the failure of such a voluntary approach could result in the need for government regulation.[5]

(b) OECD Declaration on Transborder Data Flows

Subsequent to the adoption of the OECD Guidelines on the Protection of Privacy and Transborder Flows of Personal Data, the OECD conducted further studies on issues relating to transborder data flow, including some of the economic issues that arise with respect to the transfer of information from one jurisdiction to another. In April, 1985, the member countries of the OECD adopted a further Declaration on Transborder Data Flows that expressed the wish of such countries to promote access to information, to avoid the creation of unjustified barriers to the exchange of information, and to develop common approaches and consultative mechanisms to ensure

5 See M. Rankin, "TBDF: Protectionism, Privacy, Politics and the Pacific Rim" (1985), 2 C.C.L.R. 114 at 115.

"harmonized solutions" to issues relating to transborder data flows. The following is a text of the Declaration adopted by member countries.

DECLARATION ON TRANSBORDER DATA FLOWS
(Adopted by the Governments of OECD
Member countries on 11th April 1985)

Rapid technological developments in the field of information, computers and communications are leading to significant structural changes in the economies of Member countries. Flows of computerised data and information are an important consequence of technological advances and are playing an increasing role in national economies. With the growing economic interdependence of Member countries, these flows acquire an international dimension, known as Transborder Data Flows. It is therefore appropriate for the OECD to pay attention to policy issues connected with these transborder data flows.

This declaration is intended to make clear the general spirit in which Member countries will address these issues.

In view of the above, the GOVERNMENTS OF OECD MEMBER COUNTRIES:

Acknowledging that computerised data and information now circulate, by and large, freely on an international scale;

Considering the OECD Guidelines on the Protection of Privacy and Transborder Flows of Personal Data and the significant progress that has been achieved in the area of privacy protection at national and international levels;

Recognising the diversity of participants in transborder data flows, such as commercial and non-commercial organizations, individuals and governments, and recognising the wide variety of computerised data and information, traded or exchanged across national borders, such as data and information related to trading activities, intra-corporate flows, computerised information services and scientific and technological exchanges;

Recognising the growing importance of transborder data flows and the benefits that can be derived from transborder data flows; and recognising that the ability of Member countries to reap such benefits may vary;

Recognising that investment and trade in this field cannot but benefit from transparency and stability of policies, regulations and practices;

Recognising that national policies which affect transborder data flows reflect a range of social and economic goals, and that governments may adopt different means to achieve their policy goals;

Aware of the social and economic benefits resulting from access to a variety of sources of information and of efficient and effective information services;

Recognising that Member countries have a common interest in facilitating transborder data flows, and in reconciling different policy objectives in this field;

Having due regard to their national laws do hereby DECLARE THEIR INTENTION TO:

(a) Promote access to data and information and related services, and avoid the creation of unjustified barriers to the international exchange of data and information;

(b) Seek transparency in regulations and policies relating to information, computer and communications services affecting transborder data flows;

(c) Develop common approaches for dealing with issues related to transborder data flows and, when appropriate, develop harmonized solutions;

(d) Consider possible implications for other countries when dealing with issues related to transborder data flows.

Bearing in mind the intention expressed above, and taking into account the work being carried out in other international fora, the GOVERNMENTS OF OECD MEMBER COUNTRIES,

Agree that further work should be undertaken and that such work should concentrate at the outset on issues emerging from the following types of transborder data flows:

(i) Flows of data accompanying international trade;

(ii) Marketed computer services and computerised information services; and

(iii) Intra-corporate data flows.

The GOVERNMENTS OF OECD MEMBER COUNTRIES AGREED to *co-operate* and *consult* with each other in carrying out this important work, and in furthering the objectives of this Declaration.

The foregoing indicates that in contrast to the OECD Guidelines, which emphasize a concern for the protection of personal privacy, the 1985 OECD Declaration refers to the economic and social benefits resulting from access to information and the common interest of Member countries in facilitating the free flow of information from one country to another. The Declaration also emphasizes the need to examine issues emerging from flows of data relating to international trade, computerized services and intra-corporate data flows. It remains to be seen whether the principles set out in the Declaration will be reflected in the removal of any barriers to the free flow of information among Member countries or to countries which are not members of the OECD.

(c) Council of Europe Convention for the Protection of Individuals With Regard to the Automatic Processing of Personal Data

The Council of Europe Convention has been signed by all member countries but is binding only on those countries by which the Convention has been ratified. The Convention does not provide for any sanctions for non-compliance nor does it provide individual citizens of any signatory country with any direct recourse.

The provisions relating to transborder data flows are set forth in Article 12 of the Convention, which applies to the transfer across national borders, by whatever medium, of personal data undergoing automatic processing or collected for such purpose. Paragraph 2 of Article 12 provides that any party shall not, for the sole purpose of the protection of privacy, prohibit or subject to special authorization transborder flows of personal data going to the territory of another Party. The reference to "sole purpose" in paragraph 2 would indicate that the imposition of restrictions on flows of personal data for purposes other than the protection of privacy, such as relating to national security or economic sovereignty, would not be a contravention of the Convention. Paragraph 3 sets forth certain circumstances in which a Party may derogate from the provisions of Paragraph 2, namely, where regulations have been adopted by such Party for certain categories of personal data and another Party has failed to adopt regulations to provide equivalent protection; and where the transfer is to be made to a non-contracting state through the territory of another Party to circumvent the regulations of the Party from which the transfer originates.

Articles 5-8 of the Convention set out the basic principles for the protection of data, covering such matters as the quality of data; the security of data against destruction, loss, unauthorized access, alteration or disclosure; the rights of data subjects to gain access to data; and the establishment of certain categories of data which are entitled to special protection such as data relating to religious beliefs. Articles 13-17 provide for co-operation and mutual assistance among contracting countries, while Articles 18-20 provide for the creation of a consultative committee which is to meet every two years to formulate ways of implementing or improving the Convention. The consultative committee will express opinions at the request of a member state concerning the application of the Convention and reports to the Council of Ministers.[6]

The Convention allows for accession by non-member countries and permits non-member countries as well as Council of Europe members which have not ratified the Convention to appoint an observer to the consultative committee.

3. DATA PROTECTION LEGISLATION OF OTHER COUNTRIES

As of September, 1985, data protection laws have been enacted by some thirteen countries, namely, Austria, Canada, Denmark, France, Germany, Iceland, Israel, Luxemburg, New Zealand, Norway, Sweden, the United Kingdom and the United States. In addition such legislation was under active consideration by an additional fifteen countries, namely, Australia, Belgium, Brazil, Finland, Greece, Ireland, Italy, Japan, Mexico, Netherlands, Portugal, Spain, Turkey, Switzerland and Yugoslavia.[7]

In general, data protection laws incorporate a number of common principles designed to protect the rights of "data subjects" in respect of the collection of personal data. Among these principles are that there should be prohibitions against the collection of certain types of data (for example, information about race or religious beliefs); data should be collected only for specified purposes, and in such a manner as to ensure that it is accurate and relevant; data should be stored only for the purposes specified at the time of collection and only as long as required for such purposes; data should be used only for the purposes for which it was originally collected; personal data should not be disclosed to third parties without the consent of the data subjects or except as authorized by law; data subjects should have the right to learn what information is kept about them and to cause any necessary corrections to be made; and an official should be charged with responsibility for ensuring compliance with the legislation.[8] At the same time, data protection legislation may differ in certain key areas, such as whether it applies to only the public or both the public and private sectors, whether it applies to legal persons as well as natural persons, and whether it applies to information kept in printed or manual form as well as that held in computerized form.[9]

The data protection laws of various countries also differ in the mechanisms established to enable data subjects to enforce their rights, and the nature and scope of the controls established by governments on the

6 See N. Frome, "Data Protection Law in the EEC", Paper presented at Pacific Rim Symposium on Technology and the Law, Singapore, October, 1985.

7 See N. Frome, *ibid.*; S. Kirk, "Privacy and Data Protection: Implication for Multinationals", Paper presented at the Pacific Rim Symposium on Technology on the Law, Singapore, October, 1985.

8 See S. Kirk, *ibid.*, at 3-4.

9 *Ibid.*, at 4.

collection, use and dissemination of information and on transborder flows of data. For example, unlike Canada and the United States, a number of European countries have enacted legislation to provide for the registration of data banks kept by the private sector and for the imposition of controls on such data banks and the transfer of information to other countries. An example of such legislation is the United Kingdom Data Protection Act, 1984, which received Royal Assent on July 12, 1984. The United Kingdom Act is to come into force in four stages culminating with the proclamation of the provisions relating to enforcement, de-registration and transfer prohibition notices which are to take effect in mid-1987.

The United Kingdom Act sets out eight principles of data protection, dealing with the manner in which data may be obtained, the purposes for which it may be used, the manner in which it is to be retained and kept up to date and accurate, and the rights of data subjects to obtain access to data and to have such data corrected or erased. For purposes of the Act, the word "data" is defined to mean "information recorded in a form in which it can be processed by equipment operating automatically in response to instructions given for that purpose."[10] Accordingly, the Act would not appear to apply to information which is kept in manual or printed form, although such data may be capable of being processed automatically by the use of an optical character reader. The Act is further restricted to apply only to "personal data", that is, data "consisting of information which relates to a living individual who can be identified from the information (or from that and other information in the possession of the data user)."[11]

The Data Protection Act is expressed to apply to any "data user" as well as to any "computer bureau". A data user is defined to mean any person (including a corporation, partnership or government department) who holds data. A person is deemed to hold data if: (a) the data form part of a collection of data processed or intended to be processed on behalf of that person; (b) that person controls the contents and use of the data; and (c) the data are in a form in which they have been or are intended to be processed or in a form into which they have been converted after being so processed and with a view to being further so processed on a subsequent occasion.[12] A person is deemed to be carrying on the business of a computer bureau if he provides other persons with services in respect of data; a person is deemed to provide such services if as agent for other persons he causes data held by them to be processed by equipment operating automatically or he allows other persons the use of equipment in his possession for the processing of data.

10 Data Protection Act, United Kingdom, 1984, subs. 1(2).

11 *Ibid.*, subs. 1(3).

12 *Ibid.*, subs. 1(5).

The United Kingdom Act provides for a number of exemptions with respect to the application of the Act to various categories of data. These exemptions fall into four categories, that is, total exemptions from the Act, exemptions from the provisions of the Act relating to registration of data users and the obligation to provide data subjects with access, exemptions from the provisions relating to subject access, and exemptions from the non-disclosure provisions.[13]

The United Kingdom Act also provides for the establishment of the office of the Data Protection Registrar. The Registrar is required to maintain a register of data users and persons carrying on the business of a computer bureau. The Registrar is also given powers of enforcing compliance with the provisions of the Act and may refuse to register a data user, and may issue enforcement, de-registration or transfer prohibition notices. Such orders may be appealed to the Data Protection Tribunal and appeals from the Tribunal on points of law may be made to a court.

The Act provides for certain offences in respect of the failure to register under the Act, and the failure of a person registered under the Act to hold, use, obtain, disclose or transfer personal data in accordance with the provisions of the Act. The Act also provides for civil remedies whereby any individual may be compensated for damages arising from the inaccuracy of any data or the loss or unauthorized disclosure of personal data. Moreover, a data subject may apply to a court for the correction or erasure of inaccurate data.

The United Kingdom Act authorizes the Registrar to serve a "transfer prohibition notice" in any case where data are to be transferred to a country which has not ratified the Council of Europe Convention and the Registrar is satisfied that the transfer is likely to contravene or lead to a contravention of the data protection principles set out in the Act.[14] In the case of a transfer of data to countries bound by the Convention, the Registrar may issue a prohibition notice only if he is satisfied that a further transfer to a non-Convention country is intended and the further transfer is likely to contravene or lead to the contravention of the data protection principles set out in the Act.[15]

In general, the United Kingdom Act does not apply to data users holding data outside the United Kingdom or to a person carrying on business as a computer bureau outside the United Kingdom. However, section 39 of the Act provides that data are deemed to be held where the data user exercises control over the content and use of data, while a bureau is deemed to carry on its business where it processes data or

13 See Millard, *Legal Protection of Computer Programs and Data* (Toronto: Carswell, 1985), 189-192.

14 Data Protection Act, subs. 12(2).

15 *Ibid.*, subs. 12(3).

allows a data user to use its equipment. Accordingly, if a person in the United Kingdom exercises control over the contents of a data bank located in another country, or if it causes or permits data processing equipment to be used in the United Kingdom, such a person will be subject to the provisions of the Act. Moreover, under section 39(3), where a person not resident in the United Kingdom exercises control over the processing of data in the United Kingdom through a servant or agent, the servant or agent is deemed to exercise the control or provide the services as a principal. Accordingly, the servant or agent in the United Kingdom will be subject to the provisions of the Act.

The data protection laws enacted in the United Kingdom and other European countries have been criticized on the grounds that they are difficult and cumbersome to administer, they place heavy burdens on those involved in data use which may be harmless, and they may be easily contravened through unauthorized transfers of data which cannot be detected. Such laws also lead to increased costs for the public sector as a result of the establishment of new government bureaucracies, and for the private sector as a result of the need to register data banks and to obtain licences or permits to transfer data from one country to another.[16]

The United Kingdom Data Protection Act and corresponding legislation enacted by other European countries are significant for Canada in two respects. First, such legislation may represent a model on the basis of which legislation could be enacted by the federal Parliament or provincial legislatures to regulate the holding of data by the private sector. Second, in the absence of such legislation in Canada, the Registrar under the United Kingdom Act or the corresponding authorities in other jurisdictions may find that in some circumstances, the transfer of data to Canada will result in non-compliance with the data protection principles set out in the Act. Accordingly, the failure of the voluntary approach to compliance with the OECD Guidelines announced by the Canadian Government could affect the transfer of data to Canada from the United Kingdom and other European countries.

---

16 See D.P. Farnsworth, "Data Privacy or Data Protection and Transborder or Transnational Data Flow, An American's View of European Legislation" (1983), 11 Int. Business Lawyer 114.

14

Legal Issues Relating to Electronic Funds Transfer Systems

1. INTRODUCTION

The term electronic funds transfer ("EFT") refers to any transaction involving the transfer of money, or of credits and debits, from one institution to another, or between an institution and a customer, by electronic means. The term may also be used to include various automated banking services and transactions that are ancillary to the transfers of funds. These include cash withdrawals and deposits, credit card authorizations, cheque verification, billing operations, point of sale payments, clearing house services, balance inquiries, shifting of funds from one account to another, direct payroll deposits and pre-authorized periodic payments.[1]

The trend to what has been described as a "cashless society" raises a number of legal issues. Some of these issues include when a banking transaction is deemed to be completed, the allocation of risk in the event of error or loss, the exposure of financial institutions or customers to loss as a result of the insolvency of an institution, the need for security in the means of communications, the risk of loss of personal privacy resulting from the collection of extensive information relating to cashless transactions, and the allocation of the cost of effecting an electronic funds transfer. To date, no major changes have been made to Canadian laws and regulations relating to banking and bills of exchange to deal with such issues, and such laws reflect in large measure a paper-based payment system.

2. THE EVOLUTION OF AUTOMATED BANKING AND ELECTRONIC FUNDS TRANSFER SYSTEMS

Banking practice in Canada during the past quarter century has been marked by the increasing automation of many banking functions. A major step in such evolution was the introduction of the MICR (magnetic ink character readers) system whereby every cheque drawn on any financial institution in Canada is encoded with a series of numbers representing

1 Soma, *Computer Technology and the Law* (McGraw-Hill, 1983) at 328.

the institution, the branch and the account. This system has permitted the increasing mechanization of cheque handling and processing. During the 1960s and early 1970s, there was also an increasing automation of various other information handling tasks such as the proofing and clearing of cheques, posting of credits and debits to customers' accounts, term deposit accounting, preparation of general ledgers for each day's transactions, preparation of branch returns, foreign exchange transactions, teller cash controls, money orders, letters of credit, travellers' cheques and securities handling.[2]

An important step in the evolution of automated banking in Canada involved the establishment of regional branch data preparation centres. Such centres received cheques and other documents from bank branches and carried out the necessary processing to determine what credit and debit adjustments should be made to customers' accounts. During the 1970's, the various national banks also established on-line banking networks which enabled branches to have direct access to deposit accounting, enquiry and other bank services. Financial institutions have also developed a number of new automated customer services, such as cash management, account reconciliation service, payroll service, credit authorizations, financial analysis, OCR data preparation, credit evaluation and pre-authorized payment and direct deposit plans.

An important component of the modern EFT system involves automated teller machines ("ATM's"). ATM's, which may be located at bank branches or at retail outlets, enable customers to gain direct access to bank accounts to obtain funds or to initiate other transactions. An ATM network may operate either off-line (that is, the terminal merely records each transaction on tape or other machine-readable media which are then physically transferred to the bank for processing), or on-line (whereby the terminal is directly linked to the bank's computer which processes each transaction instantaneously). The operation of an ATM requires the allocation to each customer of a personal identification number ("PIN") which is used in conjunction with a pass card to gain access to an account. It has been estimated that more than one-third of all bank customers in Canada rely on automated banking machines to do 80 per cent of their everyday banking.[3]

ATM services in Canada were initially established by financial institutions independently, so that the machines of one institution could not be accessed by customers of other institutions. Regional shared networks have been established involving credit unions, trust companies and banks, and in February, 1986, the first national shared network was

2 See "Current Status of Bank Automation" in *Branching Out* (Computers/Communications Task Force, Department of Communications, Supply and Services Canada, 1972).

3 "How to use your bank card", *The Toronto Star*, May 23, 1985, p. A15, col. 1.

established between the Bank of Nova Scotia and the National Bank of Canada. In June, 1986, Interac Association, a company established by Canada's major banks, trust companies and credit unions, commenced operation of a national shared ATM network giving customers access to some 2,500 banking machines throughout the country, with another 1,000 machines being added by the end of September, 1986.[4]

An essential component of a fully integrated EFT system involves the establishment of a method for completing the transfer of funds electronically at the point of sale ("POS"). The two essential elements of an EFT/POS system are a POS terminal located in a retail outlet, and a debit card which can be used to transmit instructions to debit a customer's account and credit the retailer's account at the time that a transaction is completed at the retail outlet. A debit card is to be distinguished from a credit card in that the former provides for an immediate debit to be made to the customer's bank account, whereas the latter provides for the debit to be made to a separate credit card account with a subsequent funds transfer to be made in favour of the card issuer.

In 1984, Telecom Canada announced its intention to design a nationwide system for electronic funds transfers at retail points of sale. Telecom Canada will commence testing of such a network in 1987 in co-operation with its member telephone companies, using a combination of the Datapac public data transfer network and special dedicated private lines.[5]

An EFT/POS system raises a number of legal, regulatory and business issues, such as who should bear the costs of establishing and operating the required network of computer terminals and communications facilities linking financial institutions, retailers and consumers, how to overcome consumer resistance as a result of the loss of advantages associated with other payment mechanisms such as delays in cheque clearances and credit card payments, and how to establish mechanisms to ensure the security of the system and the protection of privacy.[6]

It will be seen that EFT's represent a part of a wide range of enhanced banking services offered to individual and corporate customers. In addition to the actual transfers of funds, services include automated letters of credit, cash management services, inter-company networks to enable companies to buy, sell, order and issue invoices for goods on a paperless basis, and automated clearing houses whereby computer centres allow banks to make

4 "Cash to flow across Canada via machines", *The Toronto Star*, June 3, 1986, p. E3, col. 5-6.

5 See "Trial set for Funds Transfer Plan", *The Globe and Mail*, Feb. 25, 1986, p. B5, col. 1-3.

6 See "Banks, Retailers Tussle over ATMs in stores," *The Globe and Mail*, February 25, 1986, p. B5, col. 1-5; "Major Snag postponing Debit Cards", *The Globe and Mail*, March 28, 1986, p. B1, col. 1.

payments and to effect settlements. EFT systems represent a means of reducing the operating costs associated with information handling, of improving the productivity and record keeping of financial institutions and of increasing the range of services offered to customers.

3. NATIONAL AND INTERNATIONAL ELECTRONIC FUNDS TRANSFER NETWORKS

An essential element of an effective EFT system is a network to provide for the transfer of funds and of messages relating to funds transfers within and between financial institutions. Several national and international interbank networks have been established to provide for the transfer of funds and to act as settlement agents. The major networks are summarized below.

(a) FedWire

The Federal Reserve Wire System ("FedWire") was initially an interbank telegraphic communications network, and commenced on-line electronic transfers of funds in 1973. It is the major wholesale electronic payment system in the United States. The network connects all Federal Reserve Districts in the United States with member banks, other banks and offices of the United States Department of the Treasury. FedWire provides various forms of terminal links, such as computer links, dedicated leased lines and data terminals. Unlike other electronic payment networks, FedWire provides for immediate irrevocable payments, rather than same day or next day settlement service.

(b) BankWire

BankWire is a non-profit industry co-operative consisting of United States and some Canadian depository institutions. BankWire offers an on-line fund transfer service, together with various information services, such as electronic mail, delivery of return items, delivery and receipt of daily statements and forwarding of error compensation claims. BankWire settlement takes place at the end of each day. Member institutions may establish bilateral arrangements to set a net debit cap on the value of transfers that may be made between them.

(c) CHIPS (Clearing House Interbank Payments System)

CHIPS is a private sector automated clearing house operating in the United States. It is primarily an international payments mechanism and makes transfers on behalf of member banks. It presently offers same day settlement service. CHIPS has established a system to reduce the risk of fraud and exposure through implementation of bilateral credit limits.

(d) SWIFT (Society for Worldwide Interbank Financial Telecommunications)

SWIFT, a non-profit society of international banks located in Brussels, is a bank-owned telecommunications network for financial messages and instructions. The network transfers correspondent requests for funds transfers, as well as various other types of financial information and administrative messages, such as foreign exchange confirmations, loan deposit instructions and securities transactions and authorizations. The system offers member banks various types of verification and other controls as well as user security features. However, it does not offer settlement services. Subject to strict limitations, the SWIFT network assumes financial liability for system delayed messages.

4. LEGAL ISSUES RELATING TO ELECTRONIC FUND TRANSFERS

The following is a summary of some of the major legal issues that arise under Canadian law in connection with EFT systems.

(a) EFT's and the Bills of Exchange Act

Under the definition set out in subsection 17(1) of the Bills of Exchange Act,[7] a bill of exchange must be in writing and be signed by the person giving it. The various provisions of the Bills of Exchange Act dealing with acceptance, delivery, negotiation and presentment, and conferring certain rights on a holder in due course, are all predicated on the existence of a paper-based payment system. Under the Act, a stop payment order may be made on a cheque until such time as the cheque is presented for payment.

In the absence of any provision in the Bills of Exchange Act dealing with electronic transfers, the rights and liabilities of the parties to such a transfer must be determined based on any contractual arrangements made among the participants. Until such time as amendments are made to the Bills of Exchange Act to recognize EFT transactions, it is likely that financial institutions, corporations and individual customers will wish to continue using cheques in many transactions so that they can exercise the remedies set out in the Act.

7 R.S.C. 1970, c. B-5.

(b) Bank Act

The Canadian Payments Association Act was enacted as Part IV of the Banks and Banking Law Revision Act, 1980.[8] The Act provides for the establishment of the Canadian Payments Association, to consist of the Bank of Canada, every bank, every savings bank governed by the Quebec Savings Banks Act, and other persons entitled under the Act to be members. Such other persons include trust and loan companies and other persons that accept deposits transferable by order to a third party.

The objects of the Association are to establish and operate a national clearings and settlement system and to plan the evolution of the national payment system.[9] The Association is granted various powers to carry out its objects, including the power to arrange for the exchange of payment items at such places in Canada as the Association considers appropriate.

Section 72 of the Canadian Payments Association Act provides for the adoption of by-laws respecting, *inter alia*, clearing arrangements and settlements and related matters. Such by-laws are not effective until approved by the Governor-in-Council.[10] In addition to the by-laws, the Board of the Association may make such rules respecting clearing arrangements and the settlement of payment items as it considers necessary.[11]

Section 14.02 of the Clearing By-law passed by the Board of the Canadian Payments Association provides for the approval of electronic items for clearing. Electronic items are defined as payment orders recorded on magnetic tape or transmitted in any electronic message medium capable of being reproduced in writing.

The Canadian Payments Association has also established the Automated Clearing and Settlement System ("ACSS"). ACSS is an inter-institutional accounting system linking the data centres of participating institutions to one another and to their various head offices for the compiling, transmission and reconciliation of clearing exchange logs.[12]

The Bank Act also imposes restrictions on the automated services that may be offered by Canadian chartered banks. Paragraph 174(2)(*j*) of the Act provides that except as authorized by the Act and in accordance with the terms and conditions prescribed by the regulations, a bank shall not engage in Canada "in providing data processing services other than such banking related data processing services . . . as are prescribed by the regulations . . .". This provision is intended to restrict only the data

8 S.C. 1980-81-82-83, c. 40.

9 *Ibid.*, s. 58.

10 *Ibid.*, subs. 72(2).

11 *Ibid.*, s. 73.

12 See G. David, "Electronic Funds Transfer — Technological Developments and Legal Issues: Part I" (1985), 2 C.C.L.R. 65 at 68-69.

processing services that may be offered by banks to third parties, and not the banks' own data processing activities, which are related to the regular banking services that may be provided to customers. The banking related data processing services that are permitted by the regulations include the classification and reporting of payments and deposits, payment item account reconciliation, processing of pre-authorized debits and credits, various payroll services, various internal processing services that are an integral part of banking operations, various services for farmers, accounts receivable services to a bank factoring subsidiary and services for related companies.[13]

(c) Security: Risk of Fraud and Other Criminal Abuses

The widespread use of ATM's raises concerns about the security of an EFT system and the ability of financial institutions and customers to prevent any unauthorized interference with the system. Among the security problems that may be encountered are the use of counterfeit cards, the use of listening devices, the use of "spoofers", that is, devices capable of intercepting a communications line between an ATM and a bank to direct the ATM to empty itself of cash, and "skimmers", that is, devices to permit the transfer of a PIN from an original card to a counterfeit.[14]

The need to ensure the security of an EFT system raises both technological and legal issues. Although financial institutions should take adequate internal security measures to prevent any interference with an EFT system, they have little control over the use of pass words, access codes or cards used to gain access to an ATM or ETF/POS system. As a matter of law, the newly enacted provisions of the Criminal Code relating to unauthorized use of computer systems and interference with computer data and programs, as well as the provisions relating to fraud and theft, may be relied upon where the security measures are breached.

(d) Liability for Errors or Losses in Use of an EFT System

The use of an EFT system may lead to loss from a variety of causes, such as the forwarding of funds to or from the wrong account, the transmission of the wrong amount, or delays in transmission. Other issues which must be addressed in the use of an EFT system include the time at which a customer must be informed of credits or debits being made to his account, the manner in which such notice is to be provided, and

13 See "Banking and Related Data Processing Services Regulations", SOR/81-424, Canada Gazette, Part II, Vol. 115, No. 11 (May 28, 1981).

14 See Soma, *supra*, footnote 1, at 31.

the obligation of the customer to inform the bank of any errors made in any EFT transmission or debits or credits.

At present, the rights and obligations of the parties to an EFT transaction in Canada with respect to such matters are governed by the terms of any applicable contracts. In general, where a contract between two financial institutions, or between an institution and its customer, limits or excludes an institution's liability for errors or delays in the transfers of funds or messages, such limitation clauses will be binding.[15]

There are also no statutory rules in Canada governing the documentation that may be necessary to authenticate an EFT transaction. For example, should a bank be entitled to debit a customer's account when the customer denies having given an EFT instructions? In the United States, the Electronic Fund Transfer Act[16] requires documentary receipts to be given to an individual customer of each electronic transfer initiated from a terminal. Although not required to do so, most Canadian financial institutions provide such information on a voluntary basis.

(e) Privacy Concerns

The operation of an integrated EFT system requires the collection of extensive information by financial institutions and private corporations. For example, the administration of pre-authorized payment and direct deposit plans for salaries, dividends, life insurance and mortgage payments and other periodic payments requires information to be kept concerning the persons to whom or on whose behalf such payments are to be made. In the case of POS terminals which involve debit cards, records must be kept of such matters as the time, place, date and identity of the parties to any transaction. Records are also generally kept of the details of transactions made by customers using ATM machines.

The existing Canadian laws with respect to privacy and data protection were reviewed in detail in previous chapters of this Part. It was noted that no Canadian laws dealing specifically with the establishment of data banks by the private sector have been enacted. However, Canadian banks are bound at common law to preserve the confidentiality of information acquired by them concerning their customers.[17]

15 See *B.G. Linton Const. Ltd. v. C.N.R.* (1974), 49 D.L.R. (3d) 548 (S.C.C.). For a full description of the master Interbank Indemnity Agreement made among Canadian banks governing pre-authorized payment plans and suggested revisions to the Agreement, see Crawford & Falconbridge, *Banking and Bills of Exchange*, 8th ed. (Canada Law Book, 1986), at 919-924.

16 U.S.C. Title 15, 1978.

17 *Tournier v. Nat. Prov. & Union Bank of England*, [1924] 1 K.B. 461 (C.A.); see Crawford & Falconbridge, *supra*, footnote 15, at 802-816. Banks may also be subject to fiduciary

(f) Evidentiary Issues

The Canadian law of evidence relating to computer produced documents was reviewed in detail in Chapter 8, *supra.* It was noted that special provisions are contained in the federal and provincial Evidence Acts with respect to the admissibility of bank records. Several cases have considered the application of these provisions to computer produced records. The issues which arise from such provisions, such as when a record is deemed to be an original and whether foundation evidence must be adduced to admit computer records, were reviewed in detail in Chapter 2.

(g) Risk of Insolvency of Financial Institution

A major concern with respect to electronic fund transfer systems is the possibility that a financial institution will become insolvent at a time that it has negative balances through the use of a wire transfer system. If the institution does not have sufficient funds to cover settlement commitments, other institutions which are in a credit position may not be able to collect from the first. This raises the risk of a domino effect whereby other banks would find it impossible to settle their balances.

It has been estimated that the value of negative balances across the various wire transfer payment systems used in the United States could exceed $100 billion dollars at the middle of any day.[18] Commencing in March, 1986, the United States Federal Reserve requested United States banks using FedWire to establish voluntary limits on the amount by which they would over-draw their accounts at the Federal Reserve when using such wire transfer system.[19]

(h) Competition Law

Section 33 of the Competition Act[20] was enacted in 1986 to take the place of section 309 of the Bank Act. Subsection 33(1) makes it an offence for a bank to enter into any agreement or arrangement with another bank with respect to the rate of interest, the amount of any charge for service, the kind of service, or the persons for whom any service is to be provided. This provision is subject to certain exemptions set out in subsection 33(2), including an exemption for agreements for the devel-

---

duties to their customers in some circumstances: see *Standard Invts. Ltd. v. C.I.B.C.* (1985), 52 O.R. (2d) 473 (C.A.).

18 See "U.S. Moves to Restrict Electronic Banking Systems", *The Globe and Mail*, Jan. 8, 1986, p. B20, col. 1-6 (reprinted from New York Times Service).

19 *Ibid.*

20 S.C. 1986, c. 26.

opment and utilization of systems and the utilization of common facilities. These would appear to include shared ATM networks. The new Competition Act also has the effect of transferring responsibility for administration of competition law insofar as it relates to banks from the Inspector General to the Director of Investigation and Research whose office is established under the Competition Act.

5. CONCLUSIONS

The foregoing represents only an overview of some of the major legal issues arising from the increased use of electronic fund transfer systems. As a matter of technology, EFT systems could be established in the near future to replace in large part paper-based payment systems. However, a number of legal and business issues remain to be resolved before a fully integrated EFT network is established in Canada. The resolution of these issues presents some of the most interesting challenges in the field of computer law to be addressed in Canada in the coming years.

Part V

CONTRACTING FOR COMPUTER PRODUCTS AND SERVICES

15

The Making of Computer Contracts: An Overview

1. INTRODUCTION

The term "computer contract", as used in this text, refers to any commercial arrangement between two or more parties relating to the development, supply and/or use of computer products or services. The specific chapters that follow will review the provisions generally found in the most common types of contracts in the computer field. They are not intended to represent an exhaustive analysis of either the types of contracts that may be made or the scope of the provisions that may be inserted in any contract. No attempt is made, for example, to set out the kinds of recitals or definitions that should be included in any contract. The final form of any contract will depend on many factors, such as the purposes for which any products and services are being acquired, the acquisition price and the relative bargaining strengths of the parties.

Before reviewing specific contracts relating to various types of computer products and services, some of the issues that arise in connection with the process of negotiating and settling computer contracts are considered. These issues include the nature of a computer contract and how it is to be distinguished from other commercial arrangements, some of the special problems that are likely to be encountered in preparing and settling such an agreement, the role of a lawyer in the negotiation and settlement of the contract, and some of the practical problems that may be encountered by a purchaser or its counsel in dealing with different types of computer suppliers. The final section of this chapter will review the provisions that may be included in a general procurement or master agreement relating to the acquisition and implementation of a computer system.

2. THE UNIQUE CHARACTER OF A COMPUTER CONTRACT

A computer contract is a highly specialized agreement, and its negotiation and settlement require some knowledge by a lawyer of computer terminology and the functions of computers. The lawyer must also be aware of some special clauses that must be inserted in a contract of this nature and the objectives of the client in the contracting process. Finally, the lawyer must be aware of the pitfalls that a client may encounter in the process of acquiring a computer system, particularly if the products in question have not been fully developed or if their implementation is to take place over an extended period of time.

The difficulties that may be encountered in contracting for computer products and services are illustrated by the case of *Peter MacGregor Ltd. v. Fortrex Systems Corp.*[1] In that case, the Court was faced with the task of ascertaining the terms of a contract between the supplier and a customer of computer hardware and software, and in particular, whether there was a contractual obligation for the completion of the system by a particular date. The parties had executed a number of sales orders as well as various formal agreements, many of which had conflicting provisions as well as references to various other documents. The state of the contractual documents was of such confusion that Krever J. was moved to comment that if such state of affairs was "typical in the case of the purchase and sale of computer hardware and software, [it] seems to an outsider to be a less than orthodox way of conducting business."[2] The collapse of the project in that case proved costly to the user who not only failed to recover some $83,000 paid to the supplier but also suffered substantial losses in respect of costs of financing, employees' time and loss of savings in its business. The supplier, on the other hand, incurred losses of opportunities and such damage to its reputation that it was driven out of the Toronto market.

The *Fortrex* case indicates the importance of dealing in a computer contract with issues which are common to other commercial relationships. For example, the contract must set out clearly what duties are to be performed and by whom, the time of performance, the terms of payment, the circumstances for the termination of the rights and obligations of the parties, the warranties to be provided by one party to the other, and the limitations of liability and exclusionary clauses.

3. THE TIMING OF THE MAKING OF A CONTRACT

The terms of a computer acquisition contract should be settled as soon as possible after a user decides to acquire products from a particular supplier. By that time, the user should have carried out a comprehensive review of its requirements and conducted a complete investigation of potential suppliers, including an examination of the operation of any system

---

1 Ont. H.C.J., August 31, 1983, summarized at 21 A.C.W.S. (2d) 384; varied on appeal Ont. C.A., November 5, 1985, summarized at 34 A.C.W.S. (2d) 408.

2 *Ibid.*, at 1.

being considered at other sites. A user's bargaining position will be greatly weakened if it has acquired test products, commenced conversion of data or otherwise changed its business position in reliance on a particular supplier. On the other hand, the bargaining position of the supplier may also be weakened if it has expended substantial sums in reviewing a potential customer's business needs or in developing new programs without a firm contractual commitment from the user.

The user may issue a Request for Proposals to various potential suppliers which may include a draft contract in the form desired by the user. Alternatively, the supplier itself may wish to submit its form of contract with any quotation submitted to a user. If the draft is comprehensive, it will cause the parties to focus their attention on issues which must be addressed in order to complete the implementation of a computer system.

In view of the dynamic character of the process of implementing a computer system, the contract should be reviewed periodically and, if necessary, revised to reflect changes in circumstances. In some cases, it will be impossible to finalize certain terms at the time the initial contract is executed. In such event, the contract should set out a time limit for such issues to be dealt with and set out the consequences if no agreement on such issues is reached. The indefinite postponement of the resolution of key contractual issues may lead to the kind of disastrous consequences exemplified in the *Fortrex* case.

4. THE ROLE OF A LAWYER IN THE MAKING OF A COMPUTER CONTRACT

A lawyer may be required to perform a number of tasks in acting for a client in the computer field. For example, he may be required to prepare either a standard form contract to be used to supply products and services to a number of users, or a specialized contract to deal with one particular relationship. In other cases, the lawyer will be required to review and comment on a contract prepared by counsel for the other party. The lawyer may also be responsible for negotiating the terms of the contract or assisting senior management in the negotiations.

A client should be made aware of the matters which are beyond the scope of the lawyer's retainer but for which the client must receive advice or assistance. For example, a lawyer generally would not have the necessary expertise to advise a client as to the kind of computer system it should acquire in order to meet its business needs. Similarly, the lawyer should not be expected to assume responsibility for writing detailed specifications for a computer system, or even for ensuring the adequacy or completeness from a technical or business standpoint of specifications negotiated by the parties. At the same time, the user's lawyer should make his client aware of the need for such specifications as representing a definition of the product to be acquired from a supplier. The user's lawyer may also be able to identify areas in which the client may need advice from an outside computer consultant before making a formal commitment to a particular supplier.

With respect to the actual negotiation of a contract, the role of a lawyer will vary according to circumstances. If the size of the transaction and the financial circumstances of the client are a major restraint, a lawyer may be asked simply to provide comments on a standard form contract, with the client assuming responsibility for the actual negotiation of any changes. In other cases, the lawyer may participate in an early negotiating session and in a final session, with intermediate negotiations being carried out by the management and data processing personnel of the respective parties. In still more complex transactions, a lawyer may be expected to lead the negotiations on behalf of the client, although in the case of a user, such negotiations must be carried out in conjunction with the personnel responsible for the management and supervision of the computer installation.

The actual tactics of negotiation adopted by a lawyer will depend on his personal style as well as the personalities of the other parties involved. Although a confrontational and uncompromising approach may create a favourable superficial impression, it may be less effective than an approach in which the lawyers for the respective parties identify areas of possible compromise. Whatever role the lawyer plays in the actual negotiation of a contract, he should ensure that the contract is ultimately expressed in language which is comprehensible to a non-computer expert, including a judge who may be called upon to interpret the contract. In all cases, the lawyer must be aware of the ultimate objective: that is, the execution of a contract which reflects both parties' reasonable interests, which can be realistically performed, and which deals with foreseeable contingencies so as to minimize the possibility of any dispute and the disruption of each party's business that inevitably results.

5. DEALING WITH COMPUTER VENDORS: NEGOTIATING CHANGES TO STANDARD CONTRACTS

Computer suppliers are often unwilling to make substantial changes to their standard contracts. As a result, a user may find it impossible to ensure that the contract includes various provisions that it may wish for the protection of its rights, such as some of the clauses suggested in following chapters of this text. For example, a supplier may be unwilling to commit itself to a fixed delivery date, to give the purchaser the right to conduct acceptance testing, or to provide various warranties of performance or functionality. In many cases, suppliers will be unwilling to depart significantly from limitation or exclusionary clauses, in view of the inherent nature of a computer system and the extensive loss that may result from its failure.

In some cases, the supplier's sales representatives may make verbal promises with respect to certain matters which are not subsequently incorporated in the formal contract. The vendor's representatives may also promise to use all available resources to solve problems encountered by the user without regard to the terms of the contract. While such promises may provide some comfort to a client, they will be of little comfort for a user's solicitor. In this respect, regardless of the promises made by a supplier's representative or its size or reputation, the supplier is likely to insist on strict adherence to the terms of a contract if a dispute arises that the parties are unable to resolve by negotiation. Accordingly, every effort should be made to ensure that any promises made by a supplier on which the user wishes to rely are reduced to writing and are incorporated in the formal agreement.[3]

While larger computer suppliers may be unwilling to modify their standard contracts, a very different problem may be encountered in dealing with a small corporation, partnership or sole proprietorship. Such a supplier may be willing to accept whatever form of contract is suggested by the purchaser or his solicitor. However, the supplier may lack the resources to carry out the contract or the financial ability to meet any claim for damages arising from any breach of contract. In representing either a purchaser or supplier of computer products or services, a lawyer should ensure that his client is aware of the need for a proper investigation of the ability of the other party to perform the contract and the practical difficulties that may arise in enforcing what is otherwise a well-drafted and conceived contract. This will be of particular importance for a client acquiring computer hardware or software from a small vendor. In some cases, the user may wish to obtain guarantees of performance from the vendor's principals or even a performance bond to ensure satisfactory completion of the contract.

6. CONTRACTING FOR THE ACQUISITION OF A COMPUTER "SYSTEM"

Subsequent chapters of this text will review various types of contracts relating to specific computer products and services. However, a computer user rarely wishes to contract for a single product or service in isolation from any other. For example, computer equipment which is leased or purchased will be of little value without licences to use operating system software as well as various application programs to perform specific tasks.

---

3 For a discussion of the rationale underlying certain standard contractual terms and conditions from the perspective of the computer supplier, see J.H. Safer, "Computer Industry Contracting Practices", paper presented at Conference on Computer Law for the Practitioner, Canadian Bar Association for Ontario, Toronto, November 28, 1986.

In many cases, the objective of a user will be to acquire an integrated computer system: that is, a configuration of computer products and services designed to enable the user to perform a number of inter-related business functions. These functions may include centralized and/or networked word processing, accounting and other data processing functions, facsimile and electronic communications of voice, text and data, including on-line access to remote data bases.

The components of a computer system may include computer equipment, operating system software (including communications and utility programs), application programs (which may be customized or custom-developed for the user's particular needs), end user documentation to assist the user in the operation of the system, source code and technical specifications to enable the user to maintain the system, and a variety of services such as education and training, assistance in data conversion and continuing maintenance and support services.

Corresponding to the various components of a computer system which a user may require are a number of legal relationships. These may include: an agreement for the sale or lease of computer equipment; the grant of a licence or sub-licence to use operating system software; an agreement to develop custom application software together with a licence to use such software, an agreement to provide services in connection with the provisions of the system such as education and training and assistance in data conversion, an agreement to provide on-going maintenance and support services, and an agreement to place source code and other proprietary information relating to the supplier's programs into escrow for release to the user in certain defined circumstances.

In some cases, the rights and duties of the parties involved in these relationships may be set forth in a comprehensive general procurement agreement. Alternatively, such rights and duties may be set forth in separate agreements which are attached to and incorporated by reference in a head agreement. In either case, the user should ensure that it is adequately protected in the event of the failure of the computer system as a whole, which may be difficult or impossible to attribute to defects in individual components of the system.

The relationships involved in the acquisition of a computer system will be further complicated if the user contracts with more than one supplier. In such a case, the failure of one supplier to perform his obligations may result in the failure of the entire system. Moreover, relationships with a number of suppliers make it difficult to allocate responsibility for any "system failure" to any one supplier. In a "turn-key" computer installation, on the other hand, the user will be able to look to a single supplier to assume over-all responsibility for the installation of the system, although specific products such as personal computers or printers and services relating to such products may be provided by third parties.

The specific provisions that should be included in the various agreements or sub-agreements involved in the acquisition of a computer system are discussed in detail in subsequent chapters. The following are some of the provisions that should be included or referenced in any master agreement to ensure that the user acquires an integrated system, and that its implementation is carried out in a manner to minimize disruption of the user's business.

(a) Definition of System: Request for Proposals and Functional Specifications

Before entering into an acquisition agreement, a full description of the components and features of the system and the manner in which the system is to be implemented should be prepared. Such a description should be expressed in non-technical language and may be incorporated in a Request for Proposals issued to potential bidders or in the Proposal submitted by the supplier. The description of the system may include detailed published specifications for any hardware and standard operating system software, as well as an outline of the "functional specifications" for any application programs which are to be custom-developed or modified from existing programs. Such specifications are generally prepared by a user but may be developed in conjunction with a potential supplier. In any event, such specifications should be completed prior to the execution of a formal agreement and should be incorporated as a schedule to such agreement.

The system acquisition agreement should also include detailed technical specifications for any programs which are to be developed or customized from existing programs, or provide for the preparation of such specifications by a certain date. Such specifications describe in greater detail and in technical language the manner in which the programs are to operate and perform. The agreement must set out a procedure for review and acceptance of the technical specifications where they do not already exist.[4]

(b) Implementation Schedule

The agreement should contain an implementation schedule setting out the projected date for the completion of each task involved in the development and installation of the computer system. The schedule will include dates for such tasks as the development of technical specifications, completion of various modules of custom software, delivery and installation

4 For a further discussion of functional and technical specifications, see Software Development Agreements, Chap. 17, *infra*.

of computer equipment and operating systems programs, conversion of the user's existing data, installation of all application programs, provision of operator training, delivery of all user guides and other program documentation, acceptance testing of the system, final system acceptance and the commencement of maintenance and support services.

The agreement should set forth the remedies available to the user if the supplier fails to meet the milestones set forth in the implementation schedule. These remedies may include liquidated damages if the supplier fails to meet its obligations in a timely manner, and the right to terminate the agreement if default continues beyond a certain time period.

In general, the implementation schedule should provide for the various components of the system to be delivered and installed contemporaneously and for the supplier to provide a single certificate of installation. This will enable the user to carry out acceptance testing of the entire system to ensure that it operates as an integrated network.

If the user accepts delivery of only a part of the system, then in the absence of an express provision in the contract, he may be deemed to have accepted and be bound to pay for the partial system, even if subsequent components of the system are not developed, installed or accepted. Delivery of only part of the system may also create uncertainty as to the date on which acceptance testing is required to be commenced as well as the date for commencement of any applicable warranty period. Accordingly, the contract must clearly indicate whether the obligation of the supplier is to provide an integrated computer system as such, as opposed to specific hardware and software products and services.

In some cases, the supplier and user may intend for different components of the system to be completed and installed sequentially. For example, custom-developed software may be developed in a series of modules intended to perform different functions. If various components of the system are to be installed at different dates, the agreement should indicate whether each component is subject to independent acceptance testing. The agreement should also set out the consequences if one part of the system is installed and/or accepted but the supplier fails to provide subsequent components of the system.

The user may wish to have the right to return any part of the system already installed if the supplier fails to provide the entire system. In other cases, the user may be willing to accept some initial hardware and software products even if it does not receive the complete system. Still another alternative is for the parties to treat such initial components as being under a temporary lease which may be terminated if the complete system cannot be completed and accepted by a certain date. In any event, it is essential for the parties to address such issues in the formal agreement.

(c) Acceptance Testing and Acceptance

A user of a computer system which incorporates a number of products will wish to have an adequate opportunity to test the system before it is deemed to be accepted. It is also in the interests of both parties to provide for acknowledgment of acceptance to be provided by a certain date. Computer users should recognize, however, that suppliers are often unwilling to provide a lengthy acceptance testing period. Users may be willing to forego the opportunity to carry out lengthy testing if the system in question has been successfully installed and is fully operational in companies carrying on a business similar to that of the users.[5]

The purpose of acceptance testing is to enable the user to determine whether the various components of a system operate in an integrated fashion in accordance with the functional and technical specifications incorporated in the agreement. Acceptance testing should commence only after all components of the system (or at least such components as are necessary for the system to operate as an integrated unit) have been installed, all operator training has been provided, and the user has received all user documentation describing the operating conditions of the system.

Unless the parties are prepared to accept general language such as that the system must operate to the "reasonable" satisfaction of the user, the agreement should set out the acceptance test criteria to be met by the system. Such criteria may be developed as part of the technical specifications of the system. The agreement should also provide for the development of a comprehensive set of acceptance test data to enable the user to determine whether the system operates in conformity with the specifications.

The acceptance test criteria may consist of the ability of the system to perform specified functions, to handle certain volumes of data and to respond within specified response times. In some cases, the system will be deemed not to pass acceptance testing if there are malfunctions in the operation of the system of sufficient severity to cause total system failure, or if the system is incapable of performing certain functions, on a number of occasions or for a period of time in excess of permitted limits.

The acquisition agreement should set out the number of days during which the system must operate in accordance with the acceptance test criteria, and the time period following the commencement of testing by which the test criteria must be met. The supplier should agree to correct promptly any malfunctions detected during acceptance testing. The time period for the supplier to respond to a call for assistance may be specified.

The agreement should set out the remedies available to the user if

5 See Chaps. 16 and 17 for a further discussion of clauses to be included in computer contracts relating to acceptance and acceptance testing, *infra*.

the system fails to meet the acceptance test criteria. These remedies will generally include the right to return all parts of the system that have been installed and to receive a refund of all or a substantial part of the purchase price.

The agreement may also contain a provision whereby the system will be deemed to be accepted unless the user notifies the supplier of its rejection of the system within a certain number of days following the expiration of the acceptance testing period. Accordingly, it is essential for the user to provide the supplier with timely notice of any system malfunction or total system failure.

(d) Guarantees of Performance

In appropriate cases, a user will wish to obtain guarantees of performance of the obligations of the system supplier from a third party, such as a principal shareholder or an affiliated company. The guarantor would agree that upon default by the supplier, it would assume the supplier's obligation to complete the development of the computer system and to support the system after installation. The guarantee of performance may be supported by a performance bond or a letter of credit to ensure that the user will be able to recover any moneys paid as a deposit on account of the purchase price or will be able to finance the completion of the development and installation of the system.

(e) Payments and Allocation of Purchase Price

The purchase price payable for an integrated computer system may be allocated according to the value of individual components of the system. However, the agreement should provide that the aggregate purchase price is to be paid for the complete computer system, and that if any part of the system is not developed, installed and accepted, the user has no obligation to pay any part of the purchase price. In the absence of such a clause, the supplier may argue that it is entitled to receive payment of that part of the price in respect of any part of the system that is delivered and installed at the user's premises.

(f) Warranties of Performance and Operation of System

A user acquiring a computer system will wish to receive a number of warranties such as those set forth in a hardware acquisition agreement or a software development agreement which are discussed in detail in subsequent chapters. Of particular concern will be warranties that the various components of the system operate in conjunction with one another as an integrated system, that the system is compatible with the user's existing information system, and that it is capable of expansion to permit the addition of peripheral devices such as printers and terminals at a later date. The user will wish to obtain a warranty that the system will perform the functions and meet or exceed the requirements set forth in any functional specifications incorporated in the agreement.

The foregoing are some key provisions that should be included in a computer system acquisition agreement that relate specifically to the integration of the system. The agreement should contain various additional provisions, such as those found in a hardware acquisition agreement, a software development agreement, a software licence, and hardware maintenance and software support agreements, all of which are reviewed in detail in the following chapters of this Part.

16

Agreements Relating to the Acquisition and Servicing of Computer Hardware

1. COMPUTER HARDWARE ACQUISITION AGREEMENTS

Computer hardware may be acquired either by purchase or by lease. In the case of a purchase, title to the equipment passes to the customer, whereas in the case of a lease or rental agreement, the customer acquires the right to use the equipment for a fixed term at agreed-upon lease or rental rates. Both agreements will contain a number of common provisions, although a lease agreement will normally contain additional provisions relating to the commencement and termination of the lease, and the respective obligations of the parties during the lease term and upon its expiration. Although the terms lease and rental are generally used in this chapter interchangeably, as a matter of industry practice, the term "rental" means a short-term lease (generally of less than 12 months duration), and the term "lease" means an agreement of a longer term.

The final form of a hardware acquisition agreement will depend upon a number of factors, including the kind of hardware being acquired (that is, a mainframe computer system, a minicomputer or a microcomputer), the supplier from whom it is to be acquired (for example, a manufacturer, an authorized dealer or distributor, or a "value-added" dealer or remarketer who adds to or enhances a basic hardware package for a particular customer), whether the hardware is to be provided as part of an integrated computer system, the uses to which the hardware is to be put (for example, to meet the over-all business needs of the user, to perform specific business tasks such as accounting or word processing, or for use as a home computer) and the price to be paid by the user.

The following sections will review the provisions that should be considered for inclusion in a hardware acquisition agreement. The reader is cautioned that some of these provisions will be found only in complex agreements involving the acquisition of large and expensive computer systems. In other cases, a user may find that it will have to pay a higher price to obtain the legal rights and remedies that would ordinarily be given only in the case of a turn-key computer installation.

(a) Description of Equipment

The agreement should contain a complete description of the products being acquired, including the model name and series number by which the products are commonly identified, the number of units and, where the agreement involves specifically identified equipment, the serial numbers. The products may be described in separate schedules which are attached to the agreement. A general procurement or master lease agreement may also provide for the execution of subsequent orders for additional equipment. In such event, the agreement should set out the procedures and timing for the execution, acceptance and fulfilment of all orders.

(b) Term

A master lease agreement will generally commence upon the date of execution and will apply to any orders made at any time thereafter. The actual lease term of any equipment may commence on the date on which the equipment is delivered or installed, or the date on which the equipment is accepted or is deemed to be accepted. The supplier should ensure that any master lease agreement will remain in effect for a period of time corresponding to the maximum lease for any units of equipment thereunder.

A computer equipment lease may provide for its automatic renewal unless either party gives notice of termination within a certain period of time prior to expiration. If the renewal of the lease is automatic, the lease renewal term (which may be on a month-to-month or year-to-year basis) as well as the lease rate for the renewal term should be clearly specified.

A computer lease will also normally contain a clause providing for the right of termination by the lessor if the lessee fails to make any lease payments or fails to carry out its other material obligations under the agreement. The lessee will wish to receive notice of any default and be given an opportunity to cure such default before the termination is effective.

A clause providing for the termination of a lease for default by the lessee will make it difficult for the lessee to withhold lease payments on the grounds that the lessor has failed to carry out its obligations under the lease agreement. In *MGI Computer Corp. v. Systemhouse Ltd.*,[1] it was held that the plaintiffs were not entitled to an injunction to restrain the removal of its leased computer system where the plaintiffs had withheld lease payments because of the failure of the defendant to supply custom software and to train the plaintiffs' employees. Although the balance of

1 (1985), 4 C.P.R. (3d) 457 (Ont. H.C.).

convenience favoured the plaintiffs in that the loss of their computer would have a serious effect on their business, they had failed to show that the threatened harm was not adequately compensable in damages.

(c) Delivery

An acquisition agreement should indicate the date on which the computer products are to be delivered at the user's premises. Although a supplier may be unwilling to be bound by a firm delivery date, he should be required to use his best efforts to meet any scheduled date. In addition, the purchaser should have the right to cancel an order, and if appropriate to receive liquidated damages, if delivery is delayed beyond a certain time period. The agreement may provide as an alternative that if the ordered equipment is not delivered by the scheduled date, the user will be provided with alternative equipment which may be made available on a short term rental basis.

(d) Site Preparation

A hardware acquisition agreement should require the supplier to provide site specifications setting out the environmental requirements for the equipment. The specifications will include the conditions to be met with respect to such matters as electrical requirements, temperature, humidity, vibration, dust, static electricity, weight and physical design. The purchaser will be required to make any necessary changes at its own expense to ensure that the installation site complies with all such conditions. The agreement should provide for the supplier to inspect the installation site and certify that it meets the requirements of the site specifications prior to delivery of the equipment.

The user should be aware of any special requirements that may have to be met with respect to its electrical system. The modifications of its system to ensure a reliable and uninterrupted supply of electrical power, particularly in older buildings, may involve considerable expense. On the other hand, the failure to make such modifications may seriously impair the operability of the computer hardware and software which the user proposes to acquire.

(e) Title

A computer equipment supplier will wish to reserve title to the equipment until the purchase price is paid in full. In the case of a purchase or a lease finance agreement, the vendor will also wish to reserve a purchase money security interest to secure payment of the purchase price. In such an arrangement, labels should be affixed identifying the equipment as belonging to the lessor and the lessee should be required to maintain such labels on the equipment.

(f) Risk of Loss, Insurance

Standard contracts of computer suppliers generally provide that risk of loss or damage to the equipment will pass to the purchaser upon delivery to the carrier or at the customer's premises. In such event, the purchaser may be required to insure the equipment, with the vendor shown as an additional named insured, until payment of the purchase price.

In the case of a lease agreement, the lessee will be required to maintain insurance during the lease term or as long as the equipment is in the possession of the lessee. Any insurance should cover loss from such risks as fire, theft and accident as well as property damage and public liability.

(g) Installation

A computer acquisition agreement should set out the respective obligations of the parties in connection with the installation of the equipment. In general, the supplier will be responsible for such tasks as setting up the equipment, cable connections, performance of standard diagnostic tests and provision of operator training. The agreement should provide that installation will be deemed to take place only after the vendor has certified that all such procedures have been completed, that all technical specifications, user guides and manuals have been provided and that the equipment is operable. If certain procedures relating to installation are the user's responsibility, the certificate should indicate such procedures have been completed. Since the completion of installation will be a condition of the commencement of acceptance testing and any warranty period, the user should ensure that it receives a properly dated certificate of installation.

(h) Acceptance and Acceptance Testing

Provincial sale of goods legislation in Canada generally provides that unless the parties to a contract for the sale of goods agree otherwise, the goods are deemed to be accepted on the earlier of the dates on which the buyer intimates to the seller that it has accepted the goods; on which it does any act in relation to the goods after delivery that is inconsistent with ownership by the seller; or on which it retains the goods after the lapse of a reasonable time without intimating to the seller that it has rejected the goods.[2]

2 See, e.g., Sale of Goods Act, R.S.O. 1980, c. 462, s. 34.

In *Staiman Steel Ltd. v. Franki Canada Ltd.*,[3] it was noted that it was "apparent that the courts have adopted a strict interpretation [of this section] favourable to the seller."[4] However, in the case of *Burroughs Business Machines Ltd. v. Feed-Rite Mills (1962) Ltd.*,[5] it was held that the purchaser of a computer system had not "accepted" the system, notwithstanding the passage of some seven months from the time that it made payment conditional upon performance. Such cases indicate the importance of dealing expressly in a purchase or lease agreement with the manner of acceptance of computer equipment.

Computer suppliers' standard contracts generally provide that the equipment will be deemed to be accepted upon delivery, or if installation is the supplier's responsibility, upon the completion of its standard test procedures indicating that the products are operable. Such standard contracts may also provide that if installation is delayed by the purchaser, acceptance shall be deemed to occur within a certain time period following delivery. In other cases, the purchaser will be required to notify the vendor promptly of any defects, failing which the equipment will be deemed to be accepted.

The user will wish to have as long a period of time as possible before it is required to provide notice of acceptance or rejection. The agreement should specify the precise time by which such notice is to be provided to avoid any dispute as to the length of any "reasonable" period of time before acceptance is deemed to occur. The agreement should also indicate whether express notice of acceptance must be provided by the user and whether failure to provide such notice is deemed to constitute either acceptance or rejection.

For larger computer installations as well as newly released computer products, the user will wish to have the opportunity to conduct its own acceptance tests. Such testing will be of particular importance if the user is acquiring an integrated computer system consisting of hardware and software components provided by different suppliers, or if the operation of such components as an integrated network has not been successfully demonstrated at other computer installations. Acceptance testing should give the purchaser the opportunity to use the equipment in an operational environment approximating that in which the equipment will normally operate. The equipment should be deemed to pass acceptance testing only if it is shown to operate in conjunction with associated software in accordance with any specifications provided by the manufacturer or incorporated by reference in the agreement. The test criteria should be designed to enable the purchaser to determine that computer products have

3 (1985), 53 O.R. (2d) 93 (Ont. C.A.).
4 *Ibid.*, at 106.
5 (1973), 42 D.L.R. (3d) 303 (Man. C.A.).

the various features and perform the functions represented by the vendor and that all hardware and software components operate in conjunction with one another as an integrated system.

The acceptance test criteria should also enable the user to test the "reliability" of the hardware system, that is, to determine that it is capable of performing without serious malfunction or defects in operation. Reliability may be measured based on the capability of the equipment to operate at an agreed-upon "uptime effectiveness level". Uptime effectiveness level may be defined as the period of time during the purchaser's ordinary business hours that the products are capable of handling the user's normal data processing requirements and of performing their normal computing functions in accordance with all applicable specifications. The agreement may provide that the computer products will be deemed to be rejected if they require any service calls in excess of a specified number or if they cannot be made operable within a certain number of hours following any such call.

The acquisition agreement should clearly indicate the date on which acceptance testing is to commence (generally the date on which installation is completed), the number of consecutive business days that the computer products must be shown to operate satisfactorily, and the period of time within which the acceptance test criteria must be met. If the computer products fail to meet the test criteria within the agreed upon time period, the user will wish to have the right to reject such products.

(i) Charges

The total purchase price or lease or rental rates should be specified. The agreement should also clearly indicate who is responsible for other charges such as duty and federal and provincial sales taxes. Hardware suppliers' standard contracts generally require the purchaser/lessee to be responsible for all taxes, other than those based on the supplier's net income. However, the prices quoted may in fact include applicable duty and federal sales taxes. All taxes which are the responsibility of the purchaser or lessee and which the supplier is required to collect should be clearly identified in the supplier's invoice.

If a hardware acquisition agreement does not set forth a fixed delivery date, it may provide that quoted prices are effective only for a specified period of time and that such prices may be increased if delivery is delayed beyond that period. A purchaser or lessee should ensure that it is not liable for any such increase if the delay in delivery is caused by the supplier.

The agreement should also set out who is responsible for other costs such as delivery and installation charges, the cost of supplies such as cable, disks and cassettes required for the installation and operation of the equipment and accompanying systems software, and the cost of operator training and user manuals.

The agreement should also set out any volume discounts to which the purchaser or lessee is entitled. Such discounts may be based either on the aggregate number of units or the dollar value of the products acquired. In some cases, the user will wish to have the benefit of volume discounts based on the aggregate acquisitions made by all affiliated and subsidiary companies within the corporate group. In such event, all orders should be issued pursuant to a master purchase or lease agreement. Alternatively, if the companies in the user's corporate group are located in different jurisdictions, a separate agreement may be made between each user and each local supplier. In the latter case, each agreement should contain a clause to provide that for purposes of calculating volume discounts, each purchase or lease is deemed to be made pursuant to a master agreement between the respective parties' parent companies.

In the case of a lease agreement, the lessee may wish to have the right to purchase the equipment either during the term or upon the expiration of the lease. The agreement may contain a formula for calculating the purchase price, to be based on the fair market value of the equipment either at the time that the lease is made or at the time that the option is exercised. If a purchase option is exercisable for an amount less than the fair market value of the equipment, the lease agreement will generally be regarded as a security agreement for purposes of personal property security legislation, requiring the registration of a financing statement to perfect the lessor's security interest.

(j) Payment Terms

The acquisition agreement should indicate the date on which any invoice for the equipment is to be issued and the date on which such invoice becomes due and payable. A computer supplier's standard contract may provide for invoices to be issued upon the delivery or installation of the equipment. However, the purchaser should ensure that payment is not required to be made until the equipment has been accepted in accordance with any applicable acceptance test criteria.

The agreement should also set out the remedies available to the supplier in the event of non-payment or failure to pay in a timely manner. These remedies should include the right to receive interest on over-due payments at an agreed upon annual rate, and the right to terminate the agreement and to recover possession of the purchased or leased equipment.

(k) Training and Documentation

The acquisition agreement should indicate the nature and extent of any training to be provided to the user's personnel, who is to be responsible for the costs of such training and where it is to be provided. The agreement should also describe the user guides and operating manuals to which the user is entitled. Such manuals should be in sufficient detail and be provided in sufficient number to enable the user's personnel to operate the equipment without the supplier's on-going assistance. The user may also wish to receive technical specifications describing the operating conditions of the equipment. Such specifications should be in sufficient detail to enable the user to provide its own maintenance service or to obtain such services from a third party supplier. The agreement should entitle the user to make copies of operating manuals for its personnel and to receive revisions to operating manuals and technical specifications which are published by the manufacturer or which are necessary to reflect any upgrades to the equipment.

(l) Data Conversion

A customer acquiring new computer equipment may have to convert its data to a form suitable for use in conjunction with the new equipment. If the user requires any assistance from the supplier in connection with the conversion, such assistance should be specified in the agreement. In some cases, a user may wish to have the right to cancel the agreement or to delay acquisition of a computer system if the data cannot be successfully converted or if conversion is unduly delayed.

(m) Upgrades and Add-Ons

The acquisition agreement should indicate any right of the customer to replace equipment with upgraded products or to add additional units of equipment to the initial configuration. For upgrades, the agreement should indicate the amount of any credits to which the user will be entitled with respect to replaced equipment. Such credits may be determined as a percentage of the original purchase or lease price according to the number of months that have elapsed from the date of acquisition. Alternatively, such credits may be determined based on the market value of the equipment at the time of any trade-in. The agreement may also specify the price and the terms that are to govern the acquisition of any equipment that the user is entitled to add to the system at a later date.

(n) Alterations and Attachments

A lease agreement will normally contain restrictions on the right of the lessee to alter the equipment or to attach equipment of other manufacturers. Such alterations or attachments will be permitted only if they can be removed without damage to the original equipment at the lessee's cost at the end of the lease term. A user may also lose the benefit of any warranties or the right to receive maintenance services if any alterations or attachments are made to the equipment without the supplier's consent.

(o) Representations and Warranties of Supplier and User

A purchaser of computer equipment will wish to obtain a warranty that the supplier has the right to convey title to the purchased equipment free and clear of all liens, encumbrances, security interest and other charges. Similarly, a lessee will wish to obtain a covenant as to its right of quiet enjoyment, that is, that it will have the right of possession and use of the equipment without interference at any time during the lease term. Both the supplier and user may wish to obtain representations that the other party has the corporate power and authority and the financial capacity to enter into the agreement and to perform its obligations thereunder.

A user will also wish to receive various representations and warranties as to the performance and operating conditions of the hardware. These will include warranties as to the equipment's reliability, that is, that the equipment will be in good operating condition, free of defects in material and workmanship, for a specified period of time following acceptance. These representations may also relate to such matters as the functionality of the hardware (that is, the tasks which it is capable of performing), its capacity (that is, the amount of information which the equipment is capable of processing, or in the case of storage devices, which they are capable of storing), and response time (that is, the period of time required to process a certain amount of information). The user also may wish to receive warranties of compatibility (that is, that the computer hardware will operate with any accompanying operating systems software and application programs and with computer equipment making up the user's existing configuration or which the user proposes to acquire in the future).

A user will wish to ensure that the various representations and warranties survive acceptance of the computer products. The user will wish to ensure that certain warranties (for example, warranties of title) continue during the useful life of the products or during the term of any lease agreement. The supplier, on the other hand, will wish to limit the time during which the representations and warranties, particularly those relating to the functionality or reliability of the system, continue in force as well as the remedies available to the user for any breach. These remedies may include the repair or replacement of defective parts, with any liability for damages being excluded or limited to the amount of the purchase price or the amount of lease payments made for a specified time. The user, on the other hand, will wish to have the right to rescind the agreement and to receive a refund of all moneys paid to the supplier.

The acquisition agreement should indicate whether any service work required in connection with the supplier's warranties is to be performed at the user's installation facility or at the supplier's premises. If warranty service is to be performed at the user's facility, re-location of the equipment to a site outside the supplier's local service travel area may affect the availability of such service or result in extra charges being payable.

Finally, the supplier will normally insist on certain exclusions with respect to its obligation to provide warranty service. For example, such service may not cover defects resulting from improper maintenance, unauthorized modification or misuse of the equipment, use of the equipment in conjunction with any hardware or software products supplied by a person other than the supplier or operation of the equipment outside the specified environmental conditions. Additional exclusionary and limitation clauses which the supplier will wish to include in the agreement are reviewed in the final chapter of this Part.

(p) Patent and Copyright Indemnity

Most computer hardware suppliers are willing to defend the customer from any claim that the equipment or accompanying systems software infringe certain proprietary rights of any third person and to indemnify the customer against any loss, damage and expense arising from any such claim. The customer should ensure that it is entitled to be indemnified against any loss resulting from any claim or allegation of infringement even if no actual infringement is proved. In addition, the indemnity should extend to any claim of infringement of any proprietary right, including any claim based on copyright, patent or trade secret.

The obligation to defend and indemnify will normally be subject to certain conditions, including that of the supplier being promptly notified of any claim and that it have control of the defence of any action and all negotiations for its settlement or compromise. The supplier should covenant that if the use of any product is found to infringe the right of any person and is enjoined, or if the supplier believes in its reasonable opinion that such use is likely to be enjoined, it will procure for the customer the right to continue to use the product, will modify the product such that its use becomes non-infringing, or will provide a replacement product of approximately the same value and capable of performing the same functions. As a further alternative, the user may be required to return an infringing product upon repayment of all or part of the purchase price, or in the case of a lease agreement, upon cancellation of any further lease payments. A computer user may find this alternative unacceptable if the computer system being acquired is indispensable to the conduct of its business.

The indemnity clause will generally not apply with respect to any claim arising from: (a) the use of the supplier's equipment with hardware or software products supplied by a third person; (b) the manufacture or design of the equipment in conformity with specifications or instructions furnished by the customer; or (c) any unauthorized modifications to the products or their use in a manner different from that for which they were designed, as specified in any applicable specifications of the manufacturer. These exclusions will generally be acceptable to the customer except where he is relying on the supplier to provide an integrated computer system with components supplied by different manufacturers.

(q) Maintenance Service

A customer will wish to have the right to receive maintenance service during the expected useful life of the equipment, or in the case of a lease, the maximum lease term. The right to receive maintenance service may be set out in the acquisition agreement or in a separate agreement made with the equipment supplier or with the manufacturer or other third party. The maintenance agreement should deal with the term during which maintenance service is to be provided, the amount of the maintenance fee to be paid annually and any limit on the increase in such fee during the second and each subsequent year of the maintenance period. If these provisions are set forth in a separate maintenance agreement, such agreement should be incorporated by reference in the purchase or lease agreement.

In the case of a lease agreement, the lessee will generally be required to keep the equipment under maintenance during the entire lease term. Accordingly, the lessor or any third party maintenance supplier should be committed to providing such service during the full lease term, including any possible extension.

The detailed terms and conditions normally found in a maintenance agreement for computer equipment are reviewed in the following section of this chapter.

(r) Software Licence

A purchaser or lessee of computer equipment will find such equipment useless unless it is accompanied by the necessary operating systems software. Such software will generally be provided on a licence rather than purchase basis. The terms of the software licence may be included as part of the acquisition agreement or may be set out in a separate agreement. In many cases, the licence will be granted by a third party licensor. In most cases, the licence will be non-transferable and non-exclusive, will entitle the purchaser to use the software only on certain designated equipment, will restrict the number of copies of the software that may be made and the purposes for which such copies may be used, will require the vendor's proprietary legend to be inserted on all such copies, will provide that all software is the confidential information of the licensor with consequent restrictions on disclosure to third parties and will require the user to return or destroy any copies at the end of the licence term. The customer should ensure that its right to use any operating system software is for a perpetual term, or at least for such period of time corresponding to the expected useful life of the hardware products.

Other provisions to be included in a software licence agreement are discussed in detail in Chapter 17.

(s) Other Provisions

In addition to the foregoing, a computer acquisition agreement may also contain various other "standard" terms and conditions, which are reviewed in detail in Chapter 19.

2. HARDWARE MAINTENANCE AGREEMENTS

To ensure that computer equipment remains in good operating condition, a purchaser or lessee will wish to enter into a hardware maintenance agreement. The agreement should be negotiated and executed concurrently with the acquisition agreement since otherwise the purchaser or lessee may find it impossible to obtain maintenance service on acceptable terms.

If computer equipment is acquired directly from the manufacturer or an affiliated company, the maintenance agreement will generally be made with the supplier. If the equipment is supplied by an OEM supplier, a finance lease company, an authorized distributor or a value-added retailer, a separate agreement may be made with the manufacturer who is otherwise not in a contractual relationship with the customer. In still other cases, a maintenance agreement may be made with a third party service organization.

The following is a summary of the principal terms and conditions normally found in a maintenance agreement or incorporated as part of a purchase or lease agreement between an equipment supplier and the user. Although these provisions relate specifically to computer equipment, they may also extend to an integrated computer system consisting of a configuration of hardware and software products. Some of the special provisions normally found in a software support agreement are discussed in detail in a subsequent chapter.

(a) Equipment Covered

The equipment covered by the maintenance agreement should be clearly identified either in the agreement or a separate schedule. The agreement should provide that upon any additional equipment being acquired by the user pursuant to the original purchase or lease agreement or a supplemental agreement, such hardware will also become subject to the maintenance agreement on the same terms and conditions as the original equipment.

If the agreement is intended to cover any used equipment, the maintenance agreement may provide for the equipment to be inspected to ensure that it is eligible for maintenance service. The equipment may also be required to be refurbished or reconditioned in order to meet the supplier's standards. Before entering into any commitment to acquire used equipment, a purchaser or lessee should ensure that such equipment will be eligible for maintenance service.

(b) Types of Service

The two principal kinds of services normally provided under a maintenance agreement are preventative maintenance and remedial maintenance.

Preventative maintenance consists of routine check-ups and services such as cleaning, adjusting, lubrication and testing designed to prevent equipment failure and to extend the useful life of the equipment. Such maintenance will ordinarily be conducted at regularly scheduled intervals determined by the maintenance supplier. A customer may wish to exercise some control over the times at which preventative maintenance is performed to minimize any disruption of its business.

Remedial maintenance service consists of repair service and replacement parts provided in response to a service call to remedy a specific defect or malfunction in the operation of the equipment. The agreement should indicate the different kinds of remedial service that will be provided, that is, telephone assistance, diagnostic service through the use of modems and other telecommunications facilities, and on-site assistance. The agreement may set out the procedures to be followed by the maintenance supplier to ensure the operation of the equipment on a temporary basis pending a permanent solution.

Other services that may be provided under a maintenance agreement include the provision of engineering changes to improve the operation of the equipment, periodic review of the user's environmental conditions to ensure that they comply with the site specifications, and the provision of operating manuals describing the proper procedures for the use of the equipment.

(c) Period of Service Availability

The agreement should indicate the period of time during each business day that the supplier's personnel will be available to respond to any call for service. The customer may have an option to acquire extended service outside the basic maintenance period upon payment of a premium in excess of the regular charge. Alternatively, services outside the principal maintenance period may be made available at the supplier's standard or overtime hourly rates. The option chosen by the user will depend on its expectation of the reliability of the system and its need for a functioning computer system at times other than normal business hours.

(d) Response Time to Call for Maintenance Service

The agreement should set out the maximum number of hours which is to lapse between a call for maintenance service and a response by the supplier. The response time may depend on the distance between the supplier's service centre and the location of the equipment. Once maintenance service commences, the user will wish to provide for the supplier to continue such service until the products are operational even if this involves work outside the normal coverage period. The agreement may provide for service to be interrupted if additional parts or other resources are required.

The agreement may provide for an escalation procedure whereby if a solution is not found by the maintenance supplier's personnel who are assigned to answer a first call for assistance, other personnel may be called in to work on the problem. In such event, the agreement should set out the names or titles of persons who are to respond to a call for assistance at each stage of the escalation procedure and the time following the initial call for assistance that the second or any subsequent stage of such procedures comes into effect.

(e) Maintenance Fees

The maintenance agreement should set out the fees payable on a monthly or other periodic basis for the services to be provided during the basic or principal period of maintenance availability (often referred to as the "PPM"). The additional charges payable for work performed outside the PPM should also be specified.

The agreement should set out any additional costs for which the user may be liable. Such costs may include the travel time of the supplier's personnel if any service is performed outside the PPM. The user may also be required to pay costs of travel and transportation for any services, particularly if the user's computer installation is located outside the supplier's service centre. If the user is responsible for travel time or costs of travel, the agreement should set out the address of the supplier's service centre and provide that no change shall be made to such address without the user's consent. Conversely, the supplier will wish to ensure that if the equipment covered by the maintenance agreement is moved outside the supplier's service centre, it has the right to discontinue the maintenance agreement.

Other charges for which the user may be liable include the costs of telephone or telecommunications services as well as the costs of any training provided to the user's personnel.

(f) Term

A maintenance agreement will normally commence upon the expiration of any applicable warranty period since maintenance will be provided free of charge during such period. A supplier's standard form of maintenance agreement generally provides that it is made on a renewable annual basis with either party being entitled to terminate the agreement upon providing notice within a specified time period (normally 30 or 90 days) prior to the expiration of the initial year or any subsequent renewal term.

The user will wish to ensure that the maintenance supplier is bound to provide service after the expiration of the first year, since it may be difficult to obtain such service for used equipment from a third party service organization. Accordingly, the user should ensure that it will have the right to receive maintenance service for a period of time corresponding to either the estimated useful life or the lease term of the equipment. The supplier may not wish to be committed to firm maintenance fees for a term of several years, but may agree to certain limits on any increase in such fees during the second and any subsequent year of the term of the agreement. Such limits may be expressed in fixed percentage amounts or may be based on the Consumer Price Index as published by Statistics Canada.

While a purchaser will wish to ensure that it has the right to receive maintenance service during a minimum period, it will also wish to have the right to terminate the agreement in the event that it ceases to own or lease the equipment. In addition, it may wish to provide for the assignment of the maintenance agreement in conjunction with any re-sale or sub-lease of the equipment.

(g) Exclusions from Agreement

The agreement will normally set out a number of exclusions from any liability on the part of the maintenance supplier. For example, the supplier will not be responsible for the repair of damage or replacement of defective parts arising from such acts as unauthorized modifications or attachments to the equipment, the provision of maintenance by a third party without the supplier's consent, improper use of the equipment, accident, operation of the equipment at times other than those specified in the agreement, or failure to operate the equipment under the environmental conditions specified by the equipment manufacturer or maintenance supplier. Maintenance service will also normally exclude the costs of supplies such as disks, tapes and papers used in the operation of the equipment.

The maintenance agreement will also normally exclude any service necessitated by malfunctions arising from power fluctuations. Accordingly, the customer should ensure that its electrical system is upgraded or that any necessary power streamer is attached to ensure a steady and uninterrupted flow of electrical current. This will reduce the risk of any loss of information stored in the computer, or any other malfunctions resulting from fluctuations in the electrical current.

Other common exclusions include the costs of relocating the equipment to another site, the performance of any electrical work external to the equipment, and the cost of refinishing or reconditioning which becomes necessary in the event that normal maintenance service cannot keep the equipment in good working condition. If the latter exclusion is set forth in the maintenance agreement, the customer should obtain a warranty from the supplier that the equipment will not require refurbishing or reconditioning for a specified period of time, or in the case of a lease, during the lease term.

The agreement should provide for the user to notify the maintenance supplier of any proposed modifications or attachments to be made to the computer equipment. The supplier should be required to advise the user whether such modifications or attachments will interfere with the operation of the equipment, in which event it will wish to have the right to discontinue maintenance service.

If the user is dependent upon the maintenance supplier for operating supplies such as tapes, disks or cartridges, the supplier should be required to provide such supplies during the term of the maintenance agreement. However, the supplier's obligation to provide maintenance service should not be affected if such supplies and accessories are obtained from any third party.

(h) Ownership of Parts

Where maintenance services are provided for customer-owned equipment, any used parts which are replaced by the maintenance supplier will normally belong to the supplier. In the case of leased equipment, both new and replaced parts will belong to the lessor.

(i) Customer's Responsibilities

The user should be required to provide the maintenance supplier with access to the products covered by the agreement, adequate working space and the right to use such equipment and facilities as are necessary to enable the supplier to perform the various services. The customer may also be required to follow certain operator procedures to ensure that the equipment is kept in operating condition. Finally, the agreement will normally provide that a representative of the customer must be available at the site at all times that service is being provided by the supplier.

(j) Standards of Performance

The maintenance agreement should indicate the standard of performance to be met by the maintenance supplier. In general, the supplier should be required to provide such services and replacement parts as are necessary to ensure that the equipment remains in "good operating condition" throughout the term of the agreement. The supplier should be required to replace all defective parts with new parts or their equivalent. The agreement may also include covenants for the equipment to be maintained to perform in accordance with the manufacturer's specifications and with other representations and warranties set forth in the purchase agreement.

A customer may also wish to obtain a warranty from the maintenance supplier as to the minimum periods of time during which the equipment will be fully operational. If the equipment is "down" or not fully operable for a period of time in excess of the maximum allowable period set forth in the agreement, the customer should have the right to receive a credit against the monthly maintenance fee. The maintenance supplier should also covenant and agree that all services under the agreement will be carried out promptly in a workman-like manner by skilled and competent personnel.

In conclusion, if a dispute arises between an equipment supplier who is also providing maintenance service and a user, the resolution of the dispute may be complicated by the user's need to receive continuing maintenance service. For example, the supplier may threaten to cease maintenance service if the user fails to make regular lease payments. The user may be able to obtain an injunction to restrain the supplier from ceasing to provide maintenance service pending the resolution of any such dispute.

17

Agreements Relating to Computer Software

1. SOFTWARE LICENCE AGREEMENTS

(a) Introduction

A software licence agreement is a contract whereby the owner of the rights to a computer program makes a copy of that program available for use by another person. Because the licensor retains ownership, he can enter into agreements with a number of users. The agreement imposes various restrictions on the user's rights to the programs so as to preserve the licensor's proprietary rights and to maintain the value of the product.

Inherent in the concept of software licensing is the extension by contractual arrangements of the protection afforded to computer programs by intellectual property laws. As discussed elsewhere in this text, copyright law is of limited value in the protection of computer programs insofar as it protects only the expression of information rather than actual ideas. Moreover, copyright extends only to works that come within certain recognized categories (for example, literary works), and it grants protection only in respect of certain uses (for example, reproduction, adaptation or translation).

Trade secrecy law may also be of limited assistance in protecting computer programs insofar as it requires that information be treated at all times as confidential by the holder. Reliance on trade secrecy law also requires the existence of a confidential relationship between the proprietor of the programs and any person to whom the programs are disclosed. Software licences may reinforce the protection of programs that may be available under trade secrecy law by clarifying the status of software as a trade secret. Moreover, even in respect of software which is not truly confidential, a software licence sets out the permitted uses that may be made of the licensor's proprietary information.

Software licensing, then, may be regarded as a means for superimposing an additional layer of protection for computer programs beyond that provided by statute or trade secrecy law. In addition to limiting the rights of the licensee to use the programs, a software licence also reduces the risks of the programs falling into the hands of a third party. At the same time, questions may arise as to the enforceability of a software licence agreement against a third party who has not expressly agreed to the licence and who acquires the software without notice of its terms.

Software proprietors or distributors who market their programs to a number of users generally rely on a standard form of licence agreement. In the case of mass-marketed software distributed for use with micro-computers, potential users generally will have no choice but to accept the terms of such licences if they wish to acquire the product. For programs licensed to business users for substantial licence fees, on the other hand, the licensor will often be prepared to negotiate changes to reflect the user's particular business needs.

The following section of this chapter will review the types of provisions normally found in a software licence agreement. Some problems that arise where programs are licensed through distributors, or are "mass-marketed" for the microcomputer market, are also discussed in this section. The terms that should be inserted in various other contracts relating to the supply and distribution of computer programs are discussed in detail in the following chapter of this text.

(b) The Key Provisions Of A Software Licence Agreement

(i) *Description of Licensed Programs and Documentation*

The agreement should set out a complete description of the licensed programs, including a reference to any trade marks by which the programs are identified, and the particular release or version of the programs to be provided. The description of the programs may be by reference to functional specifications which are incorporated as a schedule to the agreement. The agreement should indicate whether the programs are to be provided only in object code form (that is, in a form which can be executed directly by the computer) or whether source code (that is, the "human-readable" program written in a "high-level" programming language) is also to be provided. There should also be a description of any documentation, including user guides and operating manuals, which are to be provided with the programs.

The licence should also extend to any modifications, enhancements and new releases which are provided from time to time by the licensor, including those provided under the licence agreement or any accompanying support services agreement.

(ii) *Scope of Rights to Use Programs*

A software licence agreement will set out the terms on which the licensee is entitled to use the programs. Most standard agreements will grant the licensee only a non-exclusive, non-transferable and non-assignable licence to use the programs on its computer system in the processing of its own data. The agreement may set out the kinds of uses that are permitted under the licence agreement, such as the "copying", storage, transmission and display of the programs in connection with their execution on the equipment for which they are licensed, together with associated peripheral devices such as terminals, printers or modems.

Use of the programs may be restricted to certain units or types of computer equipment located at a specified address. The relocation of the equipment to another computer facility may be subject to the licensor's consent. If the licensee wishes to use the programs on additional or upgraded equipment or at multiple locations, an additional licence fee may be payable.

The licence agreement will usually contain restrictions on the number of copies of the computer programs and accompanying program materials such as user manuals that may be made, the uses that may be made of such copies and where such copies may be kept. The licensee should be required to affix the software owner's copyright and other proprietary legend to all copies. The user will generally be required not to provide such copies to any third party.

A software licence agreement may also require the licensee not to permit other persons to obtain access to the programs, whether directly at the user's premises or through telecommunications facilities. Such unauthorized access would enable third parties to obtain the benefit of the programs without being subject to the restrictions of the licence agreement. In particular, the licensor will wish to restrict downloading of the software to other computer systems. The non-disclosure provisions that may be found in a software licence agreement are discussed in greater detail below.

A software user must carefully consider whether any conditions imposed on the right to use the software are too restrictive. For example, the user may require the right to use the programs, or to sub-license their use, for processing data belonging to affiliated or subsidiary companies. The user may also require additional copies of the software for back-up purposes, and the right to use such copies on alternative equipment located at a back-up facility. The licensee will also wish to have the right to transfer use of the program to any substitute computer equipment which it subsequently acquires for use at its premises. The licensee will also wish to have the right to assign its rights to the licensed programs (particularly operating system programs which are essential to the operation of certain computer equipment) in conjunction with any sale or other disposition of the equipment on which the programs are used.

A user will also wish to review the restrictions in the licence agreement as to its right to use the licensed programs only on a designated machine or at a designated site. A licensee may be able to negotiate a "site licence" entitling it to make multiple copies of the licensed programs for use on a number of microcomputers located at one address. In some cases, the licence will entitle the user to use the programs on a "file server" which is connected with a number of microprocessors through a local area network. Similarly, if the user has a number of branch offices, it may wish to have the right to make copies of the programs for use at such branch offices, or to permit on-line access from such branch offices to the programs installed at the user's principal computer facility. In other cases, a user may negotiate a basic licence fee for the first installation of the programs, and discounted fees for additional copies to be used at its principal computer facility or at branch offices.

(iii) *Modifications and Enhancements to Software*

The licence agreement will normally contain restrictions on the licensee's rights to modify the programs, to develop enhancements and adaptations, to convert the programs for use on other computer systems, to merge them with other programs, and to de-compile or disassemble the programs (that is, "unravel" the object code to re-create the underlying source code program). In some cases, the licensee will be given the right to modify the programs for limited purposes necessary for the conduct of its business, such as to reflect changes in applicable legislation. In other cases, the licensee will be permitted to modify the licensed programs or merge them with other programs provided that they are restored to their original condition upon termination of the licence. The agreement should indicate whether any modifications to the licensed programs are to belong to the licensor or the licensee.

The conferral of a general right of adaptation will require the user to have access to the source code which the licensor will often be unwilling to provide except on payment of substantially greater licence fees. An unrestricted right of adaptation may also lead to the licensee developing new software packages which may incorporate elements of the original programs but are otherwise treated as proprietary to the licensee. Such new programs could then be marketed in competition with the licensor's programs. Accordingly, a licensor will generally impose a number of restrictions on the licensee's rights to modify or to enhance the licensed programs, or permit such modifications and enhancements only in limited circumstances such as where the licensor becomes bankrupt, ceases to carry on business or ceases to provide support services to the licensee.

(iv) *Term of Licence*

A software licence may be granted for either a fixed time period or a perpetual term. In either case, the licence will be subject to earlier termination if the user is in default under the agreement or upon the occurrence of certain other events. If the licence is for a limited term, an additional fee may be payable upon its renewal. The licence agreement should indicate whether such renewal takes place automatically or is subject to notice being provided by either party, and should set out the fees that are payable for the renewal term.

(v) *Delivery and Installation of Programs*

The software licence agreement should specify where the computer programs are to be installed and the date of delivery and installation. The licensor should be required to deliver all components of the software package contemporaneously and the agreement should provide for the right of the licensee to cancel the agreement if delivery is not made in a timely manner.

Any responsibilities of the licensor for loading the programs on the user's computer and for running various diagnostic tests to ensure that the programs are operable should be set out. If the licensor is responsible for installation, it should be required to certify that the various steps involved in installation of the programs have been taken. The date of such certification will determine the commencement of any acceptance testing and warranty period, if applicable.

(vi) *Testing and Acceptance of Programs*

Apart from any standard tests that a licensor may run as part of the installation procedure, a software user may also wish to conduct its own tests to determine that the computer programs perform their intended functions and otherwise comply with the specifications which are incorporated as part of the agreement. Acceptance testing will also enable the user to determine whether the licensed programs can be integrated as part of its existing computer configuration. Such testing generally will involve using the programs in situations which approximate the user's actual operating environment, including extreme situations which may not be frequently encountered.

The contract should indicate the date following installation that acceptance testing is to commence, the period of time during which the programs must operate in conformity with the test criteria and the final date on which the acceptance test criteria are to be met. The contract should require the supplier to correct any "bugs" or errors, and to remedy any non-conformities with the specifications, discovered during acceptance testing. The licensee should also be required not to unreasonably refuse the software once it is demonstrated as meeting the acceptance test criteria.

As in the case of an acquisition agreement for computer hardware, it is essential that a licence for computer programs set out clearly the manner and time at which acceptance is to take place. The licensor may wish to provide that the programs will be deemed to be accepted within a certain number of days following the expiration of the acceptance testing period unless written notice of rejection is provided by the licensee within such period.

(vii) *Prices and Payment*

The software licence agreement must specify the amount of the licence fee, the manner in which it is to be paid and the date of payment. For standard software packages, the licensor may wish to receive the entire licence fee upon the execution of the licence agreement. In other cases, the licensor will generally wish to receive at least a portion of the fee upon execution of the agreement or upon the delivery or installation of the programs. The licensee, on the other hand, will wish not to be bound to pay for the programs, or at least a portion of the licence fee, until the programs have been accepted, particularly if acceptance is conditional upon conformity with certain acceptance test criteria. If the programs are not accepted or are subsequently returned because of a breach of warranty, the licensee will wish to have the right to receive a full refund of any moneys paid prior to the date of refusal or the date on which they are returned.

A licence agreement may provide for annual or other periodic royalty payments in addition to or in lieu of a one-time licence fee. Periodic payments may also be required to be made on account of maintenance or support services provided by the licensor. In either case, the licensee will wish to have a price protection clause to limit the amount of any increases during the second and any subsequent year of the term of the licence agreement. In many cases, the licensor will be prepared to accept limits on such increases for a fixed period of time (for example, five years). Such limits may be expressed as fixed percentage amounts or may be based on annual increases in the cost of living, as measured by the Consumer Price Index published by Statistics Canada.

Other services for which charges may be imposed under a licence agreement include installation, training, documentation, testing and educational materials. The licensee should ensure that all such charges are clearly identified and that the agreement provides that no other charges are payable for the products and services to be provided thereunder.

The agreement should set out the times at which invoices are to be issued and when they are to be paid. The agreement will normally provide for interest to be paid on over-due amounts and will permit the licensor to terminate the licence or any support services if the licence fee or support services fee is not paid within a certain time period following the due date.

The agreement will also normally provide for any taxes payable in respect of the licensing of the software or its use by the licensee to be the licensee's responsibility. The licensee should be aware that if the licensor is located outside Canada, it may be required to remit withholding tax in respect of the licence fees to the Canadian Government under section 212(1)(*d*) of the Income Tax Act. Accordingly, the agreement should indicate whether the amount of the withholding tax is to be deducted from the fees payable to the licensor[1]

(viii) *Software Documentation*

The software licence agreement should clearly describe all "user" documentation such as user guides and operating manuals which are to be provided to the licensee. The agreement should provide for the user guides and operating manuals to be in sufficient detail to enable the user's personnel to use the products without recourse to the licensee.

If the licensee is to be responsible for its own "support services", it will wish to receive various types of "source" documentation including printed source code listings, machine-readable source code and technical specifications. Such documentation should be in sufficient detail to enable skilled programmers to maintain the programs and to develop enhancements.

In many cases, a software licensor will be unwilling to provide the source code to its licensees. In such an event, a source code escrow agreement may be required to set out the circumstances and method by which the licensee is entitled to obtain access to the source code. The terms of such an agreement are discussed in greater detail in section 3 of this chapter, *infra*.

Where corrected versions, modifications or new releases of the computer programs are developed and provided to the user, the licensee should be entitled to receive corresponding revisions to the program documentation.

(ix) *Training*

The licence agreement should also specify the nature and extent of the training to be provided to the user's personnel. If the licensor is to assist the licensee in converting data into a form where it may be used with the licensed programs, the nature and extent of such assistance should be specified.

---

1 See Chap. 20, *infra*.

(x) *Ownership of Software*

A software licence generally will provide for the licensor to retain all proprietary rights to the software, subject only to the licensee's rights of use set forth in the agreement. Accordingly, the licensee will be required to acknowledge the licensor's proprietary rights, and not to remove any trade mark, copyright, trade secret or other ownership notice inserted on the software. Any such proprietary legend should be inserted in human readable form on any storage media in which the programs are contained and in machine-readable form in the programs themselves.

(xi) *Confidentiality and Non-Disclosure Provisions*

The software licensor may claim that the licensed programs as well as the source code and documentation relating to the programs contain confidential information which it wishes to be protected as its trade secrets. Accordingly, the licence agreement will generally require the licensee not to disclose or distribute the programs to any third party, other than its employees or other persons such as auditors or consultants to whom disclosure is necessary to enable the licensee to use the programs. The licensee should be required to take reasonable measures to protect any confidential information, including such measures as it would be reasonably expected to take to protect its own confidential information. In particular, any employees or other persons who are permitted access to any confidential information should be required to covenant not to copy or disclose such information or use it for any purpose except as set forth in the agreement. The licensor may wish to obtain direct covenants in its favour from such persons so that the confidentiality obligations may be enforced directly by the licensor. The licensee should also be required to maintain any notice of confidentiality affixed to any programs and documentation so that notice of the licensor's claim of confidentiality will be provided to third parties.

(xii) *Warranties of Title and Performance*

A licensee will wish to obtain a representation and warranty from the licensor that it has title to the software and the right to grant the rights set forth in the agreement, and that the grant of such rights will not infringe any copyright, patent, trade secret or other proprietary right of any other person. Such a representation and warranty is of particular importance where the software is developed by a software house that may have a number of employees employed on a short term basis or that may have contracted with third parties for the development of the software. In the latter case, the licensor will not have clear title to the software unless it has obtained written assignments of copyright from the parties who contributed to its development. The licensee may choose in some cases not to rely exclusively on the licensor's representations, but instead will carry out its own investigation to determine whether the licensor appears to have the rights to the software that it claims.

The licensee will also wish to obtain certain warranties of performance, including the following: (a) that the programs will operate with the hardware configuration for which they are intended, and in the case of application programs, in conjunction with operating system software supplied by the licensor or the equipment manufacturer; (b) that the programs will perform the various functions and have the features set forth in any functional and technical specifications which are incorporated in the agreement; and (c) that for a specified period of time following acceptance, the programs will be substantially error-free, and that the supplier will use its best efforts to correct defects discovered in the operation of the programs as promptly as possible.

A licensor may be unwilling to give all of these warranties, particularly if the software is provided at a low cost or if the licensor is not familiar with the licensee's particular business. In some cases, the only warranties that the licensor will be prepared to give are warranties of title and limited warranties of performance and freedom from defects. For example, the licensor may warrant that for a limited period of time (for example, 90 days) following the delivery or installation of the programs, the programs will perform in accordance with specifications and that the media on which the programs are contained are not defective. The nature and extent of the warranties to be provided by a licensor will depend on such factors as the nature of the business relationship and the price to be paid for the programs and must be carefully negotiated in order to protect both parties' interests.

(xiii) *Patent and Copyright Indemnity*

A software licence agreement will normally contain a clause whereby the licensor agrees to defend and indemnify the licensee against any claim that the software infringes any patent, copyright or other proprietary rights of a third party. The contents of such a clause are discussed in detail in Chapter 16 of this Part dealing with computer hardware acquisition agreements.

(xiv) *Termination of Licence*

The licensor will generally wish to provide for its right to terminate the licence if the licensee breaches certain provisions of the agreement, particularly those relating to restrictions on use, confidentiality or payment.

The licence agreement may also provide for its termination if any of the conditions of the licensee's rights to use the programs cease to be met, for example, if the licensee ceases to own the computer hardware for which use of the programs is licensed. The licensor may also wish to have a right of termination if the licensee ceases to carry on business, takes any action for winding up or liquidation or becomes bankrupt.

Termination of the licence may seriously impair the ability of the user to conduct business. Accordingly, any right of termination on the part of the licensor should not take effect until notice of termination is given to the licensee and the licensee fails to cure the default within the notice period.

Upon termination of the licence for any reason, the user should be required to remove the programs from its computer, and return or destroy all copies of the programs and documentation within its possession or control. However, the agreement should provide that such termination will not affect the obligations of the licensee not to disclose any confidential information or use such information provided to it.

(xv) *Other Provisions*

The licensee may also wish to have the right to obtain maintenance and support services from the licensor. It will also wish to have the right to obtain access to the source code in appropriate cases. The terms normally found in a support services agreement or an escrow deposit arrangement are discussed in detail in following sections of this chapter. Provisions for software support may be included as part of a licence agreement, and if set forth in a separate agreement, reference to such agreement should be made in the software licence.

A licensor will also wish to include in the agreement various clauses to limit or exclude its liability for any defects in the licensed programs or for any losses arising from the licensee's use of the programs. These clauses, together with various other "standard" provisions that may be included in a software licence, are discussed in detail in the final chapter of this Part.

(c) Marketing Through Distributors

If a software proprietor wishes to market its software to a number of users, particularly at an international level, it would be impracticable to negotiate licences directly with all end users. In such situations, the owner may enter into arrangements for the software to be licensed through a network of distributors and dealers. The distributors and dealers may have the authority either to grant licences directly to end users or to solicit customers who will then enter into agreements with the software proprietor.

If a distributor is granted the right to enter into licence agreements directly with end users, difficulties may arise if the distributor ceases to exist or to carry on business. In the absence of any contractual relationship between the software owner as head licensor and the sub-licensee, the owner would be unable to enforce the restrictions in the licence agreement directly against the sub-licensee. Accordingly, any sub-licence agreement between a distributor and end user should indicate that it is made pursuant to the terms of a head licence agreement between the owner and distributor. It should further state that in accordance with the terms of the head licence agreement, all rights of the distributor to the licensed programs may revert in certain circumstances to the head licensor. The sub-licensee should covenant that upon receiving notice from the head licensor of such reversion, it will continue to be bound by all provisions of the sub-licence agreement as if it had been made directly with the head licensor. The circumstances in which such reversion will occur should be set forth in the agreement made between the software proprietor and the distributor, as discussed in detail in section 5 of this chapter, *infra.*

(d) Mass Distribution of Software: The "Shrink-Wrap" Licence

The "traditional" software licence is executed between a software supplier and a business user after the terms are negotiated and settled by the parties. Even if the supplier's standard form contract is used, the user will generally have an adequate opportunity to review the contract before the licensed programs are provided. Execution of the contract will signify the user's acceptance of the terms governing its rights to use the programs.

Where software packages are marketed through retail outlets to a large number of consumer or business users, it may be impracticable to obtain a signed licence agreement from each user. As a result, it has become increasingly common for software intended for use on microprocessors to be marketed through "tear-me-open" or "shrink-wrap" licences. Under such an arrangement, the diskette or cassette containing the computer program is marketed through a retail outlet or a distributor for a fixed price. The document containing the terms of the software licence is enclosed with the diskette or cassette in a transparent wrapping so that the terms of the licence agreement or a notation referring to such terms can be read before being opened. A prominent notation will ordinarily be inserted indicating that the user will be deemed to accept the terms of the licence agreement by breaking the seal around the software package or by using the program. The user may be entitled for a limited period of time following the date of acquisition to return the package and to receive a refund of the licence fee, provided that the package has not been opened.

A shrink-wrap licence will contain restrictions on the licensee's rights to use the software similar to those set forth in any other licence agreement. In particular, it will provide for the grant of a non-exclusive and non-transferable licence to use the program on one computer, will impose restrictions against copying, will contain prohibitions against the transfer of possession of any copy of the program (or permit such transfer only if the transferee agrees to be bound by the terms of the licence agreement), and will prohibit modifications of the program or its merger with other programs. The agreement may contain limited warranties that for a specified time following the date of acquisition, the licensor will replace the program package if the medium on which the program is stored is defective, or if the program fails to perform in accordance with the specifications set forth in any documentation. The agreement will generally provide for the exclusion of all other representations, warranties and conditions as well as the exclusion of any liability on the part of the licensor for damages arising from any defects in the program or the licensee's use of the program.

The marketing of software by shrink-wrap licences raises questions of enforceability because of uncertainty both as to the nature of the transaction whereby rights to the software are granted as well as to whether the terms of the licence are accepted as a binding contract by the licensee. Uncertainty may also arise as to whether the distributor or retail outlet through which the programs are licensed has the actual or apparent authority of the head licensor to make an offer on the terms of the licence agreement to any end user. Although the enforceability of such arrangements has not been considered as yet by a Canadian court, the principles of law that apply have been considered in other circumstances.

With respect to the nature of the transaction, a shrink-wrap licence generally contains an express provision to exclude any transfer of ownership of any intellectual property rights in the program itself. However, it is possible that the transaction will be characterized as either a sale of a copy of the program, or a sale of the medium on which the program is stored accompanied by a licence to use the program itself. The characterization of the transaction in either manner may be inconsistent with certain restrictions set out in the licence agreement, such as a prohibition against the transfer of possession of a copy of the program without the consent of the licensor. The characterization of the transaction as a sale of a copy of the program may also be inconsistent with an obligation to return or destroy the program if the licence to use it is terminated.

The characterization of a shrink wrap licence as a sale will also have implications for the rights of the user and the obligations of the supplier under copyright and consumer sales legislation. For example, under section 117 of the U.S. Copyright Act, as enacted in 1980, it is not an infringement of copyright for the owner of a *copy* of a computer program to make another copy, provided such copy is an essential step in the use of the program or is made for archive purposes only. In addition, under the "first sale" doctrine which is embodied in section 109 of the U.S. Copyright Act, the owner of a copy of a copyright work is entitled to sell or otherwise dispose of such copy without the authority of the copyright owner.

With respect to consumer protection legislation, most such provincial legislation in Canada prohibits the exclusion of certain statutory warranties in the case of a "consumer sale". For example, section 34 of the Consumer Protection Act of Ontario[2] provides that the various implied warranties and conditions set forth in the Sale of Goods Act[3] may not be waived in the case of a consumer sale. These warranties and conditions include warranties of title and right to quiet possession, merchantability and fitness for the purpose for which the goods are being acquired which was made known to the seller.[4]

Other questions of enforceability may arise in connection with shrink-wrap licences because of uncertainty as to whether the terms set out in the contract are actually assented to by the licensee. In order to bind the licensee, it may be necessary to show acceptance of the terms of the licence at the time the licence fee is paid insofar as the fee represents the consideration for the licensee entering into the agreement.

In order to bind a party to the terms of a document which he does not sign, he must be shown to have accepted the terms of such document by his conduct. In a number of so-called "ticket cases", the courts have had to consider whether a customer who is handed a ticket by a railway, steamship, parking garage or cloakroom is bound by the terms set forth on the reverse of the ticket. In order to be bound by such terms, it must be shown that the customer either knew of such terms or that the other party "did what was reasonably sufficient to give notice of conditions printed on the back of a ticket."[5]

In the case of *Thornton v. Shoe Lane Parking*[6] it was held that a customer at a parking garage was bound by the terms of a notice posted at the entrance of the garage since the notice constituted the offer. However, since acceptance took place when he put his money in the machine, the contract was then concluded so that the terms of the ticket issued by the machine did not constitute part of the contract. In any event, Lord Denning noted that the customer would be bound by the exempting condition set forth in the ticket only if he knew that the ticket was issued subject to such condition or that the company did what was reasonably sufficient

2 R.S.O. 1980, c. 87.
3 R.S.O. 1980, c. 462.
4 See Chap. 19, *infra*.
5 *Hood v. Anchor Line (Henderson Bros.) Ltd.*, [1918] A.C. 837 at 842 (H.L.) *per* Lord Finlay, L.C.
6 [1971] 2 Q.B. 163 (C.A.).

to give notice of the condition. It was noted the exempting condition in that case was so wide that the court should not hold the customer bound by it "unless it is drawn to his attention in the most explicit way."[7]

The other members of the Court in the *Thornton* case did not express any view as to when the contract was concluded. However, they were in agreement with Lord Denning that the defendant failed to do what was reasonably sufficient to bring notice of the exempting condition to the attention of the customer. Megaw L.J. noted that if a contract contains a condition or restriction that is usual, it may be sufficient to prove only that the intention to attach *some* condition has been fairly brought to the notice of the other party. However, if the condition involves a restriction that is unusual in that class of contracts, it must be shown that the intention to attach an unusual condition *of that particular nature* was fairly brought to the notice of the other party. The plaintiff was not bound by an exemption when the first attempt to bring notice of the condition to the plaintiff's attention was at a time when it would have been "practically impossible" for him to withdraw from the contract.

In *Olley v. Marlborough Court Ltd.*,[8] Lord Denning M.R. held that in order to find that a notice limiting a hotel company's common law liability formed part of the contract, made by the hotel with its guest, the contract had to be proved strictly, which could be done in one of three ways.

> Now people who rely on a contract to exempt themselves from their common law liability must prove that contract strictly. Not only must the terms of the contract be clearly proved, but also the intention to create legal relations — the intention to be legally bound — must also be clearly proved. The best way of proving it is by a written document signed by the party to be bound. Another way is by handing him before or at the time of the contract a written notice specifying its terms and making it clear to him that the contract is on those terms. A prominent public notice which is plain for him to see when he makes the contract or an express oral stipulation would, no doubt, have the same effect. But nothing short of one of these three ways will suffice.[9]

In *Tilden Rent-a-Car Co. v. Clendenning*,[10] the Ontario Court of Appeal considered whether a customer was bound by an exemption clause contained on a reverse of a form signed by the customer in order to rent a vehicle. The terms on the reverse had not been read by the customer and no special measures were taken to draw the customer's attention to the provisions of the contract.

In considering whether the customer was bound by the exemption

7 *Ibid.*, at 170.
8 [1949] 1 K.B. 532 (C.A.).
9 *Ibid.*, at 549.
10 (1978), 18 O.R. (2d) 601 (Ont. C.A.).

clause, the Ontario Court of Appeal adopted the following test set forth in *Smith v. Hughes*:[11]

> If, whatever a man's real intention may be, he so conducts himself that a reasonable man would believe he was assenting to the terms proposed by the other party, and that other party upon that belief enters into the contract with him, the man thus conducting himself would be equally bound as if he had intended to agree to the other party's terms.

It was held in the *Clendenning* case that this test applied even where a contract was signed, so that the signatory could not be bound by stringent or onerous clauses inconsistent with the true object of the contract unless reasonable measures were taken to draw such provisions to that party's attention.

The foregoing authorities would indicate that in determining the enforceability of shrink-wrap licences, the courts will consider whether necessary or sufficient steps were taken to notify the customer of the terms of the licence agreement at the time that the contract is made. If any of the provisions of the licence agreement are considered stringent or onerous, there will be a greater onus to show that reasonable measures were taken to bring notice of those specific conditions to the customer's attention.

The courts may also take into account the fact that it is common practice in the software industry to license rather than to sell programs. However, it is questionable whether the courts would find that customers should be aware as a matter of general knowledge that programs are licensed and that their rights to programs would be subject to the various restrictions set out in the licence agreement. This is particularly the case since software is often marketed in retail outlets in a manner similar to the manner in which other products are sold. Accordingly, there may be an onus on the part of the retail outlet to show that sufficient notice of the very fact that the software is licensed rather than sold was brought to the customer's attention.

There are a number of steps that may be taken to bring sufficient notice of the terms of a shrink-wrap licence to a customer's attention. First, the outside of the software package should contain a prominent notice indicating that the program in the package is licensed rather than sold to the customer. Second, such notation should indicate that the licence contains various restrictions on the rights of the licensee to use, copy and disclose the programs as well as various limitations of warranties and disclaimers of liability. Third, a prominent notice containing a similar warning should also be posted in the retail outlet either where the software is displayed or at the point of sale. Fourth, the licensee should be given the opportunity to read the terms of the licence agreement before paying

---

11 (1871), L.R. 6 Q.B. 597 at 607.

the licence fee: this opportunity may be provided either by displaying the entire text of the licence document in the transparent wrapping of the software package, or by setting it out on a separate card provided to the customer in conjunction with the software package. Fifth, the customer should be given the opportunity to return the software package unopened and to receive a refund of the licence fee. Such a right may be regarded as either a condition subsequent to the transaction made at the time of payment of the licence fee, or as an integral part of the licence transaction which is completed upon the opening of the software package. In either case, sufficient notice of the right of the licensee to return the software package should be given before the right is lost by the opening of the software package.

In the United States, statutory measures to provide for the enforceability of shrink-wrap licences have been taken in two states, namely, Louisiana[12] and Illinois.[13]

The Louisiana Software Licence Enforcement Act defines a "licence agreement" as follows:

> (2) "License agreement" means any written document on which the word "license", either alone or in combination with other words, appears prominently at or near the top of such document in such a position of prominence so as to be readily noticeable to a person of average literacy viewing such documents.[14]

Section 1953 of the Act provides that, if certain requirements are met, any person who acquires software is "conclusively" deemed to have accepted and agreed to all terms of the licence agreement of such software. These requirements include the following:

> (1) A written legend or notice is affixed to or packaged with the software or copy thereof in such a manner that the legend or notice is clearly and conspicuously visible upon cursory examination of the software and related packaging; and
> (2) The legend or notice is prominently displayed in all capital letters and in language which is readily understandable to a person of average literacy; and

---

12 R.S. 51: 1951-1956, c. 27 (1984).

13 Ill. Senate Bill 800 (1985). The Illinois legislation is substantially similar to that of Louisiana, except that it provides for the deemed acceptance, upon compliance with the requirements of the Act, only of certain terms relating to proprietary rights, limitation of warranties and termination, and not of all terms of the licence agreement. For a further discussion of the legislation enacted by these two states and their implications for Canada, see C.I. Kyer and J.A. Vanduzer, "U.S. Shrink Wrap License Legislation — A Model for Canada", Canadian Computer Law Association Conference, Toronto, October 6-7, 1986.

14 *Supra*, footnote 12.

(3) The legend or notice states clearly that: (a) any use of the software or copy thereof will constitute acceptance of the terms of the accompanying license agreements; or, (b) any opening of a sealed package, envelope, or container in which the software or copy thereof is contained will constitute acceptance of the terms of the accompanying license agreement; and

(4) The legend or notice states clearly that anyone who receives the software or a copy thereof and does not accept and agree to the terms of the accompanying license agreement may, within a reasonable time, return the unused, unopened software or copy thereof to the party from whom it was acquired, or to some other identified party, for a full refund of any consideration paid; and

(5) The person acquiring the software or copy thereof takes such action as is stated in the legend or notice to constitute acceptance of and agreement to the terms of the accompanying license agreement.

Section 1954 provides that the terms of a licence agreement which are deemed to be accepted upon compliance with the requirements of section 1953 may include any of the following:

(1) Provisions for the retention by the licensor of title to the copy of the computer software.

(2) If title to the copy of computer software has been retained by the licensor, provisions for the prohibition of any copying of the copy of computer software for any purpose and/or limitations on the purposes for which copies of the computer software can be made and/or limitations on the number of copies of the computer software which can be made.

(3) If title to the copy of computer software has been retained by the licensor, provisions for the prohibition or limitation of rights to modify and/or adapt the copy of the computer software in any way, including without limitation prohibitions on translating, reverse engineering, decompiling, disassembling, and/or creating derivative works based on the computer software.

(4) If title to the copy of computer software has been retained by the licensor, provisions for prohibitions on further transfer, assignment, rental, sale, or other disposition of that copy or any other copies made from that copy of the computer software, provided that terms which prohibit the transfer of copy of computer software in connection with the sale or transfer by operation of law of all or substantially all of the operating assets of a licensee's business shall to that extent only not be deemed to have been accepted under R.S. 51: 1953.

(5) Provisions for the automatic termination without notice of the license agreement if any provisions of the license agreement are breached by the licensee.

It will be noted that the various provisions referred to in clauses (2), (3) and (4) of section 1954 will be deemed to be accepted only if title to the copy of the computer software has been retained by the licensor. Such a qualification would appear necessary since such provisions would otherwise conflict with the provisions of sections 109(a) and 117 of the U.S. Copyright Law.

The enforceability of the various provisions of the licence agreement is subject to the further qualification under section 1955 that such terms

> must be clearly and conspicuously stated in the license agreement in language readily understandable to the person of average literacy, and the license agreement must be attached to or packaged with the copy of the computer software in such a manner that the terms are susceptible to being readily examined before the act which is deemed to constitute acceptance occurs.

In the absence of any legislation in Canada similar to the Louisiana and Illinois Software License Enforcement Acts, the enforceability of shrink-wrap licences remains to be determined under the common law principles referred to above. Accordingly, it is incumbent upon software licensors to take various steps to ensure that notice of the terms of the licences are brought to the licensees' attention. Owners and distributors of software packages will also wish to ensure that they have the right to exercise some control over the manner in which retail outlets market their products.

2. SOFTWARE DEVELOPMENT AGREEMENTS

A user may enter into a software development agreement in any case where it requires a custom software package designed specifically for its business needs or a "customized" version of an existing package, that is, containing special enhancements or applications to meet the user's particular requirements. Such an agreement is generally combined with a licence setting out the rights of the purchaser to use the developed software package.

In other cases, a computer hardware manufacturer or a software supplier may wish to sub-contract the development of programs which it intends to market to third parties. In this case, the party contracting for the development of the package will wish to acquire full ownership rights and not just a licence to use the software in its own business. In still other cases, the purpose of a software development agreement may be to convert programs designed for use on one computer system into a form where they can be used on other computers.

A software development agreement is perhaps the most difficult agreement to negotiate and settle in the area of computer law. This difficulty arises from the fact that the parties must describe a product that is not yet in existence, and deal with the consequences of a process that may end in failure or require greater expenditure of time and money than is originally anticipated. Accordingly, the parties must enter into a contract of this nature in a spirit of co-operation and be prepared to negotiate changes as circumstances require. The performance of such a contract also requires the participation and the co-ordination of both parties. The parties to such a contract should also ensure that its provisions properly reflect the various stages involved in the development process. In addition, the user will wish to acquire such rights to the software necessary for the purposes for which it is to be used and have various provisions for the protection of its rights if the software is not developed in a satisfactory and timely manner.

The following are the major provisions relating to the development of computer software that should be agreed upon between a supplier and a user.

(a) Functional Specifications

The "functional specifications" describe in non-technical terms the specific functions of the software and basic features and capacities which the software should have. Such specifications represent a general description of the product to be developed, and accordingly constitute the basic parameters of a software development agreement. The functional specifications are generally prepared by the customer in consultation either with a computer consultant or with a potential supplier as part of the negotiations preceding the execution of a formal agreement. The specifications may also be incorporated in a Request for Proposals issued by a user to potential suppliers.

Where possible, the functional specifications should be finalized prior to the execution of a formal contract and be included as a schedule to the agreement. As an alternative, it should be a condition of the agreement that the specifications be agreed upon within a certain time period following execution of the agreement. In most cases, the satisfactory development and acceptance of the specifications will be a condition of the user's obligation to make any payment in respect of services to be performed under the agreement.

(b) Development of Technical Specifications

While the functional specifications represent a non-technical description of the features of a software package, the technical (or design) specifications represent a detailed technical outline of the manner in which the programs are to operate and to perform their functions. The technical specifications should include a description of the procedures for inputting data, the internal logic for processing data, output descriptions, file layouts and program documentation.

The development of the technical specifications may take place in two or more stages. The first stage may involve a "fact finding" process whereby the parties agree to meet to review the user's requirements and to develop a general outline of the technical specifications to be provided to the user. The second stage may involve the preparation of preliminary or draft reports for the specifications for the user's review and approval. The last stage will involve the preparation of the final reports using the draft or preliminary reports which have been approved by the user.

The software development agreement should indicate the date by which each of the steps involved in the development of the technical specifications is to be completed. It should also set out the period of time during which the user is entitled to review each version of the specifications and to request necessary modifications. If the final specifications with all necessary modifications are not developed in conformity with the functional specifications within the time set out in the contract, the consequences should be specified. These may include the right of the user to receive liquidated damages in a specified amount for each day that the supplier is in default, and the right to terminate the agreement if such default continues for a certain period of time following the scheduled date. However, since the development of the specifications requires the input and cooperation of both parties, the developer may be unwilling to accept liability for damages if the specifications are not completed to the satisfaction of the user.

Once the technical specifications are completed and accepted, they should be incorporated in the contract as representing the features and capacities of the programs to be developed. As such, they may supersede and replace the functional specifications. If additional features are subsequently required by the user, the contract will generally provide that they must be agreed to by both parties and that they will be subject to a separate charge.

(c) Development and Installation of Programs

Following the completion of the technical specifications, the software developer will commence coding of the programs, that is, the creation of a set of instructions or statements based on the specifications which can be entered into the computer to perform the intended tasks. The development of the programs will generally include pre-installation testing conducted at the supplier's premises to ensure that the programs are in conformity with the specifications and are "de-bugged" to eliminate major errors or deficiencies.

The agreement should also provide for the developed programs to be installed at the user's premises. Installation of the programs will generally include performance of test procedures using sample data to ensure that the programs operate in conformity with the specifications. The developer should also be required to deliver to the user a complete set of user documentation: that is, operating manuals and user guides describing the operating conditions of the programs in sufficient detail to enable the user's data processing personnel to use the programs without resort to the developer.

If applicable, the developer should also be required to deliver complete source documentation, that is, source code versions of the programs in both printed and machine-readable form, together with file specifications, manuals and other documentation. The user will wish to acquire not just the printed source code listings but also a machine-readable version of the source code since this will greatly facilitate the development of modifications and enhancements to the programs. The source documentation should contain a complete description of the operating conditions of the programs in sufficient detail to enable the user or a third party to provide maintenance and support services without assistance from the developer.

In some cases, the developer will be unwilling to provide the source code and other source documentation on the grounds that they represent confidential proprietary information of the developer, and that the price paid by the user is only for the use of the programs rather than the right to maintain the programs or to develop enhancements. In such cases, the user may be able to negotiate an escrow arrangement whereby it would have the right to acquire the source code if the developer goes out of business, becomes bankrupt or ceases to provide support services for the software.

Other steps that may be required to be taken in connection with the installation of the developed programs include the conversion of the user's data from a manual system to an automated system, or from one automated system to another, so that such data can be processed using the new programs. Modifications or additions may also be required to the user's existing hardware and software configuration to ensure that the new programs can operate in conjunction with such configuration. Any responsibilities of the software developer relating to the conversion of data or the modification of the user's existing configuration should be specified in the development agreement.

Upon the completion of all necessary steps relating to the delivery and installation of programs and documentation, a written certificate of installation should be provided to the user. The date of such certificate will determine the time at which acceptance testing of the programs is to commence.

(d) Acceptance Testing and Acceptance of Software

Following the installation of the programs, the customer will wish to have an opportunity to conduct its own tests to ensure that the programs are error-free and perform in accordance with the specifications.

The acceptance test procedures should indicate the kind of test data to be used and include a series of benchmark tests to be met by the software, that is, the volume of data to be handled, the expected outputs and the required response times. The test criteria should be designed by the user in consultation with the developer and should involve use of the programs in conditions which will approximate the actual operating environment.

The development agreement should indicate the time period available for the user to conduct acceptance tests. The agreement should provide for the user to give prompt notice to the supplier of any errors or defects discovered in the operation of the software. The developer, in turn, should be required to make all necessary modifications and corrections as promptly as possible. The computer programs should not be deemed to be accepted until the user provides written certification that the programs have been successfully operating on the computer hardware in conjunction with accompanying systems software for a specified number of consecutive business days.

If a software developer is required to provide a number of modules of computer programs, which may be developed and installed in sequence, the agreement should specify the manner in which such modules are to be integrated. For example, if one or more modules of programs are completed and accepted, but subsequent modules are not completed, the agreement should indicate whether the user is required to pay for those modules previously installed and accepted or whether such modules may be returned with the user receiving a refund of any moneys paid to date. In many cases, the user will wish to have a provision for "final system acceptance", whereby its obligation to pay the final instalment of the costs of development of the software will not arise until all modules have been installed and have been accepted as working satisfactorily as a complete package at the user's computer facility.

(e) Implementation Schedule

A software development agreement should contain a schedule of the dates by which the various steps involved in the development and implementation of the software are to be completed.The schedule will include targeted dates for such key events as completion of technical specifications, acceptance of specifications, delivery and installation of programs, conversion of data, training of personnel, conduct of acceptance testing, acknowledgement of acceptance and payment.

In many cases, it may be difficult to predict in advance how much time will be required for each of these steps. Accordingly, the agreement may provide for the parties to review the implementation schedule periodically during the course of the agreement. In some cases, the implementation schedule may represent only an estimate of the dates for the completion of various steps, rather than reflecting firm contractual commitments. In such a case, the agreement should set out a final date by which all steps involved in the development of the programs will be completed and the consequences of the failure to complete all steps by that date. Alternatively, the user should have the right to give a notice requiring the developer to rectify any default in meeting a scheduled date for performance of any obligation, failing which the user would be entitled to treat the agreement as being terminated.

(f) Project Co-ordination

It has been noted that the development of software for a user's particular needs requires the continued co-operation of the parties. Although the supplier generally will have prime responsibility for developing the software, the user will be required to provide information concerning its business as well as personnel to assist in the development of the software. In some cases, the parties may wish to identify the key employees of each party who are to devote all or a substantial part of their working time to the project. If such persons leave the company, they should be replaced by persons of equivalent skill and experience. Each party will wish to be notified of any proposed changes in the personnel working on the project on behalf of the other party. In some cases, the user will wish to require the developer to obtain its consent before any such changes are made.

The contract may also require each party to provide monthly or other periodic reports of the progress made in developing the software. Meetings should also be required to be held between the supplier and the user on a regular basis to report on any problems. Detailed minutes should be kept of all such meetings.

(g) Rights of Ownership

As noted in the preceding chapter, licences for standard software packages generally provide for all proprietary right to be vested in the licensor subject to the right of the licensee to use the software in the conduct of its business. In the case of custom-developed software, however, the determination of ownership rights as between the supplier and user may be more complex.

If the costs and risks of development of new computer programs are entirely for the user's account, the user may wish to obtain an assignment of all rights from the developer. This will enable the user not only to use the software in its own business, but also to market the programs to other persons. In some cases, the customer may acquire title to the software, subject to royalties being paid to the developer in respect of any licences granted to other persons. This will enable the supplier to share in any proceeds if the software package becomes a valuable and highly marketable product. Conversely, the agreement may provide for ownership rights in the programs to be vested in the developer, with the user being entitled to obtain a share of royalties upon any further sales or licences.

If the software development agreement involves modifications to an existing software package, the developer ordinarily will retain title to both the original package and the modifications. In some cases, a user may wish to restrict the persons to whom custom modifications may be licensed, for example, so as to exclude licensing of the software to the user's competitors.

In the event that the user acquires only a licence to use the computer programs supplied by the developer, the terms of such licence should be specified. For example, the user may be entitled to use the programs only on certain computer equipment located at the user's premises and only for purposes of processing its own data. The licensee's rights to use the programs may also be non-exclusive and non-transferable. The agreement may provide for the termination of the customer's rights to the programs in the event that it commits a material default under the agreement. Other restrictions that may be imposed on a user's right to use computer programs are discussed in detail in the preceding chapter.

The proprietary rights clause should set out the rights of the customer to make modifications or enhancements to the programs, and indicate who is to have title to such modifications and enhancements. The right to modify and improve the software should also extend to any accompanying software documentation, including source code versions of the programs, supplied by the developer. In some cases, the user will wish to have the right to acquire any modifications and enhancements to the programs which are developed by the supplier and which are provided to other users.

Where the programs to be developed include algorithms or routines taken from other proprietary programs belonging to the developer, the developer will wish to impose qualifications on any assignment of rights to the programs to the user. For example, the developer may grant only a royalty-free, non-exclusive licence to use such proprietary elements together with the right to grant sub-licences in connection with the licensing of the newly developed programs.

(h) Prices and Payment for Software

A software development agreement should set out the prices to be paid for the programs to be developed and the times at which such payments are to be made. The agreement may provide for a fixed sum to be paid to the developer only upon the completion and acceptance of the finished product. In other cases, where the costs and risks of development are for the user's account, the agreement will provide for the developer to be paid on a *per diem* basis so as to reflect the number of personnel and amount of time required to develop the product. In still other cases, the agreement will provide for a deposit to be paid on execution of the agreement, with additional amounts to be payable upon the occurrence of certain events, such as acceptance of the technical specifications and the programs. Where the software consists of several modules for which the specifications and programs are being delivered at different times, the user's obligation to make the final payment should not arise until all modules have been accepted. Such acceptance, in turn, should not take place until all modules have been demonstrated as being operable as an integrated system.

The payment structure should be carefully negotiated to reflect both parties' interests. On one hand, the software developer will wish to have the right to receive substantial up-front payments to meet the costs of development. The user, on the other hand, will wish to delay its obligation to make such payments until the programs are actually developed, or the developer has at least made substantial progress in meeting various milestones in the course of development. In any event, the user will wish to have some limit imposed on the total amount payable for the programs. In addition, the agreement should set out any rights of the user to receive a refund of any moneys paid to the developer if all programs are not completed and accepted by the user.

The contract should clearly indicate who is responsible for out-of-pocket expenses incurred under the agreement including costs of office space, out-of-town lodging, transportation and travel, telephone, computer time and testing expenses. In many cases, the developer will wish to be reimbursed by the user for such expenses, in which event the user should be entitled to receive proper invoices. In addition, some upper limit should be set on the amount of the expenses for which the user may be liable.

(i) Remedies for Non-Performance: Termination of Agreement

The agreement should set out the consequences of non-compliance by the developer of certain key provisions, including the obligation to deliver the specifications and to complete and install the programs by the dates set out in the implementation schedule. The agreement may provide that subject to a period of grace that may be granted to the developer, the user will be entitled to terminate the agreement if the various milestones are not complied with. The developer, on the other hand, will wish to have the right to terminate the agreement if the user does not meet its obligations, including the obligation to make payments in a timely manner.

Because the development of computer programs requires the continuing co-operation of the supplier and the user, each party may consider any breakdown in the development process to be the fault of the other party. To avoid any dispute arising as to which party is in default, and whether the other party may terminate the agreement for such default, the contract may contain a clause to provide for its termination without proof of default. For example, the contract may provide that if either party in its reasonable opinion believes that the software cannot be completed at the times and otherwise in accordance with the terms of the agreement, either party would be entitled to terminate on notice to the other party. While such a clause would avoid the need to prove default by the party to whom notice of termination is given, the developer may not accept such a clause unless he is given the right to receive some moneys for the work performed to the date of termination.

Each party will wish to have a right of termination at its option if certain events occur such as if the other party becomes bankrupt, ceases to carry on business, takes any action or becomes subject to any proceeding with a view to its liquidation, dissolution, winding up or reorganization, or becomes subject to any action or proceeding for the appointment of a trustee or receiver for any part of its property.

The agreement should clearly set forth the consequences of the termination of the agreement at any time with respect to any payments to be made and the rights to the software or documentation to be provided to the user. If the user entered into the agreement on the understanding it would receive a finished product, it may wish to have the right to receive a refund of all moneys paid if the agreement is terminated at any time prior to final system acceptance (that is, acceptance of all modules as an integrated software system). In other cases, the user may be prepared to pay a portion of the total purchase price against receipt of functional and/or technical specifications, a portion of the programs, or even working papers, computer files and other materials that have been prepared by the developer to the date of termination. In some cases, the user may wish to have the right to acquire title to such materials, or an unrestricted right and licence to use such materials in the conduct of its business and for the purpose of developing further modifications and enhancements which would belong to the user.

If the agreement is terminated after some but not all of the programs are completed by the supplier, such termination should not affect any rights or licences granted in respect of the completed programs prior to the date of termination. Moreover, if the developer has a right to terminate the agreement by reason of certain defaults by the user, such as the failure to make any payments, such termination should not affect any licences granted to the user for any part of the software that has been accepted and paid for prior to the date of termination.

(j) Warranties

The software development agreement should clearly specify the warranties to be provided to the user. Of particular importance is a warranty that the software will perform the various functions and have the capabilities and features set forth in the functional and technical specifications which are incorporated in or developed under the agreement. The user should also obtain warranties that the programs will operate on the user's existing computer configuration and will be compatible with such system together with any application programs already installed at the user's premises; that the programs will be substantially error-free and that the developer will use its best efforts to correct any "bugs" detected in the operation of the software for a period of time to be agreed upon following acceptance; and that the software to be provided by the developer shall be developed in a proper and workmanlike manner by skilled and competent personnel. The user will also wish to obtain warranties that the developer has the right to grant all rights to the software set forth in the agreement, and that the grant of such rights does not infringe any copyright, patent, trade mark, trade secret or other proprietary right of any third party.

Each of the parties should be required to provide a warranty that it has all necessary corporate power and authority to enter into the agreement, and the persons executing the agreement have been duly authorized to do so. In appropriate cases, each party should provide copies of its constating documents together with a certificate of incumbency and a certified copy of a resolution of its board of directors authorizing the execution and delivery of the agreement.

Each of the parties will wish to be indemnified against any loss, damage or expense arising from the breach or alleged breach of any warranty of the other party. At the same time, the software developer may seek to exclude certain warranties and to limit the damages for which it is liable for breach of any warranty. Some of the exclusion or limitation clauses that a developer may wish to have inserted in the agreement are discussed in detail in the final chapter of this Part.

(k) Other Provisions and Related Contracts

A software development agreement is frequently made in conjunction with an agreement for the acquisition of computer hardware and operating system programs which are to be used to run the new programs. Such an acquisition may involve a substantial additional investment by the user. The user would wish to make its obligation to accept and pay for the computer hardware conditional upon the satisfactory completion of the programs. However, this is frequently impracticable, particularly if the hardware supplier is an independent third party. In some cases, the user may be able to negotiate a short term rental for the hardware with an option to purchase or to convert to a long term lease once the programs have been completed.

As noted above, a software development agreement may also be combined with a software licence, in which event the various provisions discussed in the preceding chapter should be included in the agreement. In addition, the user may wish to have the right to obtain access to source code and other documentation for the developed programs, in which event a source code escrow agreement containing the terms set out in the following section should be executed unless the parties agree that such source materials are to be provided immediately. The user may also wish to enter into a software support agreement containing the terms discussed subsequently in this chapter, in order to receive continuing corrections and enhancements to the developed programs.

3. SOURCE CODE ESCROW AGREEMENTS

A software escrow refers to an arrangement whereby the owner of the rights to a computer program places the source code and accompanying documentation in the custody of a third party to be released in defined circumstances to a licensee or other person to whom rights to the program have been granted. Such an arrangement is most commonly made in the case of a software licence agreement where the licensor is unwilling to make the source code available directly to the licensee because of a fear of loss of confidentiality and other proprietary rights, and loss of control over any changes to the source code. It may also be made with a distributor or publisher who is given the right to distribute copies of the program to end users. The software escrow arrangement represents a means of balancing the interests of the software owner in protecting the confidentiality, ownership and integrity of its programs with the interests of the user in ensuring that it has access to the source code in order to support the software. An escrow arrangement will be generally made only for specialized or custom-developed programs, and will not be considered for inexpensive off-the-shelf software for which other alternative packages are readily available.

Software escrow arrangements may vary greatly in the duties assigned to the escrow agent, the circumstances which trigger the release of the source code and documentation from escrow, and the rights granted to the escrowed materials. In structuring an escrow agreement, account must be taken of the provisions of relevant bankruptcy and insolvency legislation, since the bankruptcy or insolvency of the software owner are important triggering events entitling a software user to obtain access to the source material.[15]

The following is a summary of the provisions normally found in an escrow arrangement made between the owner of rights to the computer software, the escrow agent and the user.

(a) Obligation to Deposit Materials

The agreement should set out a complete description of all materials to be placed in escrow by the software owner. These materials should include the source code for the programs in both printed and machine-readable form. The agreement should set out the programming language in which the human-readable source code is written and the kind of configuration on which the machine-readable code can be used. The agreement should set out the medium on which the source code is to be recorded and provide that the medium must be programmable and free of defects.

The escrow may extend to ancillary technical documentation, such as file specifications, program notes, manuals and other documents containing a complete description of the operating conditions of the programs. The source materials collectively should be required to be in sufficient detail that trained computer personnel will be able to make corrections to the software and to develop enhancements, modifications and improvements without recourse to the software owner. The escrow agent should also receive a complete copy of the object code program in the same form as it is provided to the user.

(b) Verification of Source Materials

The escrow agreement should set out a procedure to enable the user to determine that the materials placed in escrow represent complete, accurate and up-to-date source code versions of the programs in object code form which are in the possession of the user. In some cases, the user may be prepared to rely on an affidavit or certificate signed by the software supplier and delivered concurrently with the source materials. Alternatively, the user may wish to have the contents of the escrow verified either by the escrow agent or an independent third party or to have the right to

---

15 For a discussion of some of the legal problems that may arise in connection with escrow agreements, including the enforcement of such agreements upon the bankruptcy of the software owner, see H. Lawrence, "Software Escrow Agreements" (1984), 1 C.C.L.R. 203, and C.I. Kyer and D.E. Milner, "Drafting Source Code Escrow Agreements: Some Legal and Practical Considerations", paper presented at Conference on Computer Law for the Practitioner Canadian Bar Association for Ontario, Toronto, November 28, 1986.

verify such contents itself. The verification procedures may take place at the software owner's premises or at those of the user or a third party. In any event, the source materials should be examined or used only in the presence and under the supervision of a representative of the rights' owner. Moreover, all persons involved in the verification of the escrow should be required to treat all information obtained as a result of such verification as confidential. The user will wish to have recourse to the verification procedure when the materials are initially placed in escrow, periodically during the term of the agreement and on every occasion that updates are provided for the source materials.

(c) Updates to Escrow Materials

Upon any corrections or revisions being made to the programs which are the subject of the escrow, the supplier should be required to make corresponding revisions to the source materials and to deliver such materials to the escrow agent. Such materials should be accompanied by a further affidavit to the effect that they are current, complete and accurate. As noted above, a user will also wish to have recourse to any verification procedure when new versions of the source materials are provided. All versions of the source materials delivered to the escrow agent should be clearly labelled and dated and any previous version should be kept until such time as the user certifies that the revised programs are operating satisfactorily.

(d) Terms of Escrow

The source code escrow agreement should be structured so that a trustee in bankruptcy acting for the software supplier will not have the right to recover possession of the escrowed materials. Accordingly, the escrow agreement should provide that all materials delivered to the escrow agent are to be held in trust for the user. This will mean that such materials will not constitute part of the property of the software owner in the event of its bankruptcy. However, the agreement should provide for the source materials not to be released to the user until the occurrence of certain triggering events. The agreement may also provide for the residual right of the owner to recover possession of all materials if the agreement is terminated prior to the occurrence of any triggering event.

(e) Obligations and Rights of the Escrow Agent

The escrow agent should be required to receive the source materials from the software owner, to keep such materials under such environmental conditions as may be specified, to store them in a secure location with necessary physical security measures, to maintain such materials in confidence, and to release them to the software user only in accordance with the terms of the agreement. The agreement should provide for the escrow agent to be reimbursed for services rendered in respect of the escrow, and should indicate who is responsible for such fees.

(f) Release of Materials from Escrow

Among the circumstances in which a software user will wish to obtain access to the source materials are the commission by the software owner of any act of bankruptcy, the failure of the software owner to carry on business in the normal course, the appointment of a receiver or a receiver and manager for any part of the software owner's business or assets, and the initiation of any action or proceeding involving the liquidation or winding up of the rights' owner. The user will also wish to have the right to obtain access to the materials placed in escrow if the software owner ceases to provide support, i.e., to provide corrections, updates or enhancements, under the terms of any support services agreement.

(g) Rights of User upon Release of Materials from Escrow

The user must have certain rights to the source materials once they are released from escrow. These rights, which may be set out in the original licence agreement rather than the escrow agreement itself, should include a licence to use all source materials in order to diagnose any problems encountered in the operation of the software, to make corrections and to develop any enhancements, modifications or updates. The program owner will wish to restrict the user's rights so as not to grant any additional rights to the programs other than as set forth in the original licence agreement. For example, insofar as the original licence agreement permits the licensee to use the programs in the conduct of its own business, the licensee should not by virtue of acquiring access to the source materials thereby acquire the right to market or distribute the programs.

The software supplier will wish to ensure that upon the release of any materials from escrow, the intellectual property rights in the source materials remain with the supplier, subject to the rights granted to the user. However, the user will wish to obtain ownership of the *copy* of the source materials released from escrow. In addition, the user will wish to have the right to make copies of the source materials for back-up purposes and to make changes and modifications to the source code to enable it to use the programs in the conduct of its business.

(h) Method of Obtaining Access to Source Materials

The escrow agreement should set out the mechanism to be followed to enable the user to obtain the source materials. The agreement may provide for the user to deliver a certificate or affidavit to the escrow agent, with a copy to the software supplier, to the effect that an event of default has occurred. The software supplier will then have a specified time period (normally five or ten days) to advise the user and escrow agent whether it disputes the existence of such event of default. If no such notice is received within the specified time period, the escrow agent would be required to release the source materials to the user. If a notice of dispute is served, the escrow agent would be required not to release the source materials until directed to do so by both the supplier and user, or by an arbitrator appointed under the agreement.

The arbitration clause should provide for the supplier and user to agree on a single arbitrator, failing which each party will appoint one arbitrator, with the two arbitrators to appoint a third. The time periods during which such appointments are to be made should be specified. The arbitrator or arbitrators should be required to decide whether an event of default has occurred as soon as practicable after their appointment, and in any event within a specified time period. Both the supplier and the user should be required to co-operate and to provide all necessary information to assist the arbitrator or arbitrators in making a decision. The agreement should provide for the appointment of an alternate arbitrator if the first arbitrator(s) is unable to reach a decision. The agreement will normally provide that the decision of the arbitrator(s) is to be final and binding.

The user will wish to obtain access to all source materials as soon as practicable after the occurrence of any triggering event. In particular, if errors have arisen in the operation of the programs to such an extent that the programs are non-operational, the user's entire business may be jeopardized by the lack of support services. Accordingly, a user should ensure that the time periods for any final determination of its rights to receive the source materials are kept to a minimum.

(i) Liability of Escrow Agent

An escrow agent may wish to have a clause excluding any liability on his part in connection with the performance of his duties except for negligence or wilful misconduct. He may also wish to have an indemnity against any claims brought by any third party.

(j) Replacement of Escrow Agent

The agreement should provide for the replacement of the escrow agent and the appointment of a successor by mutual agreement of the software owner and user. In addition, the escrow agent will wish to have the right to resign by giving notice to the supplier and user. Before his resignation becomes effective, the escrow agent should be required to use his best efforts to find a successor who is mutually acceptable to the other parties. In the absence of such agreement, the escrow agent should have the right to appoint a reputable financial institution as its successor to assume its obligations under the escrow arrangement.

(k) Termination

The escrow agreement should contain a clause providing for its termination and the return to the software owner of all escrowed materials upon the termination of the licence agreement on which the user's rights to the licensed programs are based.

4. SOFTWARE SUPPORT AGREEMENTS

Subsequent to the installation of computer software, the user will require ongoing assistance to ensure that its operation is error-free and that new updates and revisions are provided periodically. These provisions may be included in a separate support services agreement or may be incorporated as part of a general procurement agreement, a software development agreement or a software licence. The following are the principal provisions normally found in an agreement relating to the provision of support services.

(a) Nature of Services

The agreement should contain a full description of all services to be provided by the supplier to the user. Among these services are the following.

(i) *User assistance*

The user will wish to receive on-going advice and assistance to overcome any problems encountered in the use of the software, including problems arising from operator error, program error or hardware malfunction.

(ii) *Error correction*

The supplier of support services should agree to correct or remedy any errors and malfunctions detected at any time by the user. The supplier may wish to limit its obligation to the correction of material errors or malfunctions or the exercise of its best efforts to do so. Such error correction service may include any temporary "fixes" or "workarounds" as may be necessary to provide temporary solutions to problems encountered pending a permanent solution. The error correction service should also include any permanent "fixes" which are to be incorporated in the user's programs.

(iii) *Modifications and Improvements*

The supplier should be required to provide any modifications or improvements necessary to ensure that the programs continue to perform their principal computing functions in accordance with the original specifications. Such modifications may also include changes which improve the operating characteristics or standards of performance of the programs but which do not change their basic functions or features. The user should also have the right to receive all modifications and improvements which are provided to other users.

(iv) *New Software Releases*

If a user has acquired a standard software package which is used by a number of licensees, the support agreement should provide for the user to receive any new software releases which are developed and supplied to other users. These new releases may involve modifications to improve the efficiency of the software as well as enhancements to enable the programs to perform new functions.

In addition to any "standard" releases, a user may wish to acquire other enhancements to reflect the user's particular requirements. For example, if the software has been acquired for use in a particular business such as insurance or banking, the user will wish to receive all necessary modifications to reflect changes in applicable legislation or regulations. The agreement should set out the conditions on which the user is entitled to require any "non-standard" support, including the rates to be charged and the obligation of the supplier to make necessary personnel available. The agreement may also set out an implementation procedure whereby the supplier will be required to consult with the user to review the features of any new release and to establish a plan for its implementation to minimize any disruption to its business. The agreement should provide for the new release to be adequately tested prior to its installation. The user may also wish to have an opportunity to conduct its own acceptance tests to ensure that the new release performs in accordance with the specifications and is compatible with the existing hardware and software configuration. If the new release cannot be installed satisfactorily and is not accepted by the user, the user will wish to have the right to receive continuing support for any existing version of the software.

(v) *Updates of Manuals and Documentation*

As any corrections, modifications or new software releases are provided by the supplier, the user will also require corresponding revisions to any user manuals and other documentation. If the user is entitled to receive source code for the applicable programs, it will require revised versions of the source code at the same time as it receives any updates or revisions to the object code programs.

(b) Included Services

Support services are generally provided to a user in one of three ways, as follows.

(i) *Telephone Assistance*

A user's systems manager and operators should be entitled to contact the supplier's response centre to seek advice concerning the operation of the software and assistance in identifying the causes of problems and obtaining solutions. Such telephone support may include direct telephone access by toll-free "hot line" to the supplier's central office. The agreement should set out the time for the service supplier to respond to any call for telephone assistance.

(ii) *Telecommunications Support*

In some cases, support may include diagnostic service through telecommunications facilities from the vendor's service centre to the customer. The user will be able to obtain such service only if its computer facility includes a modem for communicating with the supplier's response centre.

(iii) *On-Site Assistance*

If a user is unable to resolve software problems through telephone assistance or telecommunications support, it will require on-site assistance from the supplier.

(c) Manner of Providing Services

The agreement should set out the time period by which the supplier is required to respond to any call for assistance, and provide for such support to be continued until a solution is obtained. The agreement may also set out an escalation procedure whereby if no solution is obtained within a specified time after a call for assistance is made, the user will have the right to call on additional personnel to be provided by the supplier.

The user may be required to designate one or more individuals who are responsible for reporting any problems to the supplier. The supplier should be required to designate certain persons as being responsible for responding to any call for assistance and for co-ordinating the provision of services. The agreement should provide for periodic meetings between representatives of the supplier and the user to discuss the dissemination of information, and the utilization and installation of any new releases in such a manner as to minimize disruption to the user's business and utilization of existing software.

(d) Term of Support Services Agreement

The support services agreement will normally commence on the expiration of any applicable warranty period relating to the software and will continue for the term set out in the agreement. The agreement may provide for a specified period of notice (for example, 60 or 90 days) to be provided by either party to the other if it wishes to terminate the agreement at any time or on any anniversary date of the agreement. The standard form of support agreement often contains a provision entitling either party to terminate the agreement on 30 or 60 days notice given at any time or following the expiration of a minimum term such as one year. However, a user will ordinarily require support services during the entire useful life of the software. Accordingly, the user should seek a provision whereby the supplier will be required to provide service for a minimum period of time, such as three or five years, or such longer period of time as the supplier provides support to other customers with similar software products, unless the agreement is earlier terminated at the option of the user or by the supplier for cause (for example, failure of the user to pay support charges when due).

(e) Charges for Support Services

Charges for support services are generally payable on a monthly or other periodic basis and may be fixed annually. Where services are to be provided over an extended period of time, a supplier will be reluctant to have the charges fixed over the entire term of the contract. Accordingly, charges for support services may be subject to increase either at the end of any year of the term of the agreement or at any time on 30 or 60 days' notice by the supplier. A user will wish to have any increase in the charges for support services during the second and each subsequent year of the agreement limited to a certain fixed percentage, or a percentage amount based on the consumer price index published by Statistics Canada.

In any event, any increase in charges should not exceed any increase imposed on other customers having similar software configurations.

The basic charges set out in the support services agreement will generally not extend to services performed outside the hours of coverage set out in the agreement. Any services performed at such times may be charged at the supplier's overtime rates and in some cases may be subject to availability of personnel.

The basic charges for support services may also not include new releases reflecting enhancements or new features of the software. The user should ensure in such event that it has the right to receive such new releases at the same rates as are charged to other customers.

The agreement should indicate which party is responsible for any expenses incurred in the performance of support services. Such expenses may include costs of travel, transportation or lodging incurred by the supplier's personnel, as well as the costs of any telephone or telecommunications services provided under the agreement.

(f) Termination of Agreement

A support services agreement may contain a provision for its termination prior to the expiration of the applicable term upon the occurrence of certain events, such as the termination of the licence for the software being supported, or the disposal of the user's computer system. A supplier may also have the option to terminate support if unauthorized alterations are made to the software by the user, and the supplier determines in its discretion that such alterations interfere with the performance of the software. Support services for standard software may also be terminable if the user fails to keep the software at the current revision or release level. The user, on the other hand, may wish to receive continuing support for an earlier release of the software, particularly if the later release cannot be successfully installed on the user's computer facility. The agreement will also contain a provision for termination by either party upon the occurrence of certain events such as the bankruptcy or cessation of business of the other party.

(g) Obligations of User

The agreement will ordinarily impose certain obligations on the user as a condition of its right to obtain support services. For example, the user will be required to designate a responsible individual to act as liaison with the supplier, to provide the supplier with access to the user's computer system and to obtain any necessary modems or other devices required to make use of any support to be provided by telecommunications facilities. The customer should also be required to follow any operating procedures specified in the supplier's documentation, to establish any necessary procedures to reconstruct lost or altered files, data or programs, to have an individual present at all times that the support services are being provided at the user's premises and to take the necessary steps to safeguard any confidential information to which the supplier may have access.

(h) Exclusions

The support services agreement will normally set out certain activities which are excluded from the services provided by the supplier. For example, the supplier will generally not be responsible for the conversion of customer's data, for system design and in some cases for software enchancements other than the standard enhancements provided to all users. The supplier may also not be liable for any support services which are required as a result of any unauthorized modifications to the software or the use of the software with any hardware or software devices not supplied or approved by the supplier. Any services required to be performed in respect of any such matters may be subject to separate charges.

(i) Ownership of Rights to Modifications and Enhancements

Where new releases or modifications are provided with respect to standard software packages, the supplier will normally retain all ownership rights to the derivative products as well as the original software. Support services agreements may contain restrictions against copying, use or further disclosure of any derivative products supplied in the course of a software support agreement. In addition, the supplier will retain all proprietary rights to the programs made available in connection with the provision of support services.

(j) Standards of Rights to Performance and Covenants by Supplier

A supplier of support services should be required to provide such services as are necessary to ensure that the programs perform in accordance with any applicable specifications and documentation. The supplier should also covenant that the operation of the programs will be substantially error-free, that the programs will perform their programming instructions and that the supplier will use its best efforts to remedy all errors or malfunctions detected in the software as soon as reasonably practicable. The supplier should also agree to provide all services in a workmanlike manner by skilled and competent personnel.

The supplier should also covenant and agree not to disclose or use for its own purpose or advantage any confidential information concerning the business and affairs of the user obtained in the course of carrying out the support services. In particular, the supplier should be required to take all reasonable measures to protect the confidentiality of all such information, including entering into appropriate covenants and undertakings with any of its employees or representatives engaged in providing such services.

The various warranties and covenants by the supplier may be subject to certain exclusionary and limitation clauses such as those found in other computer contracts as discussed in detail in the final chapter of this Part.

5. SOFTWARE PUBLISHING AND DISTRIBUTION AGREEMENTS

The growth of the microcomputer industry for both business and consumer applications has resulted in a substantial increase in the software publishing business. A software publishing or distribution agreement refers to any arrangement whereby the owner of a computer program authorizes another person to market or distribute the program with accompanying program documentation to end users, either directly or through a network of dealers.

Distribution of computer programs through a distributor may be of advantage to the program owner insofar as the distributor may be in a better position than the owner to pay the initial promotional and advertising expenses. Moreover, software publishers may have a ready-made network of dealers and sub-distributors for program distribution. Software publishers may also be better able to prepare or to assist the program owner in preparing the necessary user manuals and other documentation required by end users. Finally, software distribution arrangements may greatly facilitate the marketing of software in the international marketplace since the owner may choose a distributor who is familiar with local laws and business practices in a particular country.

Prior to entering into a distribution agreement, a program owner will be required to disclose the programs and provide other relevant information to potential distributors. To protect his rights in connection with such preliminary disclosure, the owner should enter into a non-disclosure or confidentiality agreement with any interested distributor.

The non-disclosure agreement will provide for the owner to provide a complete copy of the program to the distributor for its review. Such disclosure may extend to the source code as well as any documentation describing the functions and features of the programs. The agreement should contain provisions restricting the distributor from copying any of the information, from making any further disclosure of such information to any third party, or from using such information for any purpose other than to enable the distributor to assess whether it wishes to enter into a distribution agreement. The agreement should also require the distributor to enter into appropriate covenants with its employees and other persons who are permitted access to the programs and other materials to prevent their further disclosure or use. Additional provisions that should be inserted in such an agreement are discussed in detail in the final section of this chapter, *infra.*

A software publishing agreement is only one of several arrangements that may be made by a software owner to market its products, particularly in the international marketplace. Alternative arrangements that may be established include marketing through sales representatives who have the authority only to solicit orders on behalf of the owner, marketing through franchisees which are granted not only distribution rights but also rights to trade marks and marketing plans, marketing through representative offices which may display products and otherwise support the software owner's business without establishing permanent offices in a local area, the establishment of local subsidiaries and foreign branch offices which carry on business in different jurisdictions, and the establishment of joint venture arrangements between the software owner and companies located in host countries in order to satisfy local ownership rules.[16] The arrangement selected by a software owner will depend on a number of factors, such as the extent of the potential market, the amount of capital which the software owner wishes to invest, the need to protect proprietary rights, and the need to comply with local laws with respect to such matters as taxation, local ownership and currency exchange.

Although the primary focus of this section will be distribution agreements *per se*, many of the provisions found in such agreements, as reviewed in detail below, may also be included in such other arrangements. The following is a summary of these provisions.

(a) Description of the Software Package

The agreement should set out the trade name or trade mark by which the programs have been identified, and should include a description of the main functions and features of the programs. The agreement should describe the hardware and operating system environment in which the program may be used. There should also be a description of the documentation to be provided by the owner or that may be prepared by the distributor.

The description of the software package and its operating environment should be in sufficient detail to prevent any subsequent dispute arising as to whether the owner has delivered the package contracted for, and

16 See R. Bernacchi, "International Distribution of Computer Software — Advanced Legal and Practical Issues", Paper presented at Pacific Rim Symposium on Technology and the Law, Singapore, October, 1985.

whether the distributor has made unauthorized modifications to the package without the owner's consent and outside the scope of the contract.

(b) Grant of Rights

The agreement should clearly describe the nature and scope of the rights granted to the software distributor. In some cases, there may be a complete assignment of all copyright and other proprietary rights to the distributor. More commonly, however, the distributor will be granted an exclusive licence to distribute the program within a defined geographical area. In other cases, the distributor will be granted only non-exclusive rights in a certain area, in which event the owner will be able to enter into similar arrangements with other distributors, or market the programs directly to end users.

The execution of an exclusive distribution arrangement represents a major business commitment by a software owner insofar as it may result in the owner being "locked into" a single distribution channel. Accordingly, such an arrangement should be made only after the owner has carefully considered whether the distributor has the necessary resources and know-how to market the product. The owner should also carefully consider whether certain minimum quotas should be imposed on a distributor as a condition of the continuation of the distribution arrangement.

Where a distributor is granted only a distribution licence rather than an outright assignment, the agreement should set out the scope of the licensee's rights with respect to the permitted uses that may be made of the work, the term for which distribution rights are granted, and the geographical area over which the distribution licence extends. The following are some of the considerations to be taken into account with respect to each of these matters.

(i) *Permitted Uses*

In general, the software distributor will wish to have the right to make copies of the work for its own purposes and for distribution, the right to distribute such copies either directly or through a network of dealers and sub-distributors, the right to promote and advertise the work, the right to make and use demonstration copies and to perform the work publicly in connection with such promotional activities and the right to prepare, copy and distribute modifications, adaptations, translations and enhancements of the work.

The distribution agreement should also include a provision to permit the grant of sub-licences to use the work to end users. Such sub-licences should contain various restrictions such as those found in other end-user licence agreements.[17] The program owner may wish to approve the terms of any sub-licence agreement to ensure that it contains such restrictions on disclosure, copying and use that are necessary to preserve the owner's proprietary rights, as well as various disclaimer clauses to limit the liability to which the owner may be subject as a result of any defects in the software.

(ii) *Term*

A software distributor will wish the agreement to be for a sufficient length of time that it will be able to recoup its investment in connection with the advertising and promotion of the work. In general, the term of an exclusive software distribution agreement may vary from one to five years (with possible renewal terms) depending on the expected useful life of the program and the period that the distributor will require to recoup its expenses and to make a profit. In some cases, the distribution agreement may be for a perpetual term, subject to any provisions for earlier termination.

(iii) *Geographical Area*

The geographical area for which distribution rights are granted may be restricted to one province or state or may extend to an entire country, such as Canada or the United States or the North American continent. In some cases, the distributor may be granted the right to distribute the program world-wide. However, it is unlikely the distributor will actually exercise such rights in all countries. Accordingly, the agreement should contain a provision whereby rights to the work in any area will be re-assigned to the owner if the program is not distributed in that area within a specified time period following notice given to the distributor.

If the distributor is granted exclusive rights to a particular territory, the agreement should indicate whether the product may be marketed through any dealer or sub-distributor located in that territory, regardless of the location of any end user, or whether the end users themselves must also be located within that territory. The parties should also address the issue of where a business end user is deemed to be located where it has local offices in a number of jurisdictions.[18]

(c) Delivery of the Work

The contract should set out a complete description of the components of the software package to be delivered to the distributor and the time

17 See section 1 of this Chap., *supra*.

18 See Bernacchi, *ibid.*, at 18-22.

at which such delivery is to take place. In general, the software publisher will require delivery of the following items.

(a) A complete copy of the program in object code form on a magnetic disk or other machine-readable medium in a form suitable for reproduction;
(b) A complete copy of the source code listings in printed form together with a magnetic disk containing the complete source code in machine-readable form;
(c) Copies of other source documentation for the program, including program notes, flow charts, diagrams and technical descriptions. If applicable, this documentation should be in sufficient detail to enable the publisher to make corrections or enhancements to the work; and
(d) Any documentation prepared for end users, including manuals and descriptions of the operation conditions of the program.

In some cases, the program owner may be reluctant to release the source code and other source documentation to the distributor. In such event, the parties may enter into an escrow arrangement whereby the source materials are delivered to an escrow agent for release to the distributor upon the occurrence of certain events. These events would include the failure of the program owner to comply with any contractual obligation to make corrections or to develop enhancements or adaptations to the program package. Any source code arrangement should contain the various terms and conditions referred to in section 3 of this chapter, *supra*.

Some of the documentation provided to the program distributor, such as the user guide and other end user documentation, may be available only in draft form. In such event, the agreement should set out the respective obligations of the parties to complete the development of such materials.

(d) Acceptance and Acceptance Testing

Where the product to be distributed is a new release or has not been fully developed, the software distributor may wish to test the programs prior to distribution to ensure that they perform the various functions and have the features represented by the owner. The acceptance testing period will also provide the distributor with an opportunity to detect any errors in the operation of the programs.

Acceptance of a program package by a distributor may be contingent upon the programs being shown to comply with agreed upon acceptance test criteria or may be subject to the distributor's discretion. In either case, acceptance testing should be required to be completed by a certain date and unless a contrary notice is provided to the owner, the distributor should be deemed to accept the program upon the expiration of that time period. The owner will generally be required to correct any errors and remedy any malfunctions and defects detected in the program during the acceptance testing period.

(e) Program Support

In addition to errors detected during the acceptance testing period, the program owner will also generally be required to correct errors detected by the distributor or any end user during a certain period of time following commencement of distribution. The agreement should indicate the length of time that the owner is required to provide corrections, and who is responsible for the costs of such corrections. The distributor will also wish to have the right to make corrections itself or to have corrections made by an outside consultant if they are not made by the owner. For such purposes, the distributor will require access, either directly or through a source code escrow arrangement, to the source code and accompanying program materials.

(f) Enhancements, Derivative Works and Competing Programs

The distributor may also wish to acquire distribution rights to any enhancements and improvements which are made to the original programs by the owner. Since such improvements would generally enhance the marketability of the work, it should be in the interests of both parties that the distribution agreement extend to such improvements.

The distribution agreement may also provide for the program owner to develop additional enhancements from time to time according to specifications established by the distributor. The contract should indicate the time period following a request from the distributor that the owner is required to develop and deliver such enhancements. If such enhancements are not prepared by the owner, the distributor may wish to have the right to arrange for their development by a third party, with the expenses to be deducted from any royalties payable to the owner. In some cases, however, the owner will be unwilling to permit the distributor to make any changes to the original programs on the ground that it retains all rights to any enhancements of such programs.

The contract should set out the respective rights of the parties with respect to versions of the program designed for use on other computers or with other operating system programs, translations of the program prepared for distribution in other languages, or derivative works that are based on the licensed programs. In some cases, the original distribution rights granted to the publisher may extend to such new versions, translations or derivative works with the responsibility of preparing such new works being that of either the author or the distributor. If they are the responsibility of the distributor, the costs of preparing the new works may be deducted from the royalties payable to the owner. A further alternative is that the author may retain all rights to such new works, subject to the distributor being granted a right of first refusal before distribution rights are granted to any other person.

The software distributor may also wish to obtain a covenant from the owner not to develop or market any product that is likely to compete with the subject program without its consent. The author may also wish in some circumstances to obtain a similar covenant from the distributor.

(g) Promotion and Marketing

The distribution agreement should require the distributor to use its best efforts to advertise and promote the work. The requirement to use "best efforts" will mean that the distributor will be bound to exert certain efforts and engage in various activities to promote and market the work.[19] The agreement may also provide for the distributor to expend a minimum amount in promoting and advertising the work during the initial stages of distribution. Such expenditures may be made in accordance with a marketing plan agreed upon between the owner and the distributor. The distributor may also be required to consult with the program owner with respect to the design of any packaging and other measures taken in connection with the distribution of the work. Finally, the distribution agreement may set out a minimum number of units of the work to be distributed within a specified time period, failing which all rights to the program package would revert to the owner.

(h) Rights to Trade Marks

The distributor will wish to acquire from the program owner the right to use any trade mark, trade name or design by which the work has been identified. The distributor should be required to enter into a registered user agreement to evidence its rights to any trade mark. Failure to execute such an agreement could result in the owner's right to the trade mark being lost. Each of the parties should agree not to use any such trade mark in conjunction with any other product or service.

Before being bound to use a particular trade mark in conjunction with the marketing of a computer program in any jurisdiction, the distributor should ensure that the mark is available. The distributor should also receive an indemnity from the owner with respect to any claims made

---

19 See *Massengale v. Transitron Electronic Corp.*, 385 F. 2d 83 (1st Cir., 1967); *Bloor v. Falstaff Brewing Corp.*, 454 F. Supp. 298 (S.D.N.Y., 1978); affirmed 601 F. 2d 609 (2nd Cir., 1979); *Perma Research & Dev. Co. v. Singer Co.*, 402 F. Supp. 881 (1975); affirmed 542 F. 2d 111 (2nd Cir., 1976).

as a result of such trade mark infringing the rights of another person. If the distributor determines that the trade mark proposed to be used in conjunction with the marketing of the program is not available, it should have the right to choose an alternative mark. The software owner, on the other hand, will wish to have the right to be consulted or to give its consent prior to any alternative mark being selected.

The agreement should indicate which of the parties is responsible for filing applications for the registration of trade marks in various jurisdictions and for the costs of such filings.

(i) Copyright and Other Proprietary Rights

The distributor will wish to register claims of copyright to evidence its rights to the work in copyright offices of the various jurisdictions in which such rights are granted. Such registrations will generally be in the name of the owner with the distributor shown as a licensee or assignee, although in the case of an exclusive licence, such registration may be in the distributor's name as the "owner" of the assigned rights. The owner should agree to execute and deliver any instruments necessary to record the grant of rights to the distributor, with the distributor being bound to execute and file all necessary documents upon the termination of the agreement to show the reversion of rights to the owner.

In the event that the source code and other source documentation are disclosed by the owner, the distributor will be required to treat all such materials as the owner's confidential proprietary information. Accordingly, the agreement should contain various non-disclosure provisions such as those discussed in the following section of this chapter. Each of the parties will also wish to require the other party to treat as confidential proprietary information any information concerning its business or marketing plans, methods of operation, names of customers and any products and devices used to protect the program from unauthorized duplication.

To preserve the rights of both parties, all copies of the work distributed under the distributor's authority should contain the copyright symbol in the form prescribed by the Universal Copyright Convention, together with the year of publication and the name of the copyright owner. The name of the distributor may also be inserted as the owner of exclusive rights, although the name of the author should be shown as the originator and creator of the work.

(j) End User Agreements

The program package will be marketed by the distributor pursuant to end user licence agreements. These licences may take the form of "shrink-wrap" licences if the programs are directed to the microcomputer market. In general, the programs will be distributed in machine-readable form only, and may contain a device to inhibit unauthorized copying.

The end user licence agreement should contain various disclaimers and limitations of liability on the part of both the owner and the distributor. The end user licence agreement should also contain various restrictions against the unauthorized copying, disclosure and use of the programs. The software owner will wish to to have the right to consent to the terms of the end user licence agreement to ensure that it contains adequate provisions for the protection of its proprietary rights, and for the limitation of its liability for any losses incurred by end users from their use of the programs.[20]

(k) Royalties

A software distribution agreement will generally provide for the payment of royalties to the owner calculated as a percentage of the net amounts received from licences or sales of copies of the programs. Alternatively, the royalties may be calculated based on a fixed percentage of a suggested resale price for each unit of the program that is distributed, directly or indirectly, to end users. The agreement should provide that notwithstanding any suggested retail price, the distributor or other person distributing the product under its authority is entitled to grant licences to the programs at a lower price to avoid any contravention of the price maintenance provisions of the Competition Act.[21]

Where royalties are to be calculated based on the amounts received by the distributor, the agreement should indicate any deductions in respect of which royalties are not to be paid. For example, royalties will generally not be payable on amounts received by the distributor on account of costs of handling or transportation, insurance, interest or other service charges, taxes and charges for maintenance and other services provided by the distributor. Where a program is distributed as part of a program package, the agreement may provide for the pro-rating of any amounts received based on the proportion of the total package represented by the subject program. The software distributor will also wish to provide that it will not be required to pay royalties on receivables that cannot be collected, or on account of any returned or defective copies of the product for which the distributor does not receive payment.

In some cases, royalties may be made payable on an escalating basis so that with any increase in the number of units of the program distributed, the royalty rates payable to the owner would also increase.

---

20 For a general discussion of the terms to be included in software licences and the issues relating to shrink-wrap licences, see section 1, *supra*.

21 S.C. 1986, c. 26, subs. 38(3).

The agreement should indicate how often royalties are payable and the period of time following the end of any royalty payment period that a royalty statement and the amount of the payment due and owing are to be remitted to the owner. The distributor should be required to submit a detailed statement of all sales or licences. The owner should be entitled to have the distributor's records examined by a qualified auditor to verify any royalty statement. The distributor will wish to receive proper notice of any such examination and to require any information concerning the distributor's business obtained as a result of such audit to be treated as confidential.

The agreement should require the distributor to pay interest at an agreed upon annual rate on any over-due royalty payments. The agreement may entitle the distributor to withhold royalties in the event of any dispute arising under the agreement. The agreement could provide for the payment of disputed royalties into an interest-bearing trust account pending resolution of the dispute. In addition, it may provide for the owner to post security as an alternative to the withholding of royalties by the distributor.

(1) Representations and Warranties

A software distributor will wish to receive various representations and warranties as to the functions and capabilities of the licensed programs. For example, the owner should warrant that the programs will have the features and perform the functions set out in any specifications incorporated in the agreement and that the programs will be substantially free of defects that would affect their performance for a specified period of time following acceptance. The distributor should also warrant that the programs will operate on any intended computer equipment and in conjunction with any applicable operating system software.

The owner should also provide various warranties of title, including: that it is the sole proprietor of the work; that the work is fully original to it; that no person contributed to the work other than employees acting in the course of their employment, and that if any other person such as consultants participated in its preparation, such persons have entered into written assignments of copyright in favour of the owner; that the owner has not assigned or encumbered the work or granted any licences, options or other rights to any other person; that it has not taken any other act that would affect the value of the work or impair the ability of the distributor to exercise its rights under the agreement; and that the work is not in the public domain and does not infringe any copyright, trade secret, patent or other proprietary right of any other person and does not contain any material which it is unlawful to include therein.

The owner should also provide certain warranties relating to the value and integrity of the work, including: warranties that there are no other versions of the work or modifications thereto which have been developed; that the source code and source documentation relating to the work have been kept confidential; that the work has not been previously published and that no registration of copyright has been made other than as has been disclosed to the distributor; and that the owner has disclosed all trade mark applications made for the work and that it has valid and enforceable rights to such trade marks.

If the owner is a corporation, it should represent and warrant that it has the necessary corporate power and authority to enter into the agreement, that the persons executing the agreement have been duly authorized to do so, and that it has the financial capability to perform its obligations under the agreement.

The owner must consider carefully whether it can properly make each of the representations and warranties that may be requested by the distributor. Some of these representations may be qualified as being to the best of the owner's information and belief.

The distributor should also provide certain representations and warranties relating to the performance of its obligations under the agreement, including representations that it has the necessary corporate power and authority and the financial capability to enter into the agreement and to perform its obligations thereunder; and that it has not previously published or entered into an agreement to publish a computer program that may compete with the licensed work. The owner may also seek covenants from the distributor that any modifications or enhancements to the program that it is authorized to make under the agreement will not infringe any copyright, patent, trade secret or other proprietary right of any third person and will not otherwise be in contravention of any agreement made between the distributor and any other party.

(m) Indemnities

The software distribution agreement will normally contain a provision requiring either party alleged to be in breach of any representation, warranty or covenant set forth in the agreement to defend any action or claim and to indemnify and hold the other party harmless from any such breach. The owner will also wish to be indemnified by the distributor against any liability arising from any claims brought by end users, whether in contract, tort or otherwise, as a result of their use of the programs, except to the extent such claims arise from breach of any warranty that the distributor is authorized to make on behalf of the owner.

The indemnifying party should be entitled to receive prompt notice of any claim and be entitled to assume or to control the settlement of any action. The right to assume such defence may be made subject to certain conditions such as the posting of security. Each of the parties may also wish to impose certain limits on the amount of its liability arising from any alleged breach of its representations, covenants and warranties. In the case of the owner, such a limit may be based on the amount of royalties to which it may be entitled under the agreement.

(n) Local Laws

The terms of any distribution agreement must reflect the laws of the jurisdiction in which such distribution occurs or which is chosen as the governing law of the contract. For example, the clause governing the protection of the software owner's proprietary rights must take into account the availability of various forms of legal protection, such as patent, copyright and trade secrecy law. In addition, if computer software is to be distributed in foreign countries, the distributor may require export licences or other consents under the laws of the jurisdiction from which the software is to be exported.[22] Some countries may impose foreign exchange controls which limit the rights of the software owner to receive royalties from a distributor.[23] Other countries may impose restrictions on the ability of an owner to terminate or not to renew agency or distribution arrangements.[24] Finally, competition or anti-trust legislation may impose limitations on various arrangements that may be established in connection with the distribution of software, such as tying arrangements, resale price maintenance clauses, territorial restraints, exclusive dealing arrangements and non-competition clauses.[25]

(o) Termination of Agreement

Each of the parties will wish to provide for the right to terminate the agreement upon the occurrence of certain events. For example, the owner will wish to have the right to terminate the agreement if the publisher is in breach of certain material covenants such as the obligation to remit any royalties on the due date, to distribute an agreed upon number of units of the work, and not to publish a competing computer program. Each of the parties will wish to have the right at its option to terminate the agreement if the other party is found to be in material breach of any of its representations and warranties or if the other party becomes bankrupt, ceases to carry on business, or becomes subject to any action or proceeding for the appointment of a receiver or receiver/manager for any part of its property and assets.

---

22 See, *e.g.*, Export and Import Permits Act, R.S.C. 1970, c. E-17.
23 See Bernacchi, *op. cit.*, at 40-42.
24 *Ibid.*, at 43-48.
25 *Ibid.*, at 49-53.

The agreement should provide that upon termination of the agreement for any reason, all distribution rights to the program would revert to the owner, although the distributor may be granted the right, at the option of the owner, to distribute copies of the work on hand at the time of termination. The distributor, of course, should be required to remit all royalties due and payable at the time of termination, as well as any royalties that may become payable in respect of licences granted prior to termination or arising from the distribution of copies on hand. The agreement should also provide for the return to the author of all copies of the source code and all other proprietary materials belonging to the author then in the possession or within the control of the distributor.

The agreement should provide that its termination would not affect any agreements previously made by the distributor with end users. However, in the event that such termination occurs as a result of certain acts or defaults by the distributor, such as its bankruptcy or failure to carry on business, all rights to any licence agreements made by the distributor would, at the option of the software owner, revert to the owner.

(p) Assignment and Sub-Licensing of Distribution Rights

Each of the parties may wish to impose certain restrictions against assignment of the agreement without the other party's consent, which should not be unreasonably withheld. The parties may also wish to provide for certain permitted transfers, including any transfer made by either party to an affiliated company or an assignment made in conjunction with the sale by either party of its business. The distributor will wish to grant sub-licences with respect to the rights granted to it by the owner to authorize dealers and sub-distributors to carry out distribution of the programs. The owner will wish to have the right to consent to the appointment of any dealers and sub-distributors, and require such persons to be bound by all of the covenants and restrictions of the distributor under the distribution agreement.

(q) Additional Provisions

In addition to the foregoing, the parties may wish to include other standard contractual provisions, some of which are referred to in the final chapter of this Part.

6. NON-DISCLOSURE AND CONFIDENTIALITY AGREEMENTS

In an earlier chapter of this text, the application of trade secrecy law to the protection of computer programs and related documentation was reviewed. It was noted that such protection requires information to be treated as confidential at all times by the holder. Accordingly, software developers and proprietors must not permit the disclosure or use of information except under conditions of confidentiality. The best means of achieving this objective is by entering into a non-disclosure agreement with any person to whom the programs or other information are disclosed for any reason.

One of the circumstances in which a non-disclosure agreement will be required is where a software developer, owner or licensee wishes to restrict use of confidential information by its employees. Such an agreement will also be necessary where an owner wishes to license the use of programs to licensees, authorize their distribution by a publisher, or provide copies of source code or other materials to an escrow agent. The materials that will be protected as trade secrets may include the source codes for any programs as well as flow charts, diagrams, program notes, and other materials describing the operating conditions and functions of the programs. Confidentiality obligations may also be inserted in software licences for programs distributed in object code form notwithstanding that such programs may be subject to widespread distribution to end users.

The following are some of the provisions that should be included as part of any arrangement whereby confidential information is disclosed by its holder to other persons. Such provisions may be included within a software licence, a software development or distribution agreement or an employment agreement, or may be set out in a separate non-disclosure agreement that is incorporated by reference as part of another agreement.

(a) Description of Confidential Material

The agreement should contain a complete yet concise description of the information which is to be treated as confidential. The list may include certain printed materials, information set out in machine-readable media, and in some cases information divulged orally. In the latter case, the circumstances of disclosure and nature of the information being disclosed should be set out in the agreement. The agreement should not apply to any information which is not truly confidential or which is a matter of public knowledge. All documentation and materials containing information intended to be treated as confidential should be clearly labelled as such.

(b) Reasons for Disclosure and Permitted Uses

The purpose for disclosure of the confidential materials, and the uses that may be made of such materials by the recipient, should be clearly set out. In the case of programs provided to a licensee in object code form, the permitted uses will be generally restricted to the execution, copying, storage and display of the programs on a designated computer in connection with the processing of data in the course of carrying on business. If source code or other materials are made available to a software licensee, the agreement should indicate that such materials are disclosed solely to enable the licensee to make modifications or corrections in connection with its use of the programs in object code form. Similarly, if a software owner wishes to disclose source materials to a *potential* publisher or distributor, the agreement should indicate that such disclosure is made solely to enable the recipient to exercise the rights set forth in the distribution agreement. The recipient should covenant not to use the materials for any purpose other than as set out in the agreement.

(c) Restrictions Against Further Disclosure

The disclosee should be required not to disclose any of the confidential information to any third party other than as expressly set out in the agreement or as is permitted in writing by the owner. Disclosure to third parties should be restricted to employees of the recipient who are either designated in the agreement or who have a need to know to enable the recipient to make use of the confidential information. In the case of a software licence, the licensee may wish to have the right to disclose the software to other parties, such as consultants retained to provide support services, auditors or accountants, or employees at a back-up computer facility or service bureau at which the programs may be used.

(d) Restrictions Against Copying

The confidentiality agreement should either prohibit the copying of any materials containing confidential information belonging to the owner or set out the number of copies that may be made and the purposes for which such copies may be used. The agreement should require any copies made by the recipient to contain a proprietary legend indicating that they contain confidential information belonging to the owner.

(e) Physical Security Measures

The recipient should be required to keep all confidential materials in a secure location to which access is restricted. Notices should be posted reminding employees of their obligation to preserve the confidentiality of the information. If confidential programs are to be installed on the recipient's computer system, the recipient should be required to put in place all necessary pass words and access codes to prevent unauthorized access to the computer system. The standard of protection to be followed by a recipient should be that which the recipient would reasonably be expected to follow to safeguard its own valuable confidential information. The owner may wish to provide for instructions to be given to the recipient as to the measures to be taken to protect such information.

(f) Covenants with Employees and Other Persons Permitted Access to Information

The recipient of confidential information should be required to enter into appropriate non-disclosure covenants with its employees and other persons permitted access to the software to preclude any copying, disclosure or use of such information in any manner not contemplated by the agreement. The form of such covenants may be attached to the non-disclosure agreement. Such persons should be required to comply with all instructions and operating procedures established by the recipient to safeguard the confidential information.

In some cases, the recipient will be made strictly liable for any wrongful disclosure or use of confidential information by its employees regardless of any measures taken by the recipient to prevent such disclosure or use. In other cases, the holder of confidential information may wish to have employees or other persons under the custody or control of the recipient enter into agreements directly with the owner. If supported by adequate consideration, this will permit the owner to take action directly against such persons even if the recipient is unwilling to join in such action.

(g) Exclusions

Confidential information may be released to third parties or become public knowledge for reasons other than its disclosure by a person subject to a confidentiality agreement. Accordingly, a person who enters into such an agreement will wish to provide for the termination of its obligations or for the exclusion of any liability on its part if confidential information becomes public for any reason other than the default of the recipient, if the information is rightfully received from a third party without a similar restriction and without breach of any confidentiality agreement, if authorization for further release of the information is obtained from the owner, or if disclosure of the information is required pursuant to any applicable law or order of a court or other tribunal.

(h) Rights on Termination

The recipient of materials containing confidential information should be required to return such materials and all copies upon the expiration or termination of the confidentiality agreement (or any other agreement to which the confidentiality agreement is ancillary) for any reason. However, the obligations of the recipient not to disclose or use confidential information should survive and continue in full force and effect with respect to information disclosed at any time prior to termination.

(i) Limitation of Liability

Finally, a recipient of confidential information may wish to restrict the circumstances in which it is liable for any use or disclosure of the information in a manner not permitted by the agreement, and limit the damages for which it may be liable. For example, the recipient may wish to provide that it would not be liable for any inadvertent disclosure of such information if it has taken reasonable steps to prevent such disclosure in accordance with its obligations under the agreement. Other limitation and exclusionary clauses that the user may wish to have inserted in the agreement are discussed in the final chapter of this Part.

18

Computer Data Base and Service Bureau Agreements

1. COMPUTER DATA BASES: OWNER-OPERATOR LICENCE AGREEMENTS

The widespread availability of microcomputers for business and consumer applications, together with the integration of computer and communications networks, have resulted in rapid increases in the use of computer data bases. A computer data base refers to any compilation of information stored in a computer system which may be made available for localized or remote on-line access. A data base may consist of materials which have been previously collected and printed in human-readable form and which have been converted to permit computer access. Alternatively, the data base may consist of materials which are collected for recording only in a computerized form and which are intended only for electronic distribution.

There are several agreements that may be necessary in connection with the establishment and operation of information storage and retrieval systems. First, an agreement will be necessary between the owner of copyright in the information to be included in the data base and the data base operator, setting out the terms on which the data base is to be established and access is to be provided to third parties. Second, the data base operator will wish to enter into agreements with end users setting out the terms governing such users' rights of access. Finally, the data base operator may enter into an agreement to authorize a distributor to provide access to the data base to end users.

This section will outline some of the provisions normally found in an agreement made between the data base "owner" and operator. The following section of this chapter will review provisions commonly found in an agreement to provide end users with access to a data base. An agreement made between a data base operator and a distributor will include many clauses similar to those found in a distribution agreement for computer programs, as discussed in section 5 of the preceding chapter of this text.

The "owner", as applied to a data base, refers to a person who owns copyright and other intellectual property rights in any existing compilation of information which is converted to computerized form, or any new compilation which is created expressly for computer access. The term "operator" means the person responsible for offering the data base to the general public or to a selected category of end users. In most cases, the operator will also be responsible for establishing and/or maintaining the data base.

The following are some of the terms and conditions that should be set forth in an agreement relating to the establishment, maintenance and operation of a data base to be made between the data base owner and operator.

(a) Identification of Materials to be Included in Data Base

Any original compilation of information may be the subject of a data base agreement. Such a compilation may include law reports, encyclopedias, dictionaries, newspaper or periodical reports, stock market reports, airline schedules or other specialized reporting services directed toward particular user groups. A data base may also consist of a compilation of personal information, such as information concerning customers of a particular business or industry, or credit information to be provided to third parties. In the latter case, the data base will be required to be maintained in compliance with applicable legislation relating to consumer reporting agencies.[1]

A data base agreement should clearly identify the materials to be included in the data base. In some cases, the data base will consist only of extracts rather than the full text of certain materials (for example, headnotes of law reports).

(b) Grant of Licence

The agreement should provide for the grant of a licence authorizing the operator to create, maintain and operate the data base. The licence should entitle the operator to offer access to the data base to end users, to promote and advertise the data base, and if applicable, to make copies of extracts of information contained in the data base available to end users.

(c) Conversion of Materials to Machine-Readable Form

The creation of a data base will require either the preparation of a new computerized compilation or the conversion of existing materials from printed to machine-readable form. This may involve either the creation of a new work of authorship or the selection and editing of pre-existing

---

1 See Chap. 11 of Part IV, *supra*.

materials for inclusion in the data base. Other tasks involved in the creation of a data base may include the creation of necessary computer programs, the actual conversion of text by the use of optical character readers or the keyboarding of the material into a computer and the addition of all necessary codes and formats to the machine-readable text. The data base agreement should indicate who is responsible for each of these duties.

The agreement should also set out any coding specifications which are to be followed in creating the data base and who is responsible to ensure that the text conforms with the required specifications. The coding specifications should be in sufficient detail that upon termination of the agreement, the owner of the data base could resume its operation or authorize its continued operation by another operator. The agreement should also set out the manner in which the information included in the data base is to be recorded, that is, by identifying the kind of computer tapes to be used and the speed at which the information is to be recorded.

(d) Offering of Data Base to End Users

The data base agreement should indicate the time period by which all materials are to be converted into machine-readable form and made available as a searchable data base for end users. The data base owner may also wish to exercise some control over the terms of any end user licence agreement and the persons to whom access to the data base is to be provided. The agreement may also set out the time periods at which access is to be provided to end users.

(e) Incorporation of Updates into Data Bases

The data base agreement should indicate how often the data base is to be updated or supplemented and who is responsible for such updating. If the data base consists of materials which are published periodically in printed form, all new issues of such publications should be required to be included in the data base within a certain time period following their publication. The agreement should indicate how such materials are to be provided, that is, as printed text, on tapes, disks or other machine-readable media or by electronic transmission.

In general, if the owner of the data base holds copyright in the printed publication, he will be required to deliver printed updates, while the data base operator will be responsible for converting such materials into machine-readable form and adding the necessary codes so that they can be included as part of the searchable data base. The time period within which the data base operator must complete the conversion of any updates should be set out in the agreement.

(f) Control of Accuracy of Data Base

The party responsible for each of the tasks involved in creating a data base should be bound to perform such tasks accurately and completely, and should be bound to correct any errors coming within the scope of its responsibilities. The agreement should provide for periodic testing to verify the accuracy of the data base. The costs of re-running any programs or of rebuilding the data base as a result of the errors on the part of either party should also be clearly specified.

(g) Marketing of Data Base

The agreement should set out the respective obligations of the parties in connection with the marketing and distribution of the data base to subscribers. In particular, the data base operator may be required to expend a certain amount on marketing and promotion, to announce the major features or attributes of the data base to potential customers, and to take the necessary steps to register any trade mark or trade name under which the data base is to be marketed. If any trade mark belongs to the owner, the operator should be required to enter into a registered user agreement evidencing its rights to use the trade mark during the term of the agreement.

(h) Payment of Royalties: Reports on Use of Data Base

A data base agreement will set out the right of the owner to receive royalties in respect of any use of the data base by end users. Such royalties will generally be calculated based on the aggregate amount of billings charged by the data base operator. The royalty rate will normally be a percentage of the billings to end users and may be on a sliding scale to reflect the increased profitability of the data base as original fixed costs are recovered.

The agreement should set out any uses to be made of the data base for which royalties will not be payable to the data base owner. For example, royalties will generally not be payable in respect of any use of the data base for training, demonstration, testing or marketing purposes for which no fees are paid by end users. Royalties will also not be payable in respect of any service charges or discounts or commissions paid by the data base operator to sub-distributors or dealers. The operator will also wish to exclude any royalties for uncollectable accounts or in respect of any rebates paid to end users because of malfunctions in the use of the equipment or in the operation of the communications network.

The data base agreement should not contain any provision requiring the operator to charge end users certain minimum amounts in respect of the use of the data base, since such a provision would be in contravention of subsection 38(1) of the Competition Act.[2] This section prohibits any person having the exclusive rights conferred by a trade mark or copyright to influence upward or to discourage the reduction of the price at which any other person in Canada supplies a product within Canada. A product is defined to include a service and a data base would appear to be a compilation in which copyright may subsist or in respect of which a trade mark may be used. Accordingly, this section would preclude a data base owner from requiring an operator to charge certain minimum amounts to end users. However, the agreement could provide for royalties to be payable based on suggested access fees, provided that the agreement clearly indicates that the operator is entitled to charge lower fees. There would also be no objection to including in the data base agreement a provision imposing maximum limits on the fees that may be charged to end users or limiting the amount by which such fees may be increased during the term of the agreement.

The data base operator should be required to provide the owner with a report on a monthly or other periodic basis, setting out such information as the number of users, the number of searches conducted and the aggregate amount of time that the data base was accessed by users. The report should be accompanied by a statement of all royalties payable to the owner as well as the amount of royalties due and payable. The owner will wish to have the right to have the operator's records audited.

(i) Rights and Title to Data Base

The data base agreement should set out the nature and scope of the proprietary rights of each of the parties to the information contained in the data base. In most cases, the data base operator will be granted an exclusive licence to operate the data base within a defined geographical area.

If the information in the data base is a new compilation or is created by the editing and selection of pre-existing materials, the party responsible for creating such work will own copyright, in the absence of an agreement to the contrary. Accordingly, a written assignment of copyright may be required from the data base operator or any other person who actually compiles the data base to the person on whose behalf it is created. The agreement should also provide for the filing of assignments of copyright in any appropriate copyright offices.

The agreement should also set out the rights and obligations of the parties with respect to the prosecution of any claim for the infringement of any proprietary rights in the data base. In general, the owner will be entitled to commence an action in its own name to protect its rights in

---

2 S.C. 1986, c. 26.

the data base with the operator being required to co-operate in any such action. In addition, the data base operator should be required to advise the owner when it becomes aware of any unauthorized copying or other infringement of the owner's rights in the data base.

(j) Protection of Trade Secrets

The information contained in a data base will generally not be regarded as confidential insofar as it is intended to be distributed to the public. However, other information relating to the data base, such as the programs used in its operation and subscriber lists, may be considered as confidential information of either the data base owner or operator. The parties may also wish to treat as confidential the provisions of the data base agreement itself. However, such confidentiality obligation should not preclude the operator from disclosing any information necessary to offer the data base to the public or to enable the owner to resume operation of the data base directly or through another operator upon termination of the agreement.

Each of the parties should agree not to disclose or use the other party's confidential information, and to enter into agreements and provide instructions to its employees or other persons who are permitted access to the confidential information to prevent any unauthorized disclosure or use.

(k) Warranties

Each of the data base owner and operator will wish to receive certain representations and warranties from the other party. For example, the operator will wish to obtain a representation and warranty that the owner owns copyright and other proprietary rights in any compilation of material included in the data base, and that no other person has been granted the right to operate a data base of such materials. Each of the owner and the operator will wish to receive representations that the other party has the corporate power and authority to enter into the agreement, and that the persons executing the agreement have been duly authorized to do so.

(l) Copies of the Data Base and Right of Access by Owner

A data base operator may wish to have the right to make additional copies of the data base to be kept on its premises and at a back-up facility. Such a right should be expressly set out in the data base agreement since the copyright owner has the sole right to authorize the making of any copies. The owner may also wish to receive a copy of the information compiled in the data base, as updated or modified from time to time. The owner may be required to pay any applicable copying charges as well as the cost of the media on which the information is recorded.

The agreement may also provide for the owner to obtain access to the data base on the same terms as end users, either with or without payment of applicable end user fees. In particular, the owner would wish to obtain such right of access for testing and demonstration purposes.

(m) Copyright Notice

The data base operator should be required to include in the data base a copyright symbol together with the name of the copyright holder and the year of publication. This information should be required to be inserted in such a manner that it will be prominently displayed on the terminal of any user conducting a search of the data base as well as on any document in printed form that may be obtained by a user.

(n) Term

The agreement should indicate the term during which the data base may be offered to end users by the operator. The agreement may be subject to automatic renewal at the end of the term unless either party provides notice of termination within a fixed time period prior to the date of expiration.

(o) Default and Termination

The agreement should set out certain events which may give rise to a right of termination by either party. In particular, the data base owner should be entitled to terminate the agreement if the operator fails to remit any payments when due and owing, if it ceases to offer the data base to subscribers or if service is interrupted for a certain time period.

Other events giving rise to a right of termination may include the breach by either party of any material covenant or obligation set out in the agreement, or the bankruptcy, receivership, dissolution, liquidation, or the failure to carry on business in the normal course of the other party.

(p) Indemnity

Each of the parties will wish to obtain an indemnity from the other party against any loss or damage arising from any error or omission made by the other party in performing any work or from any other breach of its obligations under the agreement. In each case, the indemnifying party should be entitled to receive prompt written notice of any claim or demand and to direct the defence and settlement of the action.

(q) Assignment and Sub-Contracting

The agreement may provide that neither party may assign, sub-license or encumber the agreement or its rights thereunder without the prior written consent of the other party. However, the data base operator will wish to have the right to enter into arrangements with distributors or sub-licensees to facilitate the marketing of the data base to end users.

(r) Other Provisions

An outline of additional contractual provisions that may be inserted in a data base agreement is set out in the final chapter of this Part.

2. COMPUTER DATA BASES — END USER AGREEMENTS

This section will review the provisions that may be set out in an agreement between a data base operator and a user who wishes to acquire the right of on-line access to the information contained in the data base. Many of these provisions may also be found in an agreement to provide for the establishment of an interactive computer system, that is, where a user has the right both to retrieve information from a computer data base and to provide information that may then be made available to other users.

The primary concern of the owner or operator of a data base will be to ensure that access to the data base is not granted to any person other than authorized users, and that such access does not place in jeopardy the proprietary rights of the owner. The user, on the other hand, will be concerned with obtaining guaranteed access to the data base at all necessary times to enable it to conduct its business.

The following is a summary of the main provisions in a data base access agreement in order to reflect the foregoing and other concerns.

(a) Description of Service

The agreement should describe the data base, including the information contained therein, to which the user is entitled to obtain access. In some cases, the agreement will permit the user to obtain access to a number of data bases offered by the operator. For example, the operator may be the electronic distributor for a series of related reporting services, such as news, financial information, stock market reports, airline schedules and weather reports. The agreement should indicate that the applicable data bases will be updated periodically and maintained to ensure that the information contained therein is current, accurate and complete.

(b) Communications Equipment

The agreement should describe the kind of equipment (including computer terminals and modems), operating system software and communications lines required to enable the user to obtain access to the data base. The agreement may provide that the specifications and operating conditions of such equipment and facilities will be established from time to time by the data base operator, in which event the user will wish to receive adequate prior notice of any changes to such specifications and operating conditions. In other cases, the agreement may provide for the customer to purchase or lease equipment and/or to license software from the operator, in which event the terms of such acquisition should be set out in the data base access agreement or in a separate agreement between the operator and the user.

(c) Scope of Services to be Provided to User

The rights of a data base user may include not only the right to obtain on-line access to the information contained in the data base, but also such rights as the receipt of a user guide describing the procedures to be followed to gain access to the data base, training to be provided to the user's personnel (in which event the number of employees to be trained and the length of training should be set out in the agreement), installation of any equipment and software to be provided in conjunction with the use of the data base, and the grant of the right to reproduce or to receive copies of any information contained in the data base or reports based on such information.

In some cases, a user may be entitled to receive a special package consisting of the right to obtain access to the operator's data base, together with all necessary terminal and communications equipment, software, training and support. If the user intends to utilize equipment and software provided by suppliers other than the operator, it should determine whether any equipment or software proposed to be used will be compatible with the operator's system.

(d) Method of Obtaining Access

The agreement should set out the procedures to be followed by the user to obtain services under the agreement. For example, a written order may have to be executed by the user for each data base to which it wishes to obtain access. The agreement should also provide for the disclosure to the user of any necessary access codes. The agreement should provide for changes to the access codes from time to time to reduce the risk of unauthorized access.

(e) Availability of Service

The agreement should set out the times at which access to the data base will be provided to the user. The operator may be unwilling to guarantee the availability of service at all times but may be prepared to set out a minimum percentage of the scheduled times at which service will be available. The operator should be required in any event to use its best efforts to provide service at the scheduled times. The agreement may also provide for a reduction of the fees otherwise payable by the user for the times that the service is not available.

The data base access agreement may also permit the operator to notify the user of changes in the times at which service is provided. The user will wish any such change to be subject to its consent and may wish to have the right to terminate the agreement if any such change is unacceptable.

(f) Confidentiality and Security of System

The user will be responsible for taking various steps to preserve the security of the operator's data base and to prevent unauthorized access. For example, the user may be required not to disclose access codes or permit the use of terminals by any person other than designated employees. The user will also be required to preserve the confidentiality of any secret information contained in the data base and to use any terminals or other equipment only in accordance with the terms of any user guide.

The access agreement should provide for the user not to copy, disclose or use for its own advantage any confidential information, such as programs or techniques, made available by the data base operator. The agreement will generally require the user to enter into appropriate covenants with its employees and to take such reasonable measures as it would take to protect its own confidential information in order to protect that of the operator. The operator should be required to label any materials intended to be treated as confidential and to provide to the user the details of any procedures that it is expected to follow to preserve the confidentiality of such materials. The user should also receive written acknowledgement from the operator that such procedures represent compliance with the terms of the user agreement. The user will also wish to consider whether any confidential information may have to be disclosed to persons other than its employees, such as computer consultants, to enable it to make full use of the services provided by the operator.

(g) Reports on Usage of System

The agreement should provide for the user to receive periodic reports of the extent of its usage of the system, which may be provided in conjunction with invoices in respect of payments due and owing to the operator. The agreement may provide for the user to notify the operator of any errors within a specified time period following receipt of each report.

The charges payable for the services obtained by the user will ordinarily be set out in a price schedule incorporated in or referred to in the agreement. The prices may be subject to change by notice given by the operator. However, if the agreement requires the user to utilize the services of the data base operator for a fixed term, or to make a minimum expenditure of access fees, the user may wish to receive a guarantee that the prices will not be changed for a minimum period of time such as one year, and that limits are imposed on subsequent increases during the remainder of the term of the agreement. If the operator is not prepared to agree to any such limits, the user should have the right to terminate the agreement upon receiving notice of any price increases.

The agreement will set out how often invoices are to be issued and the date on which they are payable, will provide for the user to be responsible for all taxes levied in respect of any charges under the agreement (other than those based on the operator's income), and will generally provide that the service charges do not include the costs of transportation or shipping of any materials or the acquisition of any equipment, programs or supplies required to enable the user to utilize the data base operator's services. The user should ensure that any agreement required to enable the user to obtain any materials, equipment or supplies necessary to utilize the operator's services is executed concurrently with the data base access agreement.

(h) Use of Information Contained in Data Bases

The agreement will normally set out restrictions on the uses that may be made of any information contained in the data base to which the user acquires access. For example, the user may be granted the right to use such information solely for its own purposes in conjunction with the conduct of its business. The user may be subject to certain restrictions as to the copying of such information in either machine-readable or human-readable form, as well as its disclosure to third parties.

(i) Warranties and Performance Standards

The user will wish to obtain warranties from the operator that it has the right to offer the data base to the user and to authorize the user to exercise the rights set out in the agreement. The user will also wish to obtain certain covenants as to the standard of performance to be met by the operator, including that it will exercise due care and utilize skilled and experienced personnel in offering the services set out in the agreement. The standard of care may be that which the operator, acting in a commercially prudent manner, would utilize in acquiring or providing information in the conduct of its own business.

The operator should agree to correct any errors in the information contained in the data base or the reports provided thereunder, as well as in any programs used to enable the user to obtain access to the data base. The agreement may require the user to notify the operator of any errors in any information or reports within a specified time period. The operator should also be required to use its best efforts to prevent any interruption of service and to restore service as promptly as possible if any interruption occurs. However, the operator may wish to have the right to make scheduled interruptions of service to add enhancements and modifications to the services or the computer system, or to maintain the system.

It is unlikely that the operator will be willing to guarantee the completeness or accuracy of the information contained in the data base, or guarantee that service will be provided without failure or interruption. However, the user may be able to obtain a commitment by the operator to use its best efforts to ensure the accuracy of any information and to prevent interruptions of service. The user may also be able to negotiate the right to receive rebates in respect of any interruption of service or in respect of any reports provided under the agreement that are incomplete or inaccurate.

In addition to the foregoing, the operator should be bound to provide all such programming and support services as may be necessary to maintain the data base and the reliability of the services to be provided to the user.

The warranties and performance standards may be subject to certain exclusionary and limitation clauses, as discussed in the final chapter of this Part.

(j) Term and Termination

The agreement should set out the term of the data base access agreement. The agreement may be for an indefinite term with either party entitled to terminate the agreement on thirty or sixty days written notice. If the user requires access to the data bases for a fixed period of time to carry on business, or if the operator wishes to obtain a guaranteed amount of income from the user in consideration of price concessions, the agreement may set out a minimum term as well as a minimum amount of usage to be made of the data base during such term.

The agreement should set out the circumstances in which either party is entitled to terminate the agreement for cause, such as if the other party becomes bankrupt, ceases to carry on business in the normal course, or becomes subject to any act or proceeding for the appointment of a receiver or a receiver/manager, or for its liquidation, dissolution or winding up, or disposes of all or a substantial part of its property. In addition to the foregoing, each party will wish to have the right to terminate the agreement if the other party is in breach of its material obligations, for example, if the data base operator ceases to offer the data base to the user or makes significant changes to the terms on which such access is provided, or if the user fails to pay all charges when due. The agreement should indicate any provisions of the agreement which are to survive termination, such as the obligation of the user to pay all accrued charges, to preserve the confidentiality of any trade secrets disclosed by the operator, and not to use any information obtained during the course of the agreement except as authorized under the terms of the agreement.

3. COMPUTER SERVICE BUREAU AGREEMENTS

A computer service bureau agreement refers to an arrangement whereby an individual, company or other entity makes use of the computer facilities of another organization in order to process its data or to obtain some other computer service. By using a service bureau, a user is able to avoid the expenses of acquiring its own computer system. On the other hand, by providing services to a number of customers, the service bureau may enjoy economies of scale and offer a wide range of computer services to many users.

The following are some for the provisions to be included in a service bureau agreement.

(a) Description of Services

The agreement should contain a full description of services to be provided to the user, such as data processing, on-line and off-line storage, shared software services, education and other training, user guides, update service, and communications service. The agreement should contain a complete list of all services that may be requested from time to time by a customer, which request is to be governed by the terms of the agreement. Any specialized or customized service designed for a customer's particular business should also be clearly described.

The agreement should also set out the manner in which the services are to be provided, for example, whether information is to be provided in printed form, on machine-readable media, or by transmission via telephone or other telecommunications facilities. If the data is to be transmitted electronically, the kind of terminal equipment to be utilized by the user should be specified. The agreement should also indicate whether the costs of any terminals or other telecommunications facilities or of any carriers to be used to deliver data to the service bureau are included in the service fees or are to be paid separately by the user.

(b) Term

The agreement should set out the term during which services are to be provided to the customer. The term may be for a fixed period such as one year, subject to automatic renewal at the end of such period or any subsequent anniversary date unless notice of termination is given by either party within a specified time period prior to expiration. Alternatively, the agreement may provide that it may be terminated by either party at any time on 30 or 60 days' notice. The user may find that such a notice period does not give it sufficient time to make alternative data processing arrangements.

(c) Changes in Equipment and Services

A service bureau may wish to have the right to make changes to the equipment, software, rules of security and standards of operation, place of performance, accessibility periods, and assignment of personnel. The user should ensure that the agreement permits such changes only insofar as the standards of performance or efficiency of the services to be provided by the service bureau are maintained or improved. In addition, the user will wish to receive adequate notice of any such changes and assistance and possible financial compensation from the service bureau if such changes require the user to change the form in which data are to be provided.

(d) Prices and Payment

The fees payable by the user for any of the services to be performed by the service bureau should be set forth in the agreement or in attached schedules. Such fees will generally be based on the length of time that the service bureau's resources are utilized or the number of transactions or volume of data processed. However, the user may be entitled to special reduced rates according to the length of time that it is willing to permit the service bureau to process its data and whether the service bureau's resources are to be utilized during off-peak hours.

In the event that a service bureau acquires equipment or other resources in order to provide customized services to a particular user, the agreement may set out certain minimum amounts of services to be obtained or minimum expenditures to be made by the user.

The service bureau may wish to have the right to change the rates payable by the customer on 30 or 60 days' notice. The customer, on the other hand, will wish to have the rates fixed during the first year of the term of the agreement and to limit any increases that may be made thereafter. If the service bureau is unwilling to agree to any such limits, the user may wish to have the right to terminate the agreement upon receiving notice of any proposed increases.

The agreement should indicate the frequency with which invoices are to be issued and the times at which such invoices are due and payable. The service bureau will wish to provide for interest at a fixed annual rate to be payable on any overdue accounts. The agreement will generally provide that any quoted charges do not include any taxes and that the user will be liable for all such taxes payable in respect of any services, materials or supplies provided to the user.

The agreement should set out any volume discounts to which the user is entitled. Such discounts will generally be based on the dollar value of the services contracted for. The user may wish such discounts to be calculated based on the services obtained by all companies in its corporate group. The availability of such discounts may also depend on the agreement being in effect for a minimum period of time following its commencement.

(e) Responsibilities of Customer

The service bureau agreement will identify certain matters that are the responsibility of the customer in respect of the services to be provided by the service bureau. For example, the customer will be responsible for the timeliness, accuracy and completeness of any data supplied to the service bureau. The agreement may also provide for the user to be responsible for the compatibility and good condition of any programs and machine-readable media to be used in conjunction with the service bureau's equipment. In such event, there should be prior consultation and agreement between the user and the service bureau as to the manner in which data is to be provided (that is, by delivery of machine-readable media or remote terminal access) and as to the kind of terminals and other equipment to be used for the transmission of data and for the provision of data processing services. The agreement may also call for certain specifications to be followed by the user as to how the data are to be coded and formatted.

The user will also be responsible for verifying the accuracy and completeness of any materials generated by the service bureau and for notifying the service bureau of any errors or deficiencies within a specified time following receipt, failing which they will be deemed to be accepted by the user. The user may also be responsible for maintaining duplicate copies of any data and programs provided to the service bureau to permit their reconstruction in the event of any loss, damage or destruction. Similarly, the agreement may provide for the customer to establish various procedures such as checkpoint and restart capabilities to safeguard its programs and data and to minimize any losses. The user may also be responsible for complying with any procedures, rules of operation and security measures established by the service bureau. The agreement should provide for the service bureau to provide the user with a copy of all such rules and procedures and to notify and consult with the user prior to making any changes.

(f) Property Rights

The service bureau agreement will generally provide that any programs, media and similar materials provided by either party under the agreement will belong exclusively to that party. However, any information or materials generated or created by the service bureau in the course of performing services on behalf of the user should belong to the user and the agreement should contain an express assignment of all rights to such materials to the user. The agreement should also provide that upon termination of the agreement or at any time at the user's request, the service bureau will return all information and materials belonging to the user. The service bureau may wish to provide that its obligation to deliver such materials is subject to the payment of all applicable charges and the performance of all obligations of the user under the agreement.

The agreement may also deal with the ownership of the rights to any ideas, concepts, techniques or know-how developed by either party in the course of carrying out the agreement. If such ideas and concepts are developed jointly by both parties, the agreement may provide for the joint ownership of all rights therein. Alternatively, such rights may be owned by one party, with the other party entitled to a royalty-free licence in respect of the use of any programs or other works created by the use of jointly developed ideas or concepts.

(g) Risk of Loss: Insurance

The service bureau agreement should set out who bears the risk of the loss of any programs, data or media belonging to one party while in transit or in the possession of the other party under the agreement. Although the service bureau may wish to provide that the user must bear all such risk of loss, such a provision should not apply to any loss arising from any negligent or wilful act or omission on the part of the service bureau or its employees, agents or representatives. The agreement may also provide for each party to take out insurance to cover any risk of loss which is its responsibility.

(h) Confidentiality

The agreement should provide for the service bureau to institute all reasonable measures to protect the security and to preserve the confidentiality of all information provided by the user. In particular, the service bureau should covenant not to permit any unauthorized person to obtain access to any part of the computer facility at which the user's data is to be processed, and to establish access codes and passwords and take such other measures as may be necessary to prevent any unauthorized access to or interference with the service bureau's computer system. The service bureau should also be required to obtain appropriate covenants of non-disclosure from any employees who perform any obligations in connection with the provision of data processing services.

To avoid any dispute as to the confidential status of any information or materials supplied by either party under the agreement, all materials which are intended to be so treated should be clearly labelled.

(i) Products Supplied By Customer

In some circumstances a user may supply its own proprietary software to the service bureau for utilization in processing the customer's data. The agreement will generally provide for the service bureau to review and test the software to determine its condition and compatibility with the bureau's computer configuration. The customer will be responsible for the costs of such testing and review. The agreement will also generally exclude any liability on the part of the service bureau for the incompleteness and inaccuracy of the information generated utilizing the user's software. However, the user should insist on the right to receive the results of any testing or review of the software conducted by the service bureau and to be informed if the software is not compatible with the service bureau's existing computer configuration.

(j) Patent and Copyright Indemnity

The user will wish to obtain an agreement from the service bureau to defend, indemnify and hold it harmless from any claim that use of the service bureau's resources constitutes infringement of any patent, copyright, trade mark, trade secret or other proprietary right of any third person. The user should ensure that the agreement to indemnify and defend extends to a claim based on an alleged breach of any proprietary right and not just copyright or patent claims. The agreement to indemnify and defend may be subject to certain conditions, such as the receipt of prompt notice of any claim, and the right of the service bureau to participate in or to assume the defence of any action or to agree to the settlement of any claim. The agreement should also provide that if the services provided to a customer are found to be in breach of the rights of any person, or if there is a reasonable likelihood of such breach, the service bureau will procure for the customer the right to continue to use such services or will make all necessary modifications to ensure that use of such services will be non-infringing. The service bureau will wish to provide for this obligation not to apply in respect of any modifications to its system in accordance with specifications provided by the customer.

(k) Back-up Facility and Data

The user may wish to ensure that the service bureau is bound to provide access to a back-up computer facility in the event that its system is not available for processing the user's data. The agreement should set out the period of time following the service bureau's facility becoming unavailable that access to the back-up facility will be provided. The agreement should also set out the period of time following processing that all input data as well as all monthly and annual master files are to be retained by the service bureau so that, if necessary, the data can be re-processed.

(l) Non-competition

Each of the parties may wish to ensure that the agreement contains certain provisions relating to non-competition. For example, the agreement may provide that neither party will hire or solicit the other's employees during the term of the agreement and for a specified period of time thereafter.

In some circumstances, the user may wish to require the service bureau not to provide services to the user's competitors. The service bureau will likely agree to such a provision only if it is providing a specialized service for which it is important that competitors not gain access to any of the customer's data. The service bureau will also wish to restrict such a clause so that it only precludes the provision of a certain kind of service rather than the exclusion of all possible dealings with a particular class of customers.

(m) Standards of Performance

The agreement should indicate the standards to be met by the service bureau in the provision of services under the agreement. In particular, the service bureau should be required to exercise "due care" in performing all services, to employ competent and skilled personnel, and to provide all services and to deliver all materials by the times specified in the agreement or to use its best efforts to do so. The service bureau should also be required to adhere to established standards in the electronic processing industry as to the standard of care to be followed in performing any services. In addition, the agreement should provide for the service bureau to make any corrections and if necessary to re-process any data as promptly as possible following notice of any errors or deficiencies being provided by the user. The agreement should also provide for the service bureau to take all reasonable measures to prevent the loss or destruction of any information provided by the user.

The agreement should set out the responsibilities of the parties with respect to any reports to be prepared by the service bureau being in compliance with applicable legislation. For example, if the service bureau's reports are to be filed with any governmental authority, the agreement should set out the standards or regulations to which the reports must comply. Alternatively, the agreement may provide for the user to notify the service bureau of such standards and regulations, failing which the service bureau will have no liability or responsibility.

The service bureau will wish to impose various qualifications to its obligations to provide services and certain limits on its liability for any breach of the performance standards. The service bureau may also seek an indemnity from the user for any claim made by any third party involving any injury or loss suffered as a result of any incorrect information supplied by the user, any default by the user in the performance of services under the agreement, or the infringement by any materials supplied by the user of any proprietary right of any other party. The clauses that the service bureau may wish to have inserted to deal with such matters are discussed in detail in the final chapter of this Part.

(n) Default and Rights of Termination

Each party may wish to have the right to terminate the agreement if the other party commits a material default of its obligations. In particular, the service bureau will wish to have the right to terminate the agreement if the user fails to pay all charges in a timely manner, while the user will wish to have such right of termination if the service bureau ceases to provide services in the manner set out in the agreement. The agreement should provide for notice of any default to be given to the defaulting party, and for the termination to be effective only if such default is not cured within the notice period.

Each party will also wish to have the right to terminate the agreement if the other party commits any act of bankruptcy, becomes subject to the appointment of a receiver or a receiver/manager for any part of its business or assets, or if the other party becomes subject to any action or proceeding for its liquidation, dissolution or winding up, or if it ceases to carry on business in the normal course.

The agreement should provide that termination of the agreement either upon its expiration or for cause, will not affect any accrued obligations as well as any obligations which are to survive termination, such as those relating to confidentiality, protection of proprietary rights and the return of any materials belonging to one party in the possession of the other party. The service bureau may wish to provide that upon termination of the agreement, it will not be bound to return any materials belonging to the user until payment of all charges and such materials will be subject to a lien until all charges are paid in full. Such a provision may make it impracticable for the user to withhold payment of any charges to offset any claim for damages it may have against the service bureau.

19

Contracting for Computer Products and Services: A Review of Some "Standard" Contractual Terms

1. INTRODUCTION

In the foregoing chapters, the terms and conditions that may be included in a number of specialized contracts relating to the provision of computer products and services were considered. The present chapter will review a number of provisions that the parties may wish to consider for inclusion for any one of these contracts. Such clauses may be characterized as "standard" or "boiler-plate" provisions insofar as they are to be found in one form or another in most computer contracts and indeed in many other commercial arrangements. However, the common reliance on such clauses does not mean that they are of less importance than other contractual provisions. To the contrary, their widespread use indicates their importance in determining the rights and obligations of the parties. Accordingly, the parties should carefully consider the implications of such provisions before their final form is settled.

The most important of the so-called "standard" provisions in a computer contract are those clauses relating to the exclusion or limitation of the liability of one or both of the contracting parties. Accordingly, the main emphasis of this chapter will be a review of these clauses together with some of the legal doctrines that may limit the enforceability or application of such clauses. The remainder of the chapter will review additional "standard" clauses that the parties should consider for inclusion in any computer contract.

2. EXCLUSIONARY, LIMITATION AND INDEMNITY CLAUSES

(a) A review of common exclusionary clauses

In entering into a contract to acquire computer products, a user will wish to obtain various representations, warranties and covenants from the supplier as to the quality of the products, their compliance with technical specifications and standards, and their capacity to perform certain functions and to meet certain business requirements. The user will also wish to obtain warranties of title and the right to enjoy quiet possession of any products acquired from the supplier. In the case of a contract for computer services, such as a hardware maintenance or software support agreement, the user will wish to obtain warranties that such services will be performed in a workmanlike manner by skilled and competent personnel so as to keep the hardware or software in good operating condition. The user will also wish to exercise its normal contractual remedies for any breach of such obligations, such as to recover damages, obtain specific performance or injunctive relief or rescind the contract and obtain a refund of any moneys paid.

By contrast, the supplier will wish to limit the scope of its obligations as well as the amount of damages for which it may be liable for breach of such obligations. The supplier may also wish to have the right to be indemnified against claims brought by third parties. Among the various limitations, exclusionary or indemnity clauses that a supplier may wish to have inserted in a contract to supply computer products and services are the following.

(i) Clauses limiting the time period during which any warranties are to run and imposing certain conditions as to the application of such warranties. For example, a warranty as to computer equipment being in good operating condition may be for a limited period of time such as 90 days. Similarly, a warranty that computer programs will perform in accordance with their specifications may be subject to certain conditions of usage, such as that the programs not be modified without authorization, that they be used on specific equipment and that they not be merged with other programs. An exclusionary clause may provide that upon the occurrence of any of these events, any warranty provided by the supplier becomes null and void.

(ii) A clause excluding all representations, warranties, conditions, promises or guarantees other than those expressly set out in the written contract. Such a clause is intended to provide for the exclusion of any obligations to which the supplier may otherwise be subject as a result of any statutory warranties or conditions, implied terms arising from any custom or usage of the trade or collateral promises or representations such as those contained in sales literature or representations made by the supplier's sales personnel.

(iii) Clauses limiting the nature and scope of the remedies available to the user as a result of any breach of the supplier's obligations. For example, the agreement may provide that the only remedy available to the user is the repair or replacement of a defective product. Alternatively, the user's remedies may be restricted to the rescission of the agreement and the repayment of any amount paid by the user. The supplier may also impose a limit on the amount of damages for which it may be liable, for example, an amount based on the purchase price for any equipment, the licence fee paid for any programs, or the fees payable during a specified time period under a service agreement. The agreement will also normally exclude any liability for certain kinds of losses incurred by the user, such as consequential, incidental or special damages, including loss of profits, lost business or loss of data.

(iv) Clauses purporting to exclude or limit any liability on the part of the supplier in respect of the breach of any independent legal duty that may be imposed upon it in connection with the performance of its obligations under the agreement. For example, the supplier may wish to exclude any liability for negligence, misconduct or any other tortious act or omission by the supplier or its agents, servants or representatives.

(v) Clauses providing for the supplier to be indemnified in respect of any claims brought by third parties. For example, a software rights owner may wish to be indemnified against any claims brought by a distributor's customers as a result of any defect or malfunction in any products supplied to the distributor or any default by the owner in the performance of its obligations.

(vi) Clauses setting out the procedures to be followed to enable the user to exercise its rights against the supplier. Such clauses may require the user to notify the supplier of any alleged breach of its obligations or to return a defective product within a specified time period. Alternatively, the user may be required to refer any dispute to arbitration before commencing any legal proceedings. The agreement may also provide for any legal proceeding to be commenced within a specified time period following the breach of any legal duty, which period may be less than the applicable period under any limitations statute.

(b) Doctrines Limiting Ability of Contracting Party to Rely on Exclusionary Clauses

It is beyond the scope of this work to consider in detail the legal doctrines that qualify or restrict the right of a contracting party to rely on exclusionary clauses drawn in its favour. However, three such doctrines which may be of particular importance in the field of computer contracting will be considered. The first such doctrine is that an exclusionary clause will be interpreted strictly against the party in whose favour it is drawn, and it must be brought clearly to the attention of the person against whom it is to be asserted. Second, an exclusionary clause may in some circumstances not apply in the case of a fundamental breach of contract. Finally, the consumer protection legislation of various Canadian provinces provides for various statutory warranties and conditions applicable to any consumer sale that may not be waived by agreement of the parties.

The following sections will consider, in turn, the implications of the foregoing doctrines for transactions involving computer products and services.

(i) *Strict Construction Doctrine*

A limitation or exemption clause, particularly in a standard form document, will generally be construed against the party in whose favour it is drawn with the "utmost strictness" and must be drawn in "clear and unambiguous terms".[1] Such a construction will be made on the basis of the *contra proferentem* doctrine, that is, that if there is any ambiguity in a term of a contract, "it must be resolved against the author if the choice is between him and the other party to the contract who did not participate in its drafting."[2] The doctrine will not apply, however, if a provision of a contract, if read in the context in which it appears, is clear and unequivocal.[3]

The courts will be particularly reluctant to interpret a clause as excluding liability for negligence unless the intention to do so is clear and unambiguous. In *Canada Steamship Lines Ltd. v. R.*,[4] the Privy Council laid down three principles to be applied in interpreting a clause of this nature:

> (1) If the clause contains language which expressly exempts the person in whose favour it is made (hereafter called "the proferens") from the consequence of the negligence of his own servants, effect must be given to that provision. . . .
> (2) If there is no express reference to negligence, the court must consider whether the words used are wide enough, in their ordinary meaning, to cover negligence on the part of the servants of the proferens. . . .
> (3) If the words used are wide enough for the above purpose, the court must then consider whether "the head of damage may be based on some ground other than that of negligence." . . . The "other ground" must not be so fanciful or remote that the proferens cannot be supposed to have desired protection against it; but subject to this qualification, . . . the existence of a possible head of damage other than that of negligence is fatal to the proferens even if the words used are prima facie wide enough to cover negligence on the part of his servants.

The principles set out in the *Canada Steamship* case were approved by

1 *Losier v. St. Paul Mercury Indemnity Co.*, [1957] O.W.N. 97 at 98 (Ont. C.A.) *per* Schroeder J.A.

2 *McClelland & Stewart Ltd. v. Mut. Life Assur. Co. of Can.*, [1981] 2 S.C.R. 6 at 15 (S.C.C.) *per* Estey J.

3 *Ibid.*, at 16.

4 [1952] A.C. 192 at 208 (P.C.).

the House of Lords in the case of *Smith v. U.M.B. Chrysler (Scotland) Ltd.*[5] However, in *Ailsa Craig Fishing Co. v. Malvern Fishing Co.*,[6] Lord Fraser noted that the principles set forth in the *Canada Steamship* case were

> not applicable in their full rigour when considering the effect of clauses merely limiting liability. Such clauses will of course be read *contra proferentem* and must be clearly expressed, but there is no reason why they should be judged by the specially exacting standards which are applied to exclusion and indemnity clauses. The reason for imposing such standards on these clauses is the inherent improbability that the other party to a contract including such a clause intended to release the proferens from a liability that would otherwise fall upon him. But there is no such high degree of improbability that he would agree to a limitation of the liability of the proferens, especially when . . . the potential losses that might be caused by the negligence of the proferens or its servants are so great in proportion to the sums that can reasonably be charged for the services contracted for. It is enough in the present case that the clause must be clear and unambiguous.[7]

It was noted further in that case that the limitation clause in question could be construed as applying to a total failure of the contract since it referred expressly to "the provision of, the purported provision of, or *failure* in provision of" the services contracted for. Moreover, the House of Lords noted that because the limitation clause in question applied to any liability "at common law", such words were sufficient to apply to negligence.

In addition to the need for any exclusionary clause to be expressed in clear and unambiguous language, the courts may also refuse to give effect to such a clause unless it is shown that adequate measures were taken to bring it to the attention of the party against whom it is to be enforced. In *J. Spurling Ltd. v. Bradshaw*,[8] Lord Denning stated that in finding whether a clause is part of a contract, "the more unreasonable a clause is, the greater the notice which must be given of it. Some clauses which I have seen would need to be printed in red ink on the face of the document with a red hand pointing to it before the notice could be held to be sufficient." Similarly, in *Thornton v. Shoe Lane Parking Ltd.*,[9] Lord Denning noted that a clause contained in a ticket issued by a machine at a parking garage was "so wide and so destructive of rights that the court should not hold any man bound by it unless it is drawn to his attention in the most explicit way."

In *Tilden Rent-A-Car Ltd. v. Clendenning*[10] it was held that a party to a standard form contract who seeks to rely on stringent and onerous

5 1978 S.C. 1 (H.L.).
6 [1983] 1 W.L.R. 964 (H.L.).
7 *Ibid.*, at 970.
8 [1956] 2 All E.R. 121 at 125 (C.A.).
9 [1971] 2 Q.B. 163 at 170 (C.A.).
10 (1978), 18 O.R. (2d) 601 at 609 (Ont. C.A.).

provisions "should not be able to do so in the absence of first having taken reasonable measures to draw such terms to the attention of the other party." It was noted that the doctrine should be applied even where the contract was signed since the signature could not be relied upon if obtained in such circumstances that it does not represent an acquiescence to unusual or onerous terms. By contrast, in the case of *Dyck v. Manitoba Snowmobile Association*,[11] the Supreme Court of Canada gave full effect to an exemption clause signed by the appellant who participated in a sporting event so as to relieve the respondents of liability for negligence on the grounds that the appellant knew or should have known of the dangerous nature of the sport.

The foregoing authorities indicate that adequate measure must be taken to ensure that an exemption or limitation clause contained in a computer contract is brought to the attention of the party against whom it is to be enforced. If such a clause is set out in a standard form contract, it should be highlighted in bold print. If it is set out on the reverse of a purchase or licence agreement, a prominent notation referring to it should be contained on the face page of the agreement signed by the customer. There should also be an express acknowledgment by the customer that he has read and understood the terms of the agreement. Finally, the purchaser should be asked to sign the document only after he has been given an adequate opportunity to review it, and he should be so advised either orally or in writing at the time the contract is proffered for signature.

(ii) *Fundamental Breach*

A further doctrine that may in some circumstances limit the application of an exemption or limitation clause is that of fundamental breach. The concept of "fundamental breach" has been described as meaning "nothing more or nothing less than the type of breach which entitles the innocent party to treat it as repudiatory and to rescind the contract."[12]

The question of the applicability of an exemption clause in the case of a fundamental breach has been the subject of considerable judicial debate in England and Canada for several decades. On one hand, it has been contended that as "a rule of law", exempting clauses only avail a party "when he is carrying out his contract in its essential respects" and not "when he is guilty of a breach which goes to the root of the contract."[13] Following that decision, the House of Lords in *Suisse Atlantique Société d'Armement Maritime S.A. v. N.V. Rotterdamsche Kolen Centrale*[14] held that

11 [1985] 1 S.C.R. 589 (S.C.C.).

12 *Keefe v. Fort* (1978), 89 D.L.R. (3d) 275 at 279 (N.S.C.A.) *per* Pace J.A.

13 *Karsales (Harrow) Ltd. v. Wallis*, [1956] 1 W.L.R. 936 at 940 *per* Lord Denning M.R.

14 [1967] 1 A.C. 361 (H.L.).

the question of the application of an exclusionary clause to a fundamental breach was a matter of construction rather than a rule of law. In spite of the decision in the *Suisse Atlantique* case, it was held in the case of *Harbutt's Plasticine Ltd. v. Wayne Tank & Pump Co.*,[15] and in subsequent decisions that the doctrine of fundamental breach could be relied on to prevent the application of an exclusionary clause where the breach brought the contract to an end.

In the case of *Photo Production Ltd. v. Securicor Transport Ltd.*,[16] the House of Lords held that the so-called "rule of law" that a fundamental breach brings a contract to an end, thereby precluding reliance on an exclusionary clause, was non-existent. Instead, it was a question of construction of the contract as a whole as to whether an exclusion clause applied to any breach of contract. It was noted that "Any persons capable of making a contract are free to enter into any contract they may choose: and providing the contract is not illegal or voidable, it is binding upon them."[17]

The decision of the House of Lords in the *Photo Production* case was approved by the Supreme Court of Canada in the case of *Beaufort Realties (1964) Inc. v. Chomedey Aluminum.*[18] The Supreme Court cited with approval the quotation by Lord Wilberforce in the *Photo Production* case of the headnote in the *Suisse Atlantique* case to the effect that "the question of whether an exception clause was applicable where there was a fundamental breach of contract was one of the true construction of the contract." Notwithstanding the acceptance of this doctrine, however, the Supreme Court found based on the facts of the *Chomedey* case, that the waiver clause could not be relied upon as a matter of construction of the contract considering the fundamental breach committed by the contractor.

In *Cathcart Inspection Services Ltd. v. Purolator Courier Ltd.*,[19] the Court applied the test set out in the *Photo Production* case to the effect that it was a matter of construction of the contract as a whole as to whether an exclusionary clause applies in the case of a fundamental breach. Grange, J.A. held that the proper test was whether it was fair and reasonable to attribute to the parties an intention for an exclusionary clause to survive a fundamental breach by a party in whose favour it is drawn. The Court also relied on the judgment of Brooke, J.A. in *Leading Investments Ltd. v. New Forest Investments Ltd.; H.W. Liebig & Co. v. Leading Investments Ltd.*[20] in which he stated that "when a contract is in issue the court should look for and be guided by the reasonable expectation of the parties so

15 [1970] 1 Q.B. 447 (C.A.).
16 [1980] A.C. 827 (H.L.).
17 *Ibid.*, at 853 *per* Lord Salmon L.J.
18 [1980] 2 S.C.R. 718 (S.C.C.).
19 (1982), 139 D.L.R. (3d) 371 (Ont. C.A.).
20 (1981), 34 O.R. (2d) 175 at 183-4 (Ont. C.A.).

long as it is compatible with their written contract." Based on these tests, Grange J.A. held that a clause providing that the carrier would not "be liable for any special, consequential or other damages for any reason whatever including delay in delivery" should not be interpreted as encompassing non-delivery.

Several Canadian cases have considered whether an exemption clause may be relied upon in the case of a contract for the supply of photocopying and computer equipment. In *Canadian-Dominion Leasing Corp. v. George A. Welch & Co.*,[21] the appellant leasing company had leased a photocopier to the respondent on the basis that the respondent selected both the equipment and the supplier. The leasing agreement provided that there were no representations, warranties and conditions, and that the respondent would make any claims solely against the supplier. The court held that notwithstanding that the equipment did not work, in all of the circumstances, including the dealings between the parties and the intent of the parties as expressed by the "no warranties" clause, it was fair and reasonable for the exclusionary clause to survive notwithstanding that the breach might be regarded as fundamental.

An exclusionary clause was also upheld in spite of a fundamental breach in similar circumstances in *Polaris Leasing Co. v. J.E.R. Associates Inc.*[22] and in *Nabel Leasing Div. of Citicorp Leasing Can. Ltd. v. Walwyn Stodgell Cochran Murray Ltd.*[23]

The foregoing review indicates that the trend of recent English decisions has been toward a more liberal interpretation of limitation clauses, with indemnity and exclusion clauses being subject to more strict scrutiny. Moreover, the courts will give effect even to exclusionary clauses, whether or not the event giving rise to the claim may be characterized as a fundamental breach, if the courts find on a true construction of the contract that the parties intended to exclude liability for such breach. At the same time, English courts have been able to rely on a new statutory test of reasonableness as limiting the ability of a defaulting party to rely on a limitation clause.[24]

To date, Canadian courts do not appear to have developed different tests as to the degree of scrutiny to be given to limitation as opposed to indemnity and exclusion clauses. However, they have adopted the doctrine set forth in the *Photo Production* case that a contract must be construed as a whole to determine what effect should be given to an exclusionary clause in the event of a fundamental breach. Canadian courts would also

21 (1981), 33 O.R. (2d) 826 (Ont. C.A.).
22 [1984] 3 W.W.R. 368 (Man. Co. Ct.).
23 (1982), 41 N.B.R. (2d) 291 (N.B.Q.B.).
24 Unfair Contract Terms Act, 1977 (U.K.), c. 50 consolidated in Sale of Goods Act, 1979 (U.K.), c. 54. For an analysis of recent English cases, see M.H. Ogilvie, Note (1984), 62 Can. Bar Rev. 389.

appear to be adopting a judicial test of reasonableness in finding whether as a matter of construction the parties intend an exclusionary clause to survive a fundamental breach. However, the failure of the courts to articulate in each case the precise grounds on which an exclusionary clause fails to meet the reasonableness test raises uncertainty as to what drafting techniques should be adopted to ensure the survival of an exemption clause.

Notwithstanding the lack of clarity as to the present state of the law, a number of measures may be taken in drafting exemption clauses to be incorporated in computer contracts. First, if it is the intention of the parties to exclude liability for negligence or in connection with any common law action, such exclusion should be expressly set out. Second, there should be an express reference to consequential, indirect or special damages, including loss of profits, loss of use, economic loss, or any loss arising from any claim by any third party. Third, the exemption clause should expressly refer to any claim arising from any default or failure of performance as well as arising from the performance or purported performance of the obligations of the parties. Fourth, it may be advisable to make express reference to fundamental breach so as to preclude a court from finding that the parties did not intend to include such breach within the ambit of the clause. Finally, in view of the judicial trend, at least in England, to apply a less strict test to a limitation as opposed to an exclusionary clause, a supplier of computer products and services may wish to consider setting out a maximum limit on the amount of damages for which it will be liable in any circumstances; such a limitation clause may be inserted in place of or in addition to an exemption clause *per se* although a court may find that such a clause creates ambiguity as to the interpretation to be given to a separate exemption clause.[25]

In conclusion, in view of a growing tendency of the courts to adopt a test of reasonableness in determining what effect should be given to an exemption clause, particularly upon the occurrence of a fundamental breach, a supplier of computer products and services should ensure that any such clause will meet such test. If a contract for the supply of computer products and services gives the purchaser an adequate opportunity to test the system before acceptance, and if a limitation clause limits the vendor's liability upon the occurrence of any default to the repayment of the purchase price, it is unlikely that a court would find that a clause excluding any other liability would be outside the reasonable expectation of the parties. However, if the contract purports to exclude any liability of any kind notwithstanding any default or failure by the supplier, the courts may attempt to construe the contract so as not to give effect to such a clause, regardless of how strained an interpretation may be necessary to achieve such objective.

---

25 *Ailsa Craig Fishing Co. v. Malvern Fishing Co.*, [1983] 1 W.L.R. 964 (H.L.).

(iii) *Consumer Transactions*

The ability of a contracting party to rely on various exempting conditions is further limited insofar as the contract involves a consumer transaction. In this connection, the consumer protection legislation of most provinces provides for certain implied conditions and warranties applicable to consumer transactions which may not be waived by the parties. The following is a summary of the principal provisions of such legislation.

In Ontario, section 34 of the Consumer Protection Act,[26] provides that any implied conditions or warranties applicable to a "sale of goods" under the Sale of Goods Act,[27] may not be waived in the case of a consumer sale. Any clause of a contract purporting to negate or vary such conditions is void and severable from the remainder of the contract. A "consumer sale" is defined to mean a sale made in the ordinary course of business to a purchaser "for his consumption or use", but does not include a sale to a purchaser for resale, to a purchaser whose purchase is in the course of carrying on business or to a corporation, partnership or association of individuals.

The implied warranties that may not be waived under the Ontario Sale of Goods Act in the case of a consumer sale are that the seller has the right to sell the goods; that the buyer will have quiet possession of the goods, that the goods will be free of any encumbrances in favour of any third party; that the goods will correspond with any description; that the goods will be fit for the purpose made known to the seller as the purpose for which they are being acquired; and that the goods will be of merchantable quality.[28]

Section 34 of the Consumer Protection Act applies only to a sale of goods. The word "goods" is defined in section 1(*g*) of the Sale of Goods Act to mean all chattels personal, other than things in action and money. A "contract of sale" is defined in section 2 of that Act to mean a contract whereby the seller transfers or agrees to transfer the property in the goods to the buyer for a money consideration.

In view of the definitions of "goods" and a "contract of sale", it is questionable whether the prohibition against waiver of the implied warranties and conditions under the Sale of Goods Act would apply in the case of a grant of a licence to computer programs which does not involve the transfer of property in any tangible goods. However, such warranties may apply in the case of a shrink-wrap licence for computer software if the transaction is characterized as a sale of a copy of the program or involves a transfer of ownership of the medium in which the programs are contained.

---

26 R.S.O. 1980, c. 87.

27 R.S.O. 1980, c. 462.

28 *Ibid.*, ss. 13-15.

The Manitoba Consumer Protection Act[29] also provides for certain warranties and conditions which are implied, notwithstanding any agreement to the contrary, in every "retail sale of goods" and every "retail hire-purchase of goods".[30] A retail sale is defined substantially the same as a consumer sale in Ontario, but excludes from the definition any sale in which the cash price exceeds $25,000.[31] The implied warranties and conditions are similar to those in the Ontario Act with an additional implied condition that the goods are new and unused unless otherwise described. The Manitoba Act also provides for an implied condition in every retail sale of services that the services shall be performed in a skilful and workmanlike manner, although such condition may be waived if expressly agreed in writing by the seller.[32] Finally, the Manitoba Act provides that every oral or written statement made by a seller regarding the quality, condition, quantity, performance or efficiency of goods or services that is contained in an advertisement or made to a buyer shall be deemed to be an express warranty respecting those goods or services.[33]

In Nova Scotia, section 20C of the Consumer Protection Act[34] provides for implied conditions and warranties applicable to every consumer sale which are similar to those in Manitoba. An additional implied condition set out in the Nova Scotia Act is that the goods shall be durable for a reasonable period of time having regard to the use to which they would normally be put and the circumstances of the sale. A consumer sale is defined to mean a contract for the sale of goods or services made in the ordinary course of business to a purchaser for his consumption or use, but does not include a sale to a purchaser for resale, to a purchaser whose purchase is in the course of carrying on business, to an association of individuals, corporation or partnership or by a trustee in bankruptcy, receiver or liquidator.[35]

The Quebec Consumer Protection Act[36] also sets out certain warranties which are deemed to be included in any contract of sale or of lease of goods or services made between a consumer and a merchant in the course of his business. A consumer is defined in the Act to mean a natural person except a merchant who obtains goods or services for the purposes of his business.[37] Among the warranties prescribed by the Act

29 R.S.M. 1970, c. C200.
30 *Ibid.*, s. 58.
31 *Ibid.*, subs. 1(*t*).
32 *Ibid.*, subs. 58(6).
33 *Ibid.*, subs. 58(8).
34 R.S.N.S. 1967, c. 53.
35 *Ibid.*, subs. 20C(1).
36 S.Q. 1978, c. 9/P 40.1.
37 *Ibid.*, subs. 1(*e*).

are: that the goods are free of any charge or encumbrance;[38] that they are fit for the purposes for which goods of that kind are normally used;[39] that they are durable in normal use for a reasonable time having regard to their price, the terms of the contract and the conditions of their use;[40] that if the goods will be of a nature requiring maintenance, replacement parts and repair service will be available for a reasonable time (unless the consumer is warned in writing before the contract is made that the merchant or manufacturer does not supply replacement parts or repair service);[41] and that the goods conform with any description set out in the contract and with any statements or advertisements made by the merchant or manufacturer.[42] A merchant or manufacturer is also bound by any written or verbal statement made by its representative.[43]

The Quebec Act also sets out certain conditions as to the form and contents of any conventional warranty. For example, any exclusions in a conventional warranty must be clearly indicated in separate and successive clauses, and the duration must be determined precisely.[44] In addition, any writing evidencing a warranty must set out: the name and address of the person offering the warranty; a description of the goods that are the subject of the warranty; whether the warranty may be transferred; the obligations of the person granting the warranty in the case of a defect in the goods or the improper carrying out of services under the warranty; the manner in which to obtain execution of the warranty; and its duration.[45] If a manufacturer's warranty is valid only if goods or services are supplied by a merchant certified by the manufacturer, any other merchant must notify the consumer in writing that the warranty is not valid, failing which he will be bound to assume the warranty at his expense.[46] No charges may be exacted for the performance of a conventional warranty, or in respect of the cost of transportation or shipping for performance of the warranty, unless such charges are clearly stated.[47]

A consumer who has entered into a contract with a merchant may exercise directly against the merchant or a manufacturer a recourse based on a latent defect in the goods or the lack of instructions necessary to protect the user against a risk or danger of which he would otherwise

38 *Ibid.*, s. 36.
39 *Ibid.*, s. 37.
40 *Ibid.*, s. 38.
41 *Ibid.*, s. 39.
42 *Ibid.*, ss. 40-41.
43 *Ibid.*, s. 42.
44 *Ibid.*, ss. 44, 46.
45 *Ibid.*, s. 45.
46 *Ibid.*, s. 47.
47 *Ibid.*, ss. 48, 49.

be unaware.[48] Such rights may also be exercised as against the manufacturer by a subsequent purchaser.[49] A consumer who has entered into a contract with a merchant may take action directly against the merchant or the manufacturer to assert a claim based on a breach of the implied warranties of fitness of goods for their normal purpose or their durability, or the availability of replacement parts; the fitness and durability warranties may also be enforced against the manufacturer by a consumer who is a subsequent purchaser of the goods.[50]

The Consumer Products Warranties Act of Saskatchewan[51] sets out certain statutory warranties deemed to be given by a retail seller of a consumer product. These warranties relate to title, freedom of the goods from any encumbrances, description, acceptable quality, fitness for any particular purpose for which the goods are being bought made known to the seller, correspondence of the goods with any sample, durability and availability of spare parts and repair facilities.[52] A "consumer" is defined under the Saskatchewan Act to mean a person who buys a "consumer product" from a retail seller unless he acquires the product for purposes of re-sale or for use in a business.[53] A consumer product is defined to include any goods ordinarily used for personal, family or household purposes.[54] A retail seller means a person who sells consumer products to a consumer in the ordinary course of business but excludes a trustee in bankruptcy, receiver, liquidator, sheriff, auctioneer or a person acting under a court order.[55]

Under the Saskatchewan Act, subject to certain qualifications, a manufacturer of a consumer product is deemed to give to consumers the same statutory warranties (other than a warranty of title) as those given by the merchant. The manufacturer is liable, however, only for his own breach of such warranties or of any express or additional written warranties.[56]

If a retail seller is sued for a breach of any statutory warranty, and the manufacturer is in breach of the same warranty, the retail seller is entitled to recover any losses that he suffers and that are the responsibility of the manufacturer.[57]

The Saskatchewan Act also contains detailed provisions with respect

48 *Ibid.*, s. 53.
49 *Ibid.*, subs. 53(4).
50 *Ibid.*, s. 54.
51 R.S.S. 1978, c. C-30.
52 *Ibid.*, s. 11.
53 *Ibid.*, subs. 2(*d*).
54 *Ibid.*, subs. 2(*e*).
55 *Ibid.*, subs. 2(1).
56 *Ibid.*, s. 13.
57 *Ibid.*, s. 16.

to the contents of any additional written warranties given by a warrantor.[58] It also prescribes the remedies available to a consumer with respect to any breach of the statutory or express warranties.[59]

The New Brunswick Consumer Product Warranty and Liability Act[60] applies to every sale or supply of a consumer product, unless made by a seller or supplier who is not a distributor of consumer products or unless made by a person acting as a trustee in bankruptcy, liquidator, receiver, sheriff or person acting under court order.[61] A consumer product is defined in the Act to mean any tangible personal property of a kind commonly used for personal, family or household purposes.[62]

In every contract for the sale of a consumer product under the New Brunswick Act, the following statements are deemed to be express warranties given by the seller to the consumer: (a) any oral statement in relation to the product unless the circumstances show that the seller does not rely, or it is unreasonable for him to rely, on the statement; (b) any written statement made to the buyer in relation to the product, whether or not the buyer relies on the statement, unless the circumstances show that it would be unreasonable for him to rely on the statement; and (c) any statement made to the public, whether the buyer relies on the statement, unless the circumstances show it would be unreasonable for the buyer to rely on the statement.[63]

A seller is deemed to make any statement made by his agent or employee unless it is proved that the agent or employee was not acting within the scope of his actual, usual or apparent authority. The seller is also deemed to make any statement in writing on the product or its container or on any label or sign attached to, in proximity to or accompanying the product, unless it is proved the statement was made by a person other than the seller and that he neither knew nor ought to have known the statement was made.[64] The Act provides that where there is a written contract, oral or other extrinsic evidence is admissible to establish an express warranty notwithstanding that it adds to, varies or contradicts the written contract.[65]

The New Brunswick Act also sets out a number of statutory warranties included in every contract for the sale of a consumer product, including warranties of title and the right of quiet possession, a warranty that the goods are unused, a warranty that the goods are fit for the purposes for

58 *Ibid.*, ss. 17-18.
59 *Ibid.*, ss. 19-30.
60 S.N.B. 1978, c. C-18.1.
61 *Ibid.*, s. 2.
62 *Ibid.*, s. 1.
63 *Ibid.*, subs. 4(1).
64 *Ibid.*, subs. 4(2).
65 *Ibid.*, s. 5.

which they are normally used as well as any particular purpose made known to the seller and a warranty that the goods will be durable for a reasonable period of time.[66]

The Act sets out various remedies available for breach of the statutory and implied warranties, including rectification of the breach, damages for any losses suffered, and rejection of the product against the refund of any moneys paid.[67] Any person who is not a party to a contract for the sale of a consumer product but who suffers a "consumer loss" (that is, a loss suffered otherwise than in a business capacity, or suffered in a business capacity on account of liability incurred for a loss not suffered in a business capacity) because of a breach by a seller of a warranty provided by the Act may recover damages against the seller if the loss was reasonably foreseeable.[68]

The New Brunswick Act also provides that except as expressly provided in sections 25 and 26, the parties to a contract for the sale of a consumer product may not exclude or restrict any warranty or remedy provided by the Act.[69] Section 25 provides that the parties to a contract for the sale of a consumer product may agree to exclude or restrict any remedy provided by the Act for breach of an express warranty, but such agreement shall be incffective to the extent that it is shown that it would not be fair or reasonable to allow reliance on such agreement, which is to be determined according to the circumstances of each case.[70] Where a buyer makes a contract to purchase a consumer product in the course of a business, the parties may agree to exclude or restrict any warranty or remedy provided by the Act but except as provided in section 25, such agreement shall be ineffective with respect to any consumer loss for which the seller would otherwise be liable.[71]

The foregoing review indicates that the consumer protection legislation of various Canadian provinces restricts the rights of the supplier of products and services to exclude certain liabilities on its part and to limit the remedies available to the consumer. Accordingly, manufacturers and vendors of various products and services directed to the consumer market must consider carefully the provisions of such legislation in structuring their marketing and distribution arrangements and in drafting contracts to be made with consumers.

---

66 *Ibid.*, ss. 8-12.
67 *Ibid.*, ss. 15-22.
68 *Ibid.*, s. 23.
69 *Ibid.*, s. 24.
70 *Ibid.*, s. 25.
71 *Ibid.*, s. 26.

3. INTEGRATION CLAUSE

An integration clause defines the documents which constitute the entire agreement between the parties. For example, a system acquisition agreement may consist of a master purchase or procurement agreement having attached to it a number of supplemental agreements dealing with certain aspects of the acquisition. These supplemental agreements may include licences for operating system software and application programs, a hardware maintenance agreement, a software support agreement, and a source code escrow agreement. An integration clause should also identify any schedules intended to form part of the agreement: for example, schedules describing the functional or technical specifications for the products to be acquired or an implementation schedule setting out the dates on which certain obligations are to be performed.

A computer supplier and user should consider carefully what documents, if any, should be incorporated by reference in the contract. A reference to a request for proposals issued by the user or to a proposal submitted by the supplier may be useful as indicating the features of a computer system or of customized software to be provided by the supplier. However, the supplier may be reluctant to have the entire document incorporated by reference in the contract since the supplier will be bound by all promises and representations contained therein.

Purchasers of computer products and services should be aware that integration clauses are intended to restrict the application of representations and promises made by sales representatives as well as claims made in sales or promotional literature. If any oral promises or any claims made in promotional literature or a proposal prepared by the supplier are important for the user, they should be reduced to writing and incorporated expressly as part of the contract documents.[72] A computer supplier's sales personnel will normally not be involved in the actual administration of an executed agreement, and the purchaser will generally not be entitled to rely on verbal assurances given by such representatives during the course of the negotiation of the agreement.

4. ASSIGNMENT

An agreement for the supply of computer products and services will generally contain a prohibition against its assignment by either party without the other party's consent. In the case of an agreement for the

72 In *Unified Systems Ltd. v. Clearwater Lobsters Ltd.*, Nfld. T.D., March 26, 1986, summarized at (1986), 38 A.C.W.S. (2d) 461, it was held that a clause in an agreement between a software developer and customer stating that the agreement superseded all proposals, previous negotiations and all other communications relating to the subject matter, meant that the terms of a prior proposal made by the developer did not form part of the contract.

licensing of computer software, the agreement may restrict the licensee from making any assignment or transfer of the software or the licensee's rights therein without the licensor's consent. The licensor may require the payment of an additional licence fee, require the execution of a satisfactory licence agreement by an assignee or impose other obligations as a condition of granting such consent.

A user may wish to limit the scope of any restrictions imposed on it with respect to the assignment of its rights under the computer agreement. For example, the user will wish to have the right to assign any licence to computer programs in conjunction with the sale of the equipment on which the programs are used, particularly if made in conjunction with the sale of its business, or other corporate reorganization. The user may also wish to have the right to transfer any leased computer equipment and to assign or sub-license the right to use computer software to other companies within the user's corporate group, that is, to any affiliate or subsidiary companies. The right to grant such assignments or licences may be subject to certain restrictions, such as that the programs be used on designated computer equipment or with certain operating system software. The user may also wish to have the right to grant a security interest in its computer system, including computer software, in order to obtain financing for the conduct of its business.[73]

A computer user will wish to consider whether the agreement should restrict the right of the supplier to assign or sub-contract its duties without the user's consent. For example, if a customer under a maintenance or support services agreement is relying on the personal skill and expertise of a particular supplier, it will wish to have the right to insist on performance personally by that supplier.

5. FORCE MAJEURE

This type of clause excuses delays or defaults in performance by either or both of the contracting parties due to some cause beyond the reasonable control of the party concerned. The majority of force majeure clauses contain detailed lists of the types of events which excuse performance such as acts of war, riots, strikes and other labour disputes. The events which are specifically listed may be included "ejusdem generis" as coming within any cause beyond the other party's reasonable control. A customer may wish to restrict the events which trigger the application of the force

73 For a discussion of the legal implications of the creation and perfection of security interests in computer software, see W. Semple, "The Legal Incidents of Computer Software and its Use as Collateral in Secured Transactions" (1982-83), 7 Can. Bus. Law J. 450, and A.R. Szibbo, "Secured Transactions in the Computer World", Paper presented at International Bar Association, Section on Business Law, 21st Biennial Conference, New York, New York, September 18, 1986.

majeure clause, or may wish to have the right to terminate a contract for which the supplier is in default even if the supplier is excused from liability for damages as a result of the force majeure clause.

6. TIME OF THE ESSENCE

A clause stipulating that time is of the essence may be of particular importance if a purchaser or lessee of computer hardware, or a customer contracting for the development of computer software, requires the hardware or software by a particular date to carry on business. Unless such a clause is inserted, or the right is granted expressly under the contract, the failure of one party to complete its obligations by a certain date may not entitle the innocent party to treat the contract as at an end. A clause stating that time is of the essence may also be important for a supplier if the user is required to notify the supplier by a certain date whether it accepts certain products or services.

7. NON-WAIVER OF DEFAULTS

This clause provides that if a party waives or does not pursue its remedies against the other party who commits one event of default, such waiver will not affect the non-defaulting party's remedies with respect to any future default of the same or any other obligation. This could be of importance to a customer which has contracted for the delivery of a computer system consisting of components to be delivered at different dates or with certain milestones to be achieved at stipulated intervals. The customer may be prepared to waive strict performance if the supplier fails to meet one deadline but on the second such occurrence it may wish to rescind the contract. Accordingly, its remedies should not be affected by its previous waiver.

8. SUCCESSORS AND ASSIGNS

Notwithstanding any prohibitions on assignment, this clause should be included to ensure that the agreement is binding on any successors or assigns to the contracting parties. This clause should also provide that notwithstanding any permitted assignment, the assigning party will not be relieved of its obligations to the other party.

9. SEVERABILITY

This clause provides that if any portion of the contract is determined to be unenforceable, only that portion of the contract will be rendered ineffective and the balance of the contract will continue in full force and effect. Such a clause would mean that if a provision purporting to exclude or limit the liability of one contracting party is found to be unenforceable, such as because of non-compliance with consumer protection legislation, the contract will be read as if such provision were omitted and the balance of the contract will remain in effect. However, the prohibited clause will be rendered unenforceable in its entirety, and will not be interpreted by a court as if it were drafted in a manner that would be acceptable.

10. NOTICES

For practical reasons it is advisable to specify the proper means of giving notice to each of the contracting parties and defining when notice is deemed to be effective. For example, the contract will generally require any notice to be in writing and to be given by personal delivery, or by telex or mail. The notice will generally be effective on the date of delivery, the date of sending a telex, or a specified number of business days following the date of mailing. The effective date of a notice may be critical when calculating times for payment, the effective period of a warranty or whether notice of a breach or possible claim against the parties has been given properly.

11. ARBITRATION

The parties to a computer agreement may wish to provide for the appointment of an arbitrator to resolve any dispute that the parties are unable to resolve in connection with the interpretation and performance of the agreement. The arbitration clause will generally set out the time period by which an arbitrator is to be appointed and provide that if the parties are unable to agree on a single arbitrator, each would be entitled to appoint its own arbitrator, and the two selected would appoint a third. The agreement should set out the place at which the arbitration is to be conducted and the rules which are to govern such arbitration, for example, the Arbitrations Act of the jurisdiction whose laws are to govern the agreement or the arbitration rules of the International Chamber of Commerce.

12. GOVERNING LAW AND FORUM

A contract should specify the laws of the jurisdiction which will govern the interpretation and enforcement of the contract, particularly when one of the contracting parties such as a computer supplier is carrying on business in several jurisdictions. The contract should also set out the jurisdiction in which any legal proceeding under the contract is to be brought.

Part VI

TAXATION ASPECTS OF COMPUTER TRANSACTIONS

20

Tax Aspects of the Transfers of Computer Products

1. INTRODUCTION

The purpose of this Part of the text is to consider the income and commodity tax aspects of the transfer, servicing and development of computer products in Canada. Chapter 20 begins with a review of some of the terms used by Canada's revenue authorities to distinguish among the different components of a computer system. There follows a brief description of the means by which computer products are transferred and a detailed analysis of the tax consequences of such transfers. The following chapter considers the tax aspects of the servicing and the development of computer products.

2. COMPONENTS OF A COMPUTER SYSTEM

Canada's revenue authorities generally consider a computer system to be composed of two major components: computer hardware and computer software.[1] Computer hardware refers to the physical components of the system including the central processing unit and peripheral equipment such as card readers and printers.[2] Computer software, on the other hand, refers to the computer programs and related procedures and technical

---

1 In para. 9 of Memorandum D10-8-13 Revenue Canada, Customs and Excise ("RCCE") refers to "firmware" which may be regarded as a third component of a computer system. To date, however, Canada's revenue authorities have not developed special rules for computer firmware.

2 Subs. 1104(2) of the Income Tax Regulations (the "ITR") defines "general-purpose electronic data processing equipment" as follows:

> [E]lectronic equipment that, in its operation, requires an internally stored computer program that
>
> (a) is executed by the equipment,
>
> (b) can be altered by the user of the equipment,
>
> (c) instructs the equipment to read and select, alter or store data from an external medium such as a card, disk or tape, and
>
> (d) depends upon the characteristics of the data being processed to determine the sequence of its execution.

It is generally accepted that most computer hardware falls within this definition.

documentation which allow the hardware to perform different functions.

All of Canada's revenue authorities distinguish between operating or systems software and applications software.[3] Systems software generally is considered to be software that is fundamental to the operation of the computer whereas applications software is software which permits the computer to perform specific functions for the benefit of the user.[4] Most provincial revenue authorities further distinguish between custom applications software and canned or prepackaged applications software. Custom software refers to programs designed and developed solely to meet the specific requirements of a user whereas canned or prepackaged software refers to programs designed for general application without the needs or requirements of a specific user in mind.[5]

3. TRANSFERS OF COMPUTER PRODUCTS

(a) Types of Transfer Agreements

Computer hardware may be purchased or leased. Where hardware is purchased, the purchase price generally is payable in a single lump sum or in two or more instalments which may be made payable at the time of delivery, installation and/or acceptance of the hardware. In many cases,

---

3 Subs. 1104(2) of the ITR provides that "computer software" includes "systems software" and a right or licence to use computer software. Systems software is defined as follows:

> [A] combination of computer programs and associated procedures, related technical documentation and data that
> (a) performs compilation, assembly, mapping, management or processing of other programs,
> (b) facilitates the functioning of a computer system by other programs,
> (c) provides service or utility functions such as media conversion, sorting, merging, system accounting, performance measurement, system diagnostics or programming aids,
> (d) provides general support functions such as data management, report generation or security control, or
> (e) provides general capability to meet widespread categories of problem solving or processing requirements where the specific attributes of the work to be performed are introduced mainly in the form of parameters, constants or descriptors rather than in program logic,
>
> and includes a right or licence to use such a combination of computer programs and associated procedures, related technical documentation and data.

4 Such functions may include payroll, inventory or accounting.

5 This distinction is legislated in s. 1 of the Social Service Tax Act, R.S.B.C. 1979, c. 388, as amended, which defines computer software as follows:

> [P]ackaged or prewritten computer software programs that are designed for general application, or the right to use those programs, and includes modifications to those programs and modifications of those programs for a particular purchaser or lessee, but does not include a computer software program that is designed and developed solely to meet the specific requirements of the purchaser or lessee.

an initial deposit will be payable at the time of execution of the purchase agreement, with the final payment being due on the date of acceptance. Where hardware is leased, lease payments are made on a periodic basis over the term of the lease.

Regardless of whether hardware is purchased or leased, the accompanying systems software generally is licensed. Where the hardware is purchased, the term of the licence to use the systems software is often perpetual (subject to earlier termination if the licensee is in breach of the terms of the licence agreement) and the licence fee is payable in a lump sum, which may be due at the same time as the hardware purchase price. Where hardware is leased, the term of the licence to use the systems software matches that of the hardware lease and the licence fee is usually payable on a periodic basis in accordance with the schedule for the hardware lease payments. Prepackaged applications software generally is licensed whereas custom applications software may be licensed to the customer for whom it is designed, or all rights to such software may be acquired outright by the customer.

(b) Transfers of Computer Products — Income Tax Implications

(i) Tax Implications for Transferor

The sale, lease or licence of computer products in Canada by a transferor resident in Canada does not give rise to any special income tax consequences to the vendor, lessor or licensor. Amounts generally are included as business income of the transferor in the year in which they become receivable although a reserve may be available to the transferor in respect of amounts not actually due until after the end of the year.[6]

(ii) Tax Implications for Transferee

A transferee of computer products will be concerned with two tax issues relating to the acquisition, namely the manner in which the costs of acquisition may be deducted, and the extent to which investment tax credits may be claimed in connection with the acquisition. As noted in greater detail below, the general investment tax credit is being phased out beginning in 1987 and the credit will be eliminated for most types of property acquired after 1988.

---

6 Para. 20(1)(*n*) of the Income Tax Act, (the "ITA") S.C. 1970-71-72, c. 63 as amended. See also subsection 20(8) of the ITA. Where used computer equipment that constitutes capital property of a taxpayer is resold, the taxpayer is required to include as income any recapture of capital cost allowance and one-half of any capital gain (see ss. 13 and 39 of the ITA).

(A) *Deduction of Costs of Acquisition.* The rules applicable to the deduction of the costs of acquisition of computer products differ for computer hardware and systems software, on the one hand, and applications software, on the other. In the former case, the rules differ according to whether the hardware is purchased or leased.

Purchase of Computer Hardware and Systems Software. A purchase of computer hardware and a perpetual licence of the underlying systems software generally result in the acquisition of a capital asset.[7] Accordingly, the purchase price of the hardware and the licence fee for the right to use the systems software are not fully deductible in the year of acquisition but rather are deductible only over time in the form of capital cost allowance ("CCA").[8] The maximum amount of CCA that may be claimed in a particular year is determined in accordance with the provisions of Part XI of the Income Tax Regulations ("ITR") and depends upon which of the classes enumerated in Schedule II of the ITR is applicable to the hardware and systems software acquired.

Most computer hardware and systems software acquired after May 25, 1976 falls under Class 10(f) of Schedule II of the ITR.[9] The following property is included in Class 10(f):

> [P]roperty not included in any other class that is . . .
>
> (f) general-purpose electronic data processing equipment and systems software therefor, including ancillary data processing equipment, acquired after May 25, 1976, but not including property that is principally or is used principally as
>
> (i) electronic process control or monitor equipment,
> (ii) electronic communications control equipment,
> (iii) systems software for a property referred to in subparagraph (i) or (ii), or
> (iv) data handling equipment unless it is ancillary to general-purpose electronic data processing equipment.

Capital cost allowance in respect of the capital cost of property included in Class 10(f) may be claimed on a declining balance basis at the rate

7 The purchase and perpetual licence result in the acquisition of an asset of enduring benefit. See the *dictum* of Viscount Cave in *Atherton v. British Insulated & Helsby Cables Ltd.* (1925), 10 Tax. Cas. 155 at 192 (C.A.).

8 Paras. 18(1)(*b*) and 20(1)(*a*) of the ITA.

9 Prior to 1983, Revenue Canada, Taxation ("RCT") took the position that a licence to use computer software constituted either an eligible capital expenditure or, if the licence was for a limited period, the acquisition of Class 14 property. The computer industry strenuously objected to this position and in 1983 the Department of Finance amended the ITR retroactive to May 25, 1976 to provide that computer software includes a right to use computer software. Accordingly, for CCA purposes a licence to use computer software is treated in the same manner as an outright purchase.

of 30 per cent per annum subject to the restriction that the maximum CCA that may be claimed in the year of acquisition is limited to one-half of the amount that may otherwise be claimed.[10] The capital cost of hardware that is purchased includes the purchase price of the hardware together with any federal or provincial sales taxes or customs duties paid thereon as well as any costs of delivery and installation.[11]

Computer hardware and systems software that are more specifically described in another of the classes enumerated in Schedule II are included in that other class. For example, computer hardware and systems software acquired after May 25, 1976 that is attached to an aircraft probably is included under Class 9(h). Computer hardware and systems software specifically excluded from Class 10(f) by subparagraphs (i) to (iv) thereof and computer hardware and systems software acquired prior to May 26, 1976 fall under Class 8(i).[12] Capital cost allowance in respect of the capital cost of Class 8(i) property may be claimed on a declining balance basis at the rate of 20 per cent per annum subject to the restriction that the maximum amount of CCA that may be claimed in the year of acquisition is limited to one-half of the amount that may otherwise be claimed.[13]

New computer hardware and the systems software therefor that would otherwise fall within Class 10(f) or Class 8(i) are excluded from such classes and are included instead in Class 29 if they are used in Canada primarily in the manufacturing or processing of goods for sale or lease.[14] Revenue Canada, Taxation ("RCT") considers that "primarily" means "chiefly" or "principally" and that otherwise qualified property used more than 50 per cent of the time in manufacturing or processing activities qualifies as Class 29 property. RCT also takes the position that direct or indirect use of a computer in manufacturing or processing includes direct manufacturing and processing applications, and ancillary activities such as maintaining inventory records, production scheduling, engineering design and production control, but does not include the maintenance of financial and accounting information such as accounts receivable and payable records,

10 For property acquired after November 12, 1981. See subs. 1100(2) of the ITR. See also subs. 1100(2.2) for circumstances where the so-called "half-year rule" under subs. 1100(2) does not apply. Where a taxation year is less than 12 months, the CCA must be prorated; see subs. 1100(3) of the ITR.

11 See D. Keith McNair, "The Meaning of Cost in Canadian Income Tax", *Canadian Tax Paper No. 69* (Canadian Tax Foundation, 1982) at 107.

12 Class 8(i) includes tangible capital property not included in another class. RCT accepts that systems software could be included in Class 8(i). *Quaere*, however, whether systems software is tangible capital property. See *Continental Commercial Systems Corp. (Telecheque Can.) v. R.* (1982), 18 B.L.R. 52 (B.C.C.A.).

13 *Op. cit.*, footnote 10.

14 Class 29 treatment also is available to certain "principal business corporations" which lease hardware to lessees who may reasonably be expected to use the equipment in the manufacturing or processing of goods for sale or lease.

general ledger accounts, payroll records, customer lists, and sales invoices and analysis.[15] If the maximum amount of capital cost allowance is claimed each year in respect of Class 29 property, the capital cost may be deducted fully over three years.[16]

Lease of Computer Hardware and Systems Software. Where computer hardware is leased, the lease payments are deductible by the lessee on a current basis and the lessor is entitled to claim CCA on the property leased.[17] Where licence fees for the underlying systems software are payable on an annual or more frequent periodic basis, these fees also will be deductible as they are payable.

Characterization of Acquisition: Purchase or Lease? The characterization of the acquisition of computer hardware as either a purchase or a lease of hardware is important in view of the different income tax considerations described above. For example, where the term of a lease is shorter than the term over which the costs of purchasing hardware may be deducted in the form of CCA, leasing allows the cost of acquiring the hardware to be deducted more quickly. Additionally, if the user of the hardware is in a low income tax bracket, leasing also may be advantageous if the CCA that may be claimed by the lessor results in lower rental charges to the lessee.[18]

Unfortunately, there is little jurisprudence that establishes the criteria for determining whether a particular transaction is a purchase or a lease for income tax purposes.[19] RCT takes the position in IT-233R that a transaction will be treated as a purchase of the underlying property rather than as a lease thereof where

(a) the lessee automatically acquires title to the property after payment of a specified amount in the form of rentals,

(b) the lessee is required to buy the property from the lessor during or at the termination of the lease or is required to guarantee that the lessor will receive the full option price from the lessee or a third party (except

15 Interpretation Bulletin IT-147R2, para. 6.

16 Para. 1100(1)(ta) of the ITR (25% in year of acquisition, 50% in second year and 25% in third year). For Class 29 property acquired prior to November 13, 1981, the capital cost may be deducted fully over two years (50% in year of acquisition, 50% in second year).

17 Unless the lessor is a "principal business corporation" the leasing property rules may apply to restrict the amount of CCA that may be claimed. See subss. 1100(15)-(20) of the ITR.

18 Neil H. Harris, "Computer Systems Canadian Taxation Issues", a paper delivered by the author to the Computer Law Association, Inc. seminar on *Distributing Computer Products to Canada: Legal & Business Considerations* October 3-4, 1983.

19 See Wallace M. Howick, "Leasing" in Corporate Management Tax Conference, 1986 (Canadian Tax Foundation: 1986).

where such guarantee is given only in respect of excessive wear and tear inflicted by the lessee),

(c) the lessee has the right during or at the expiration of the lease to acquire the property at a price which at the inception of the lease is substantially less than the probable fair market value of the property at the time or times of permitted acquisition by the lessee. An option to purchase of this nature might arise where it is exercisable within a period which is materially less than the useful life of the property with the rental payments in that period amounting to a substantial portion of the fair market value of the property at the date of inception of the lease, or

(d) the lessee has the right during or at the expiration of the lease to acquire the property at a price or under terms or conditions which at the inception of the lease is/are such that no reasonable person would fail to exercise the said option.[20]

Subsequently to the publication of IT-233R, the Canadian Institute of Chartered Accountants (CICA) published guidelines for determining for accounting purposes whether a lease should be regarded as the acquisition of an asset (a "capital lease") or as an ordinary lease.[21] These guidelines generally provide that if substantially all of the benefits and risks of ownership relating to the leased property are transferred from the lessor to the lessee, the transaction should be treated as a purchase rather than as a lease. The guidelines also set out specific rules which provide that a lease will be treated as a capital lease if the lease term is equal to 75 per cent or more of the economic life of the leased property or the present value of the lease payments is equal to at least 90 per cent of the fair market value of the property at the time of the lease.

The guidelines published by the CICA for accounting purposes are much broader than the tests enumerated in IT-233R. Revenue Canada, Taxation has not adopted the guidelines for tax purposes and, therefore, a lease may be treated as a sale for accounting purposes but as an ordinary lease for income tax purposes. It is a matter of speculation, of course, whether adoption of the guidelines by Revenue Canada, Taxation would withstand the scrutiny of the courts.[22]

Where, for income tax purposes, a lease is considered to be a purchase rather than a lease, the lessee is not entitled to deduct lease payments on a current basis but may deduct CCA according to the class in which the hardware is included. Where a lease is treated as an ordinary lease

---

20 Interpretation Bulletin IT-233R, para. 3.

21 S. 3065, *CICA Handbook* (The Canadian Institute of Chartered Accountants).

22 IT-233R is presently under review by RCT. See *R. v. Metro. Properties Co.*, [1985] 1 C.T.C. 169 (Fed. T.D.) for a discussion of the application of generally accepted accounting principles for income tax purposes. This case is currently under appeal to the Federal Court of Appeal. Compare *Oxford Shopping Centres Ltd. v. R.*, [1980] C.T.C. 7 (Fed. T.D.); affirmed on appeal [1981] C.T.C 128 (Fed. C.A.).

and the lessee subsequently purchases the property, the annual payments made by the lessee continue to be deductible in the year incurred. Where, however, the lessee purchases the property for an amount less than the fair market value of the property at the time of purchase a portion of any gain realized by the lessee on a subsequent disposition of the property may be treated as a recapture of CCA and included as income of the lessee.[23]

Applications Software. Most applications software, whether purchased or licensed, is included in Class 12(o) of Schedule II of the ITA Regulations and the cost of acquiring such software may be fully deducted over two years.[24] Applications software does not qualify for treatment under Class 29 even if it is used primarily in the manufacturing or processing of goods for sale or lease.

(B) *Investment Tax Credit.* Investment tax credits may be earned by acquiring certain types of property or by making certain qualified expenditures.[25] Investment tax credits earned in a year may be used to reduce income tax otherwise payable in the year. To the extent that investment tax credits earned in a year exceed the tax otherwise payable in the year, the excess investment tax credits may be carried back three years and forward seven years to reduce tax payable in those years. In addition, qualifying Canadian controlled private corporations and individuals are entitled to a cash refund generally equal to 40 per cent of investment tax credits earned in respect of property acquired or qualifying expenditures made after April 19, 1983 and before 1989.[26] Other taxpayers are entitled to a cash refund equal to 20 per cent of investment tax credits earned in respect of property acquired or expenditures made during the same period.[27]

The purchase of computer hardware and the licence of systems software in 1987 generates an investment tax credit equal to 5 per cent of the capital cost of the hardware and software acquired provided the property is used in certain qualifying activities, including the manufacturing and processing of goods for sale or lease.[28] For these purposes, manufac-

23 Subs. 13(5.2) of the ITA.

24 The rate of CCA that may be claimed is 100% but the half-year rule in subs. 1100(2) of the ITR applies. The capital cost of applications software acquired prior to November 13, 1981 may be deducted fully in the year of acquisition. Applications software that is more specifically described in another class falls within that other class.

25 Subss. 127(5) and (9) of the ITA.

26 S. 127.1 of the ITA. A refund equal to 100% of the investment tax credit generated by current qualifying expenditures is available to qualifying Canadian controlled private corporations and individuals.

27 Para. 127.1(2)(*b*) of the ITA.

28 Subs. 127(*a*) of the ITA. The credit is 20% if the property is acquired for use in Atlantic

turing and processing are defined to exclude data processing.[29] Accordingly, an investment tax credit generally is not available in respect of the acquisition of computer hardware and systems software by persons in the data processing business.[30] An investment tax credit generally is not available to a lessee of computer hardware[31] or in respect of the acquisition, whether by purchase or by licence, of applications software. The general investment tax credit rate will be reduced to 3 per cent for property acquired in 1988 and to nil for property acquired after 1988.

(iii) *Cross-border Transfers*

Special income tax considerations arise when computer products are transferred in Canada by a non-resident transferor.

(A) *Computer Hardware.* Where a non-resident sells computer hardware in Canada in the course of carrying on a business in Canada any profits earned by the non-resident in Canada are subject to Canadian tax under Part I of the ITA.[32] If, however, the non-resident is resident in a country with which Canada has concluded an income tax convention, the non-resident generally is exempt from Canadian tax unless the profits are reasonably attributable to a permanent establishment of the non-resident in Canada.[33] Where a non-resident sells computer hardware in Canada other than in the course of carrying on a business in Canada, the profits earned by the non-resident in Canada are not subject to Canadian tax.

Where a non-resident leases computer hardware to customers in Canada otherwise than in the course of carrying on a business in Canada, any lease payments are subject to a 25 per cent withholding tax under Part XIII of the Act. The rate of tax may be reduced, however, under an applicable income tax convention.[34] Where a non-resident leases computer

---

Canada, the Gaspé Peninsula or a prescribed offshore region and 7% (3% in 1988, nil after 1988) if the property is acquired for use in a prescribed designated region. The rate is 50% if the property qualifies as certified property for use in certain regions and 60% if the property qualifies as approved project property for use in an approved project on Cape Breton Island.

29 Para. 127(11)(*b*) of the ITA.

30 See Interpretation Bulletin IT-145R, para. 40.

31 Unless, of course, the lease is treated as a purchase under the ITA. See *op. cit.*, footnote 20.

32 See the extended meaning of carrying on business in Canada under s. 253 of the ITA. See also *Tara Exploration and Dev. Co. Ltd. v. M.N.R.*, [1970] C.T.C. 557 (Ex. Ct.); affirmed on other grounds [1972] C.T.C. 328 (S.C.C.).

33 See, *e.g.*, Article VII of the Canada-U.S. Income Tax Convention.

34 Para. 212(1)(*d*) of the Act. The rate is reduced to 10% under the Canada-U.S. Income Tax Convention (the "New Convention") and under the Canada-U.K. Income Tax Convention (1978) and to 15% under the Canada-Japan Income Tax Convention (1965).

hardware to customers in Canada in the course of carrying on a business in Canada, any profits earned by the non-resident are subject to tax under Part I of the Act and no withholding tax under Part XIII of the Act is exigible.[35] Where the profits of the non-resident are not reasonably attributable to a permanent establishment of the non-resident in Canada, the amount of Canadian tax ultimately payable under Part I of the Act may be limited to a fixed percentage of the gross lease payments made to the non-resident.[36]

(B) *Computer Software.* Interesting considerations arise in the context of licence fees paid to non-residents for the right to use either systems software or applications software. Where the licence fees are earned by the non-resident in the course of carrying on a business in Canada, the analysis set forth above with respect to lease payments made to a non-resident carrying on business in Canada should equally apply. If the fees are earned by the non-resident otherwise than in the course of carrying on a business in Canada, however, the fees may not be subject to tax under the ITA or, in the case of fees paid to a resident of the U.S., may be exempt from Canadian tax under the Canada-U.S. Income Tax Convention.

Paragraph 212(1)(*d*) of the ITA provides, in part, as follows:

> (1) Every non-resident person shall pay an income tax of 25% on every amount that a person resident in Canada pays . . . to him as . . .
>
> (*d*) rent, royalty or similar payment, including, but not so as to restrict the generality of the foregoing, any payment
>
> (*i*) for the use of or for the right to use in Canada any property, invention, trade name, patent, trade mark, design or model, plan, secret formula, process or other thing whatever . . .
>
> but not including
>
> . . . .
>
> (*vi*) a royalty or similar payment on or in respect of a copyright in respect of the production or reproduction of any literary, dramatic, musical or artistic work. . . .

As discussed in Chapter 4, it was held in *Apple Computer v. Mackintosh Computers Ltd.*[37] that a computer program is a subject matter in which copyright exists notwithstanding that the program is embodied in a silicon

The new Canada-Japan Income Tax Convention, signed in Tokyo on May 7, 1986 but not yet ratified, will reduce the rate to 10%.

35 Para. 214(13)(*c*) of the Act and s. 805 of the ITR. RCT takes the position that the Canadian resident must withhold tax under Part XIII unless a waiver is obtained under s. 805 of the ITR.

36 See e.g., Articles VII and XII of the Canada-U.S. Income Tax Convention.

37 (1986), 10 C.P.R. (3d) 1 (Fed. T.D.).

chip in the computer. Accordingly, it may be that by virtue of sub-paragraph 212(1)(*d*)(vi) of the ITA no withholding tax is exigible in respect of licence fees paid to a non-resident for the right to use computer software programs in Canada. The *Apple Canada* case has been appealed to the Federal Court of Appeal, however, and RCT has indicated in discussions with the author that, pending the appeal, RCT will abide by its position that a computer program is not a matter in which copyright subsists, and, therefore, that sub-paragraph 212(1)(*d*)(vi) does not apply to exempt licence fees paid to non-residents for the right to use computer software programs from withholding tax under Part XIII of the ITA.

Even if licence fees paid to non-residents for the right to use computer software in Canada are subject to withholding tax under paragraph 212(1)(*d*) of the ITA, fees paid to the U.S. residents may be exempt from Canadian tax under the Canada-U.S. Income Tax Convention (the "New Convention"). In *R. v. St. John Shipbuilding & Dry Dock Co.*,[38] a U.S. resident corporation granted the taxpayer a non-exclusive licence to use a computer software program in connection with the design and construction of the taxpayer's ships. The U.S. company also supplied users' manuals and programmers' manuals as well as tapes containing technical information. The taxpayer's right to use the software was unlimited and the taxpayer paid for the software in three instalments which were not related to its use of the software. The Minister of National Revenue imposed a withholding tax equal to 15 per cent of the payments and the taxpayer appealed. The Federal Court of Appeal agreed with the Minister that the payments were probably subject to withholding tax under paragraph 212(1)(*d*) of the ITA as a "payment for the right to use in Canada any property or thing". The Court then considered whether the payments were exempt from Canadian tax by virtue of the 1942 Canada-U.S. Income Tax Convention (the "Old Convention"). Article I of the Old Convention provides, in part, as follows:

> An enterprise of one of the contracting States is not subject to taxation by the other contracting State in respect of its industrial and commercial profits except in respect of such profits allocable . . . to its permanent establishment in the latter State . . .

It was clear that the payments made to the U.S. corporation constituted industrial and commercial profits of the U.S. company. The U.S. company did not, however, have a permanent establishment in Canada. The Court reasoned, therefore, that the payments were exempt from Canadian tax unless they could be considered rentals or royalties for the purposes of

38 [1980] C.T.C. 352 (Fed. C.A.).

the Old Convention.[39] Paragraph 6(a) of the Protocol to the Old Convention provides as follows:

> The term "rentals and royalties" referred to in Article II of this Convention shall include rentals or royalties arising from leasing real or immovable, or personal or moveable property or from any interest in such property, including rentals or royalties for the use of, or for the privilege of using, patents, copyrights, secret processes and formulae, goodwill, trade marks, trade brands, franchises and *other like property.* [emphasis added]

The Court concluded that what the taxpayer had received under the licence could not be regarded as the privilege of using "patents, copyrights, secret processes and formulae, goodwill, trade marks, trade brands or franchises" but could be considered as the privilege of using "other like property". The Court held, however, that the payments were not in the form of royalties or rentals because the licence was perpetual and the payments were not in any way related to use. Accordingly, the payments were exempt from withholding tax and the taxpayer's appeal was allowed.

Under the New Convention, the term royalties is defined as follows:

> The term "royalties" as used in this Article means *payments of any kind* received as a consideration for the use of, or the right to use, any copyright of literary, artistic or scientific work (including motion pictures and works on film, videotape or other means of reproduction for use in connection with television), any patent, trade mark, design or model, plan, secret formula or process, or for the use of, or the right to use, tangible personal property or for information concerning industrial, commercial or scientific experience, and, notwithstanding the provisions of Article XIII (Gains), includes gains from the alienation of any intangible property or rights described in this paragraph to the extent that such gains are contingent on the productivity, use or subsequent disposition of such property or rights. [emphasis added]

This definition differs from that under the Old Convention in two important respects. First, royalties include "payments of any kind . . .". Accordingly, lump-sum payments such as those made by the taxpayer in the *St. John Shipbuilding* case likely are included as royalties. On the other hand, the definition in the New Convention does not include the catch-all provision "and other like property". RCT takes the position that software programs are secret processes and that withholding tax is exigible in respect of software licence fees paid to U.S. residents.[40] It will be recalled, however, that in the *St. John Shipbuilding* the Court indicated that payments for the right to use software programs were not payments for the right to use "patents, copyrights, secret processes and formulae, trade brands or

39 Article II of the Old Convention provides that the term "industrial and commercial profits" shall not include, *inter alia*, rentals and royalties.

40 See Revenue Canada, Taxation Release dated January 26, 1983.

franchises." It is difficult to see how software programs could fall within any of the other types of property referred to in the definition of royalties under the New Convention. Accordingly, unless the finding in *St. John Shipbulding* is reversed in a subsequent case in view of the decision in *Apple Computer* that programs are protected by copyright, licence fees paid to a U.S. resident for the right to use software may not be subject to Canadian withholding tax.

(c) Transfers of Computer Products — Customs Duties

Goods that are imported into Canada may be subject to a customs duty imposed under the Customs Tariff.[41] The amount of duty, if any, imposed on imported goods is equal to a prescribed percentage of the value of goods for duty purposes (value for duty). The value for duty of goods imported into Canada is determined in accordance with rules prescribed under the Customs Act.[42] The Customs Tariff classifies goods within one of 2,500 to 3,000 Tariff Items and sets the rate of duty applicable to goods within each Item. The rate of duty applicable to goods within any particular Item depends upon the country of export of the goods and for this purpose, the Customs Tariff classifies countries of export within one of five categories: British Preferential ("BP"), Most Favoured Nation ("MFN"), General, General Preferential Tariff ("GPT") and United Kingdom and Ireland ("UK & Ire."). Both the United States and Japan, the two largest exporters of computer products to Canada, are classified within the MFN category.

(i) *Value for Duty*

Effective January 1, 1985, Canada adopted, by way of an amendment to the Customs Act, a new system of valuation, known as the transaction value system, for the purpose of determining the value for duty of goods imported into Canada.[43] This system replaced the fair market value system which had been used in Canada for over 100 years.

The new system provides that the primary basis for the valuation of goods imported is their "transaction value", which is generally the price paid or payable by the importer to the exporter, plus or minus certain prescribed adjustments.[44] The transaction value method is available only in respect of goods sold for export to Canada and if certain conditions

41 R.S.C. 1970, c. C-41.

42 S.C. 1986, c. 1 (proclaimed in force: November 10, 1986).

43 S.C. 1984, c. 47, s. 2.

44 *Ibid.*; subs. 37(5).

are satisfied.[45] Where the importer and exporter do not deal at arm's length, the transaction value method is available only if the importer can demonstrate that the relationship did not influence the price charged or that the transaction value of the goods meets certain "test values".[46] Where the transaction value method is not available, the value for duty must be determined by the first of the following methods which is applicable:[47]

(i) the transaction value of identical goods, if any, sold for export to Canada;
(ii) the transaction value of similar goods, if any, sold for export to Canada;
(iii) the deductive value of the goods;
(iv) the computed value of the goods; and
(v) the residual method.

Where computer hardware is purchased by a Canadian importer from an arm's length exporter, the value for duty generally is the transaction value of the hardware, that is, the price paid or payable by the importer to the exporter plus or minus the prescribed adjustments. Where the hardware is purchased from an exporter with whom the importer does not deal at arm's length (for example, where a Canadian subsidiary purchases hardware from its U.S. parent) the transaction value method may not be available and recourse to one of the alternative methods may be necessary. Where computer hardware is leased, the transaction value method is not available and one of the alternative methods must be applied. Where there is no hardware sold for export to Canada that is identical or similar to the hardware leased, the computed value method likely will be applicable in determining the value for duty of the leased hardware.

Determination of the amount of duty payable in respect of computer software imported into Canada is somewhat more cumbersome. Reference must be made not only to the value for duty provisions of the Customs Act but also to the Computer Carrier Media Remission Order.[48] The value for duty of computer software is determined under the regular valuation provisions of the Customs Act and includes the value of the carrier medium, the value of the instructions or data contained on it and the value of reproducing the instructions on the medium.[49] An importer initially is required to pay duty at the applicable rate on the value for duty but is entitled to claim, within two years of the date of importation, a remission equal to the amount of duty in respect of the value of the instructions

---

45 *Ibid.*; subs. 37(1).

46 *Ibid.*; subpara. 37(1)(*d*)(ii) and subs. 37(3).

47 *Ibid.*; s. 36. The order of consideration of (iii) and (iv) may be reversed at the written request of the importer (subs. 36(3)).

48 Financial Administration Act, R.S.C. 1970, c. F-10, s. 17, SI/85-20 dated January 31, 1985.

49 RCCE Memorandum D8-3-15.

or data contained on the medium.[50] Accordingly, where a remission is claimed, duty is payable only in respect of the aggregate of the value of the carrier medium and the value of reproducing the instructions on the medium.

In most cases, the Computer Carrier Media Remission Order will be available to reduce the duty payable on applications software, which is generally contained on media separate from computer hardware. Most systems software is not contained on separate media but forms an integral part of computer hardware. Revenue Canada, Customs and Excise, generally requires the value of systems software to be included in computing the value for duty of computer hardware.

(ii) *Tariff Classification and Rates of Duty*

Prior to 1980, there were no specific tariff items for computer hardware or computer software. Computers arguably could be imported under any one of a number of items and it was often difficult to ascertain which tariff item was most appropriate in a particular instance. To resolve the uncertainty, the Customs Tariff was amended, effective January 1, 1980, to add Tariff Item 41417 which reads as follows:[51]

> Electronic data processing machines and apparatus, peripheral equipment for use therewith including data entry, data preparation and data handling machines and apparatus; accessories and attachments for use therewith; parts of all the foregoing; none of the foregoing to include telephone and telegraph apparatus and parts thereof:
>
> Tariff 41417-1 other than the following:
>
> 41417-2 Drum storage memories disc packs; disc drives; disc files; disc cartridge cleaners; card readers; card punches; card readers/punches; paper tape readers and punches; badge readers and punches; document transport mechanisms; tape to card punches; reproducing punches; line printers and page writers; card printers; matrix printers including buffered printers; card or paper tape verifiers; collators; card sorters; process control apparatus which converts analog signals from or to digital signals, the foregoing not to include sensors; card conditioning equipment; parts of the foregoing.

Computer products that are classified under Tariff Item 41417-1 generally are subject to duty at the rate of 3.9 per cent on the value for duty while products classified under Tariff Item 41417-2 generally may

50 Where the value of the medium is invoiced separately, the remission generally is granted at the time of importation.

51 S.C. 1980-81-82-83, c. 67, subs. 7(1).

be imported duty-free.[52] Computer software contained on disks is classified under Tariff Item 41417-2 whereas computer software contained on tapes is classified under Tariff Item 41417-1. On January 20, 1986, an Order-in-Council was passed exempting from duty all imports of computer parts and semi-conductors.[53] The exemption was revoked in June 1986, but in February 1987, the Minister of Finance announced that it would be reinstated.[54]

(d) Transfers of Computer Products — Federal Sales Tax

The federal government, under the Excise Tax Act,[55] imposes a single incidence sales tax of 12 per cent on the sale price of most goods manufactured in Canada and on the duty paid value (that is, the value for duty plus customs duty) of most goods imported into Canada.[56] Where goods are manufactured and leased in Canada, the value of goods for federal sales tax (FST) purposes may be determined by the Minister of National Revenue.[57] FST is not exigible in respect of transfers to certain exempt purchasers such as hospitals and universities or in respect of transfers of certain goods such as machinery and equipment used directly in the manufacture or production of goods for sale.[58]

In general, all Canadian manufacturers must be licensed and must pay tax at the earlier of delivery or passage of title of goods manufactured by them.[59] Where goods are leased, tax is payable on the lease payments. Where goods are imported into Canada, and are not to be further manufactured, the importer generally is not required to be licensed and tax is paid at the time of importation.

A manufacturer of computer hardware should be licensed under the Excise Tax Act and should collect FST on the sale price (or lease payments) of computer hardware manufactured and sold (or leased) by him under taxable conditions. Clause 26(6)(*c*)(ii)(B) of the Excise Tax Act permits

52 The rates are as follows:

	BP	MFN	General	GPT	UK & Inc.
41417-1	Free	3.9%	25%	Free	3.9%
41417-2	Free	Free	35%	Free	Free

53 Customs Tariff s. 11, SI/86-8 dated February 5, 1986; P.C. 1986-124 dated January 16, 1986.

54 See Budget Papers released on February 18, 1987. In February 1986, the Government also announced its intention to replace the current Tariff and adopt in its place the Harmonized Commodity Description and Coding System, effective January 1, 1988.

55 R.S.C. 1970, c. E-13.

56 *Ibid.*; s. 27.

57 *Ibid.*; s. 28.

58 *Ibid.*, s. 45 and s. 29 and Schedule III.

59 *Ibid.*, subpara. 27(1)(*a*)(i) and subs. 31(1). An exemption from the licence requirement is available to small manufacturers under subs. 31(2).

transportation and installation costs to be deducted from the selling price for the purposes of determining the amount of FST payable.[60] Purchasers should ensure, therefore, that these costs are invoiced separately.

Not surprisingly, the application of FST to transfers of computer software is less straightforward. Generally, in administering the Excise Tax Act, Revenue Canada, Customs and Excise distinguishes between systems software and applications software. Where, under taxable conditions, systems software and hardware are acquired in one transaction from one manufacturer, Revenue Canada, Customs and Excise considers both the systems software and the hardware to be taxable, even if they are invoiced separately. If, however, systems software is purchased from someone other than the manufacturer of the hardware, the systems software is not subject to FST. Generally, applications software is not subject to FST unless it is acquired from the manufacturer of hardware as part of a complete package of hardware, systems software and applications software and the applications software is not invoiced separately.[61]

(e) Transfers of Computer Products — Provincial Sales Tax

All of the provinces of Canada, with the exception of Alberta, impose a retail sales tax (RST) on the purchase or lease of tangible personal property and on the provision of certain services.[62]

Provincial RST generally is not imposed on goods purchased for resale or for use outside of the province. In addition, exemptions from the tax are available to certain purchasers and in respect of certain goods or services.

Where computer hardware is sold or leased, it is subject to RST in all provinces unless an exemption applies. Where hardware is sold, RST generally is payable on the sale price, although some provinces permit deduction of charges for installation or delivery where such charges are invoiced separately. Where hardware is leased, RST is payable on the lease payments.

There is less uniformity among the provinces regarding the exigibility

---

60 See also the Sales Tax Transportation Allowance Regulations and Memorandum ET 204, April 26, 1985.

61 Revenue Canada Press Release dated January 1, 1979.

62 Nfld. Retail Sales Tax Act, S.N. 1978, c. 36.
P.E.I. Revenue Tax Act, R.S.P.E.I. 1974, c. R-14.
N.S. Health Services Tax Act, R.S.N.S. 1967, c. 126.
N.B. Social Services and Education Tax, R.S.N.B. 1973, c. S-10.
Que. Retail Sales Tax Act, R.S.Q., c. I-1.
Ont. Retail Sales Tax Act, R.S.O. 1980, c. 454.
Man. Retail Sales Tax Act, R.S.M. 1970, c. R150.
Sask. Education and Health Tax Act, R.S.S. 1978, c. E-3.
B.C. Social Service Tax Act, R.S.B.C. 1979, c. 388.

of RST on transfers of computer software. All of the provinces consider systems software to be an integral part of computer hardware and most transfers of systems software are subject to RST.[63] Prepackaged or canned applications software generally is subject to RST but a transfer of custom software generally is considered to constitute the provision of a non-taxable service and is not subject to RST. In Manitoba and Quebec, RST is not payable on any applications software, whether custom or prepackaged, that is acquired for use in a business. Newfoundland and Nova Scotia impose RST on resales of custom software.[64]

---

63 It is unclear whether a licence of systems software is taxable in Newfoundland. See Paul K. Tamaki and Carol Anne Jackson "Retail Sales Tax Update" (1986), 3 C.C.L.R. 53.

64 It is arguable that provincial retail sales is not exigible, except in British Columbia, in respect of transfers of computer software because computer software is not tangible personal property. See *Continental Commercial Systems Corp. (Telecheque Can.) v. R., supra*, footnote 12. The British Columbia Social Services Tax Act expressly subjects computer software to tax; see *supra*, footnote 5.

21

Tax Aspects of the Servicing and Development of Computer Products

This chapter reviews the tax aspects of the provision of maintenance and support services for computer hardware and software. The tax implications of the development of computer products are also considered.

1. SERVICING COMPUTER PRODUCTS

(a) Hardware Maintenance Agreements

In the typical hardware maintenance agreement the hardware supplier agrees to provide preventative and remedial services in exchange for a maintenance fee, usually payable on an annual basis. Preventative services include routine inspection and various standard services at regular intervals to ensure the computer hardware remains in good working condition. Remedial services are provided in the event the computer malfunctions. Payments for both types of services generally are deductible by the payor and constitute income to the payee.

Subject to the comments below, where maintenance services are performed in Canada by a non-resident otherwise than in the course of carrying on a business in Canada, the non-resident is generally not subject to Canadian tax on any fees earned in Canada.[1] Where the services are performed as part of a business carried on in Canada by a non-resident who is resident in a country with which Canada has concluded an income tax convention, the non-resident will not be subject to Canadian tax on service fees unless the fees are attributable to a permanent establishment of the non-resident located in Canada. Regardless of whether the activities of the non-resident are in the course of carrying on a business in Canada, section 105 of the ITR requires the Canadian payor to withhold 15 per cent of payments made to a non-resident for services performed in Canada. The non-resident is therefore put to the rather inconvenient task of having

1 It may be difficult for the non-resident who provides maintenance services in Canada to demonstrate that he is not carrying on business in Canada. In certain circumstances, paras. 212(1)(*a*) or 212(1)(*d*) may impose a withholding tax on the fee payments.

to complete a Canadian income tax return to claim a refund of the withholding tax.[2]

In most provinces, payments in respect of maintenance services will be subject to retail sales tax.

(b) Software Support Agreements

Software support agreements generally provide for the supplier to provide on-going assistance in the use of the software, the correction of errors and malfunctions, and the provision of modifications to improve the operating condition of the software as well as new releases to reflect enhancements or new features. Payments for such services generally are deductible by the payor and income of the payee. Where payments are made to non-residents, the analysis set forth above with regard to payments to non-residents under hardware maintenance agreements will apply.[3]

Where the software support services result in enhancements to the software, any payments may be considered to be in respect of the acquisition of a new capital asset and will therefore be deductible only in accordance with the provisions of Part XI of the ITR. In addition, payments made to enhance software may be subject to withholding tax under Part XIII of the Act as discussed above.

2. DEVELOPMENT OF COMPUTER PRODUCTS

(a) Computer Hardware

The development and production of computer hardware in Canada does not give rise to any special income tax considerations.

The design, construction and testing of prototype hardware generally constitutes scientific research and experimental development for the purposes of the ITA and expenditures that are attributable to these activities generally are currently deductible under section 37 of the ITA.[4] In addition, such expenditures, other than certain prescribed expenditures, generate investment tax credits.[5] Qualifying Canadian controlled private corporations are entitled to an investment tax credit equal to 35 per cent of the first $2,000,000 of qualifying expenditures in a year.[6] A general investment

2 A waiver from the withholding obligation may be obtained where the non-resident can demonstrate that he is not subject to Canadian tax because of a treaty exemption.

3 The obligation to withhold tax under s. 105 of the ITR does not apply if services are performed by a non-resident outside of Canada. *Quaere* whether software services performed over telephone lines originating in the United States are performed in Canada.

4 See Interpretation Bulletin IT-439, para. 16.

5 Subss. 127(5) and (9) of the ITA. See also s. 2902 of the ITR.

6 Subss. 127(10.1)-(10.4) and (10.6) of the ITA.

tax credit equal to 20 per cent of qualifying expenditures is available to other developers and in respect of expenditures over $2,000,000 per year incurred by Canadian controlled private corporations.[7] The rate is increased to 30 per cent when the scientific research and experimental development is carried out in Atlantic Canada or in the Gaspé peninsula.

The costs of developing and producing computer hardware for sale, other than the costs that qualify for the special treatment under section 37 of the ITA, generally should be inventoried and deducted only upon the sale of the hardware. Where computer hardware is manufactured for use in the producer's own business or for lease, the costs of development and production which are not deductible under section 37 generally should be capitalized and deducted over time in the form of CCA. As discussed above, most computer hardware is included as Class 10 or Class 29 property.

The production of computer hardware in Canada constitutes manufacturing and processing and the profits derived therefrom are eligible for the manufacturing and processing credit under section 125.1 of the ITA.

Machinery and equipment used directly by a manufacturer of computer hardware generally is exempt from federal sales tax and from provincial retail sales tax.

(b) Computer Software

There is considerable uncertainty as to whether the costs of developing computer software qualify as expenditures in respect of scientific research and experimental development and are therefore eligible for current deduction under section 37 of the ITA. Generally, Revenue Canada, Taxation, takes the position that most computer software development is not scientific research and experimental development.[8] Where, however, the development of software involves an innovative program that incorporates a technological advance in a particular area of application, such development could qualify as scientific research and experimental development. The Minister of National Revenue has recently published guidelines dealing, inter alia, with how to distinguish routine computer software development from innovative or novel work which is eligible for deduction under section 37.[9]

Where computer software having a useful life of less than one year

---

7 See footnote 5, *supra.*

8 The designing, building and testing of prototype systems software is considered to be scientific research and experimental development.

9 Information Circular No. 86-4 entitled "Scientific Research and Experimental Development," August 29, 1986. See especially Part 6.

is developed for use by the developer, the costs of development are currently deductible.[10] Where the useful life of the software is greater than one year and the software is to be used by the developer or licensed by him, the costs of development normally should be capitalized and deducted over time in the form of CCA. Where the software is developed for sale, the cost of development should be inventoried. Notwithstanding these general principles, software developers often deduct development cost currently and this practice appears to have been accepted by Revenue Canada.[11]

Computer software development does not constitute manufacturing and processing and profits therefrom are therefore not eligible for the manufacturing and processing tax credits under section 125.1 of the ITA.

10 Interpretation Bulletin IT-283R, para. 13.

11 James R. Wilson, "Recent Developments in the Taxation of Computer Technology", Paper delivered to Computer Law Update '84 Conference (Insight, June 25-26, 1984).

Index